Interacting with Geospatial Technologies

Interacting with Geospatial Technologies

MORDECHAI (MUKI) HAKLAY

Department of Civil, Environmental & Geomatic Engineering, University College London, UK

WILEY-BLACKWELL

A John Wiley & Sons, Ltd., Publication

Registered Office
John Wiley & Sons Ltd, The Atrium, Southern Gate, Chichester, West Sussex, PO19 8SQ, UK

Other Editorial Offices
9600 Garsington Road, Oxford, OX4 2DQ, UK
111 River Street, Hoboken, NJ 07030-5774, USA

For details of our global editorial offices, for customer services and for information about how to apply for permission to reuse the copyright material in this book please see our website at www.wiley.com/wiley-blackwell.

Library of Congress Cataloging-in-Publication Data

Haklay, Muki.
   Interacting with geospatial technologies / Muki Haklay.
       p. cm.
   Includes bibliographical references and index.
   ISBN 978-0-470-99824-3
 1. Geographic information systems.   2. Human-computer interaction.   I. Title.
   G70.212.H34 2010
   910.285–dc22

                                                      2009050268

A catalogue record for this book is available from the British Library.

Set in 10.5/12.5pt Garamond by Aptara Inc., New Delhi, India.
Printed in Great Britain by Antony Rowe Ltd., Chippenham, Wiltshire.

First Impression   2010

Contents

Preface

Geographical Information Systems (GIS) have been around for almost 40 years, yet the number of users remains relatively small, and the level of skills required from an operator remains high. In comparison, within two years of releasing Google Earth, 22.7 million people downloaded the application – and many used it. Why?

A significant part of the answer lies in the area of Human-Computer Interaction (HCI) – but there aren't any textbooks for a GIS developer to pick off the shelf to provide the guidelines and information needed to create a successful GIS interface. This book aims to fill this gap, as well as discussing the wider area of geospatial technologies that are currently being developed – from Portable Navigation Devices (PNDs) to Web-mapping applications.

The need for the book emerged from the realization that GIS and other applications of computerized mapping have gained popularity in recent years. Today, computer-based maps are common on the World Wide Web, mobile phones, satellite navigation systems and in various desktop computing packages. The more sophisticated packages that allow the manipulation and analysis of geographical information are used in the location decisions of new businesses, and for public service delivery in respect of planning decisions by local and central government. Many more applications exist and some estimate that several million people across the world are using GIS in their daily work (Longley *et al.*, 2005).

Yet, many applications of GIS are hard to learn and to master. This is understandable as, until quite recently, the main focus of software vendors in the area of GIS was on the delivery of basic functionality and development of methods to present and manipulate geographical information using the available computing resources. As a result, little attention was paid to usability aspects of GIS. This is evident in many public and private systems where the terminology, conceptual design and structure are all centred around the engineering of geospatial technologies and not on the needs and concepts that are familiar to the user.

There is currently no agreed definition of HCI, but a working definition of the Association for Computing Machinery Spacial Interest Group of Computer-Human Interaction (ACM SIGCHI) describes it as 'a discipline concerned with the design, evaluation and implementation of interactive computing systems for human use and with the study of major phenomena surrounding them'. Hence, HCI is concerned with enhancing the quality of interaction between humans and computer systems within the physical, organizational and social aspects of the users' environment to produce systems that are usable, safe and functional (Sharp *et al.*, 2007).

Research on these issues is based on the assumption that the needs, capabilities and preferences for the way users perform an activity should influence the design and implementation of a system in order for it to match users' requirements. Knowledge about the users and the work they need to accomplish, as well as about the technology, is required to meet this approach to systems design, which makes HCI a multidisciplinary field of research.

Research into HCI aspects of GIS started in earnest in the mid 1980s. In the mid 1990s several books were published in this area, including the books edited by Medyckyj-Scott and Hearnshaw, *Human Factors in Geographical Information Systems* (1993) and Nyerges *et al.*,

Cognitive Aspects of Human-Computer Interaction for Geographic Information Systems (1995a), although unfortunately they are no longer widely available.

Within GIS research, the basic understanding that 'spatial data is special' and therefore requires special methods of handling, manipulating and interacting is widely accepted. This is also true for aspects of HCI within GIS and therefore there is a need for specific research which deals with the design of interfaces for geographical information.

Today, HCI research in GIS covers topics such as interface design, cognitive models of spatial knowledge, digital representations of spatial knowledge and how well the digital representation of such knowledge in a GIS translates into intuitive human reasoning, the use of geographical information in the knowledge discovery process and many other areas.

This book's aim is to provide an introduction to HCI and usability aspects of Geographical Information Systems and Science and the geospatial technologies that are based on them. Its aim is also to introduce the principles of Human-Computer Interaction (HCI); to discuss the special usability aspects of geospatial technologies which designers and developers need to take into account when developing such systems; and to offer a set of tried and tested frameworks, matrices and techniques that can be used within geospatial technologies projects.

M. Haklay
London, September 2009

About the authors

Mordechai (Muki) Haklay is a Senior Lecturer in GIS at UCL. He has been carrying out research on usability and GIS since the late 1990s and published many academic papers in this area. His research focuses on usability engineering aspects of geospatial technologies.

Aaron Marcus, President, Aaron Marcus and Associates, Inc., Berkeley, California, USA. He is the first graphic designer to use computers, has published six books and 250 articles/papers, lectures worldwide, is an AIGA Fellow and member of CHI Academy.

Annu-Maaria Nivala is a usability specialist at DIGIA Oyj, Finland, a private IT company. Between 2002 and 2008 she was carrying out her PhD research at the Geoinformatics and Cartography Department of the Finnish Geodetic Institute where her research topics were related on usability perspectives for the design of interactive maps.

Antigoni Zafiri holds an undergraduate degree in Marketing followed by postgraduate studies in Environmental Management and Geographical Information Systems. During her master studies, she carried out a GIS snapshot study and her research focus on usability aspects of GIS.

Artemis Skarlatidou is an Engineering Doctorate student at UCL. Her main research interests involve usability and User Experience elements of web-based GIS. Her EngD focuses on trust aspects of web-based GIS.

Carolina Tobon is a software developer. She holds MSc and PhD degrees in GIS from UCL. During her studies, she specialised in the design and evaluation of visualization tools to support the exploration of large multivariate datasets, and published several papers in this area.

Catherine Emma (Kate) Jones is a lecturer in Human Geography at the University of Portsmouth. She completed her PhD at University College London in 2008 in the area of Health Geography. She is a specialist in GIS for collaborative research in Social and Urban Geography.

Clare Davies has researched cognition and geographic information since the early 1990s. Currently a Senior Research Scientist at the Ordnance Survey, she focuses on people's concepts and everyday language of space and place, and making geographic information reflect them.

Jessica Wardlaw is an Engineering Doctorate student at UCL. Jessica is researching usable GIS mapping techniques for the analysis and visualization tools that are used to process healthcare statistics in the UK National Health Service.

Jochen Albrecht is an Associate Professor at Hunter College, CUNY. His research concentrates since the early 1990s on the notion of process, both the computer-based representation of the geographic phenomenon as well as the management of GIS workflows.

(Lily) Chao Li holds MSc and PhD degrees in GIS from UCL. Her research interests are in spatial data mining and modelling, mobile geo-services, wayfinding and spatial cognition. Her recent book on "Location-Based Services and Geo-Information Engineering" is published by Wiley.

Robert Laurini is distinguished professor at INSA, University of Lyon, France. He is a specialist in geographic data bases and visual information systems, especially for urban and environmental planning. He published more than 150 peer-refereed papers and several books, among them "Information Systems for Urban Planning: A Hypermedia Cooperative Approach" in 2001 and "Fundamentals of Spatial Information Systems" with Derek Thompson in 1993.

Stephanie Marsh is a Usability Consultant with behavioural research consultancy Bunnyfoot. Her PhD 'Using and Evaluating HCI Techniques in Geovisualisation' was completed at City University in 2007.

How to use this book

Who should read this book

When developing this book, we thought of two groups of readers: first, GIS professionals who are developing applications in their workplace, or students at advanced levels (Masters or PhD) who need to develop a geospatial technology application as part of their research. We expect these readers to have knowledge of GIS, but not to be familiar with principles of usability engineering or Human-Computer Interaction, as these are not covered by the existing literature on GIS and cartography. The second group is of software engineers and developers without an education in geography, cartography or GIS who are developing geospatial technology applications. These readers may have come across some of the principles of user-centred design and usability engineering, as these areas receive some attention in computer science and information technology studies.

While this text can be used as a teaching text, we wrote it for the reader in mind who is self studying the material from the book. We therefore provided in each chapter case studies that illustrate the topics that are discussed in the chapter. We deliberately avoided using jargon from either HCI studies or GIScience. Where we use jargon, we made sure that it is explained.

We did not intend to cover the full area of GIScience or HCI – this is impossible within the limited space of the book, when considering that current popular textbooks on HCI can run to 800 pages, and GIS books hold over 500 and more pages. Instead, we attempted to distil the core issues that are relevant for developers of geospatial technologies, and included sections on further reading that link the chapters to the wider literature. The reference list at the end of the book is also a valuable source as it includes the most significant papers that have emerged from the research into HCI and GIS.

Structure of this book

The book is divided into three parts: Theory, Framework, and Practicalities and Technique. Chapter 1 provides an overview of HCI, GIS and then HCI for GIS. The aim of the chapter is to provide the reader with the intellectual history of these fields, and to understand how they have evolved. It also provides an overview of the topics that will be covered in the book.

The *Theory* section covers the principles and scientific theories that are fundamental to usability. The most fundamental aspect is the issue of human understanding of space, which is covered in Chapter 2. This chapter covers aspects of spatial cognition and other aspects that relate to human ability. Chapter 3 moves to the principles of cartography. These might be familiar to some readers who are coming from a geography background, although it is specifically focusing on aspects of cartography that are relevant to technology design. Chapter 4 moves from the individual to social interaction, and deals with the principles of collaborative computer systems and participatory GIS.

The *Framework* section focuses on the principles of User-Centred Design (UCD) in Chapter 5, with specific discussion of how these design philosophies relate to geospatial technology, and then moves to the practical implementation of UCD in usability engineering in Chapter 6. In this chapter we cover how usability engineering can be integrated into the software development process. For readers who are familiar with HCI literature, this section can be helpful in situating these frameworks within the specific context of geospatial technologies.

The final section of the book moves to *Practicalities and Technique* of creating usable geospatial technologies. We start in Chapter 7 with the principles of application planning, including a look at the process of task analysis, which helps the designer to understand what the user will do with the system, and covers several techniques for carrying out the application planning process. Chapter 8 covers practical aspects of cartography such as selection of colour, the process of classification of data and the organization of map layout. In Chapter 9, the basic elements of an application interface are discussed, including aspects of the graphical user interface. Chapter 10 is dedicated to usability evaluation techniques, and provides a comprehensive description of the various techniques and their role in the development of geospatial technologies. Chapter 11 deals with single user environments, which are mainly desktop systems used in organizational settings, as most GIS packages are, although this chapter also deals with the issue of mobile HCI. Finally, Chapter 12 is dedicated to web-based applications and discusses the various considerations that need to be taken into account when web-based mapping applications are being developed.

The use of examples in case studies

Each chapter in this book includes at least one case study. The aim of the case studies is to illustrate issues that are discussed in the sections in which they are integrated, so as to make the principles and the points of the discussion concrete and tangible. To provide a comprehensive overview of usability aspects we used cases where we highlight some good aspects and some bad aspects in different geospatial technologies and applications. It is very important to note from the outset that none of the criticism is aimed at singling out the developers of a specific application or case. These are examples of problems and issues that are widespread in the area of geospatial technologies, and the specific cases that are presented can be praised for many of the elements that they do include and implement. We hope that even if some examples are critical of a specific system, they will be used as a motivation for improving it. Significantly, all our examples are not designed to single out any system as the worst offender, rather to show the problems with the current state of the art.

Acknowledgements

First and foremost, to all the authors who contributed to this book, which was challenging to write. Your efforts and insights are sincerely appreciated.

Many people contributed to the development of this book. I am always in debt to Professor Angela Sasse, of UCL computer science department, who guided me to readings in HCI over a decade ago during my PhD studies, as well as Dr Paul Densham, who introduced me to the early research on HCI and GIS from the 1990s. Professor Paul Longley provided invaluable advice on early ideas for the book. I would also like to thank many colleagues and students at UCL who helped in researching HCI aspects of GIS, including Rachel Alsop, Sephi Berry, Steve Bosher, Claire Ellul, Steve Edney, Nicola Francis, Kate E. Jones, Hanif Rahemtulla, Ana Simao, Artemis Skarlatidou, Carolina Tobón, Tina Thomson, Jessica Wardlaw, Antigoni Zafiri and the participants in the studies that we have carried out.

Special thanks to researchers from across the world who provided inspiration and insights for this book, and especially to Eran Ben-Joseph, William Cartwright, Nicholas Chrisman, Jason Dykes, Max Egenhofer, Corné van Elzakker, David Fairbairn, David Forest, Reginald Golledge, Jenny Harding, Ari Isaac, Piotr Jankowski, Robert Laurini, David Mountain, Timothy Nyerges, Barbara Poore, Alexander Pucher, Adena Schutzberg, Carl Steinitz and the members of the ICA commission on Use and User Issues.

Finally, thanks to my partner Tanya Pein, who has provided consistent and loving encouragement and support during the process of conceiving and writing this book.

Section I

Theory

1 Human-computer interaction and geospatial technologies – context

Mordechai (Muki) Haklay and Artemis Skarlatidou

This book is about interaction with systems that contain and represent geographic information – information about objects and activities that occur on the face of the Earth. While Geographic Information Systems (GIS) have been around since the mid 1960s, only in the late 1980s was attention turned to the ways in which people interact with them. The recent growth of geographic information technologies (geospatial technologies) – from Internet-based applications such as Google Earth or Microsoft Virtual Earth to GPS-based navigation devices – means that today, there are more people who are exposed to and use geographic information daily than ever before. GIS, though less consumer friendly, is a routine tool in public and private sector organizations where it is being used to manage land ownership, plan public services, locate new shops and design delivery routes.

Case Study 1.1: Should you trust your Personal Navigation Device?

'A 20-year-old student's car was wrecked by a train after she followed her portable navigation device (satellite navigation or sat nav as they are known in the UK) onto a railway track. Paula Ceely, was driving her Renault Clio from Redditch, Worcestershire, to see her boyfriend at his parents' home in Carmarthenshire for the first time.

She was trying to cross the line in the dark when she heard a train horn, realized she was on the track, and the train smashed into the car.

Transport police said drivers must take care with satellite navigation.

The car was carried about half a mile (800m) down the line by the Pembroke Dock to Swansea train, although Ms Ceely escaped injury in the incident near Whitland.

A second-year student at Birmingham University, she had borrowed the sat nav from her boyfriend, Tom Finucane, 21. "**I put my complete trust in the sat nav** and it led me right into the path of a speeding train," she said "**The crossing wasn't shown on the sat nav**, there were no signs at all and it wasn't lit up to warn of an oncoming train. Obviously I had never done the journey before so I was using the sat nav – **completely dependent on it**" she said.'

(adapted from BBC, 2007, emphasis added)

Interacting with Geospatial Technologies   Mordechai (Muki) Haklay
© 2010 John Wiley & Sons, Ltd

Though clearly millions of people are using Portable Navigation Devices (PND) on a daily basis, and many of them complete their journey without any incident of the sort which Ms Ceely experienced, this story illustrates the impact of geospatial technology design. The end results of the design which did not highlight the railway crossing in its cartography almost ended in a fatal disaster.

Significantly, interaction is not the only issue here. The story highlights other aspects of geospatial technologies, which are not covered in this book. For example, there is an issue with the accuracy of the geographic information that was loaded on the device – the information is usually less accurate and comprehensive in rural areas where there are fewer inhabitants. In addition, there is the issue of the algorithm (the software code) that was used to calculate the path that guided Ms Ceely towards the railway crossing: a different algorithm design might have chosen a safer path. These are generic aspects of GIS which are covered in books dedicated to the computational aspects of these systems (Worboys and Duckham, 2004, Longley *et al.*, 2001) or the analysis aspects of geographic information (De Smith *et al.*, 2007). While references will be made to these sources, this book will not discuss these issues at length.

Another facet of Ms Ceely's story is interaction beyond the geographic domain – she was driving her car, and surely she entered the address of her boyfriend's home using the PND interface. Some of the interface is low level – such as the shape of buttons, the order of characters on the screen etc. These aspects of operating a computing device are part of general Human-Computer Interactions (HCI) issues, and there is a wide range of literature that covers the design and study of general computer systems (Sharp *et al.*, 2007, Dix *et al.*, 2004). Here, too, this book will refer to other sources, but it is aimed to be a companion to this literature with a focus on the unique aspects of geographic information and its handling.

After stating what aspects will be excluded, it is time to turn to those that are included. This section of the book covers the basic theory which is needed to understand interaction aspects of geographic information. It starts, in this chapter, with an overview of the two fields that provide the grounding for this book, and which have just been mentioned – these are the fields of Human-Computer Interaction and Geographic Information Science (GIScience). After introducing these fields, the discussion turns to the combination of HCI and GIScience and the third section provides an overview of the main strands of research and development in this area over the past two decades. The chapter concludes with some guidance to useful resources that can be used for further reading in both areas.

1.1   Human-computer interaction and usability engineering background

The first commercially available computers that appeared in the 1950s were very large and expensive machines that could only be operated by specialists who were able to program them. Technological advances however, have dramatically changed this situation by decreasing both the cost and size of computers. The advent of the silicon chip, for instance, allowed the miniaturization of circuits and the development of powerful computers with large storage capacities. This also facilitated the innovation of the personal computer (PC) in the 1970s, which in the 1980s came to be used by a wide variety of people who were not computer experts or programmers but who utilized computers for a vast range of applications. The success of the PC, however, has also been made possible by improved understanding of the ways in which humans interact with computers, since this has enabled the design of systems that support a larger user population with the broadest range of requirements.

Human-Computer Interaction (HCI or sometimes the variant CHI) appeared during the 1970s, emerging from the fields of ergonomics and 'Man Machine Interaction' that can be dated back to the 1940s. The original focus of HCI was in issues of the 'User Interface' or those aspects of the computer systems with which the end-users come into contact (like screen layout). During the mid 1980s, the term HCI was adopted and the field became broader – aiming to tackle all aspects of interaction with computers. Each of these fields illuminates a specific aspect of computer operation and use.

HCI has implications for many aspects of information systems. For example, Landauer (1995) used the lack of usability as an explanation for the productivity paradox – the fact that in spite of the continual and growing investment in computerized systems in the work-place, the productivity of the American workforce did not increase (and even decreased in some sectors). Some evidence shows that similar trends can be observed in other Western countries. This trend is identified in the period between the early 1970s and the early 1990s. Landauer's explanation for this paradox is that a lack of attention to usability when com-puterizing work processes and tasks results in wasted effort and counterproductive software products. Computers do not necessarily improve productivity – they can actually hamper it in applications that deal with more sophisticated manipulation of information rather than clear-cut calculations.

One of the core concepts that emerged from HCI research is User Centred Design, Development and Deployment or UCD (Landauer, 1995; Sharp et al., 2007; Dix et al., 2004). While User Centred Design is covered in detail in Chapter 5, it is worth noting here that it is a development philosophy that puts usefulness and usability at the centre of the process and evaluates them empirically. Put simply, within UCD, developers and designers are required to put the end-user at the centre of the design process. Thus, it requires the designer to understand what the user's work environment looks like; what tasks they are trying to accomplish; what the requirements and needs of the user are etc. In adopting the UCD concept, the likelihood of creating useful and effective systems increases.

HCI aims to create systems that provide functionality appropriate to their intended use, and which are 'good enough to satisfy all the needs and requirements of the users and other potential stakeholders' (Nielsen, 1993: 24). These people, however, may vary in their computer literacy skills, world views, cultural backgrounds or domain knowledge. Thus it is important to understand the ways in which people use computer systems in particular settings if system design is to support users in an effective and efficient manner. Furthermore, users expect computer systems to be useful for achieving their goals not only in terms of the appropriateness of the functionality they may provide, but also in terms of how well and easily such functionality can be operated (Nielsen, 1993; Preece et al., 1994).

Apart from understanding how to improve users' work processes, HCI is also concerned with understanding how people use computer systems in order to develop or improve their design. The aim is to meet users' requirements so that they can carry out their tasks safely, effectively and enjoyably (Preece et al., 1994). Usability deals with these issues and it applies to all aspects of a system's user interface, defined here as the medium through which a user interacts and communicates with the computer (Nielsen, 1993). Usability refers to the effectiveness of the interaction between humans and computer systems and it can be specified in terms of how well potential users can perform and master tasks with the system.

A system's usability can also be measured empirically in terms of its learnability, efficiency, memorability, error rate and user satisfaction (Nielsen, 1993). The ease of learning a product is measured in the time it takes a person to reach a specified level of proficiency in using it, assuming the person is representative of the intended users. Efficiency refers to the level of productivity that the user must achieve once the system has been learned. Memorability measures how easily a system is remembered, either after a period of not using it or by casual users. An error in this context is defined as 'any action that does not accomplish the desired goal' (Nielsen, 1993: 32). Counting such actions provides a measure of a system's error rate. Satisfaction refers to how pleasant the system is to use. Preece and her colleagues (1994) also mention throughput, flexibility and user attitude towards the system. Ease of use or throughput is comparable to Nielsen's efficiency and error rate as it is defined as 'the tasks accomplished by experienced users, the speed of task execution and the errors made' (Preece et al., 1994: 401). Flexibility refers to the extent to which the system can accommodate tasks or environments it was not originally planned for. Attitude is comparable to Nielsen's user satisfaction or how pleasant it is to use the system.

Usability Engineering (UE) (which is covered more fully in Chapter 6) is an approach aimed at integrating central concepts and lessons that were learned through HCI research into software design processes in a way that they can be applied in a consistent and efficient manner in software development projects.

The integration of UCD principles in the software development process is done through the creation of frameworks, techniques, and matrices that can be deployed systematically and rigorously. By developing such methods and tools, UE aims to ensure that the concept of usability is translated into measurable criteria and into a set of actions that the developer can carry out through the life cycle of the software.

Of course, since UE is reliant on cognitive models of tasks and abstract manipulation of information and since the final product will be used by a range of users with differing culture, age and education attainment, UE is not an engineering discipline where criteria and methods are rigidly defined and where predictions will work deterministically in every case. Furthermore, in the software development processes, it is unlikely that presubscribed matrices that were set at the early stages of the design process guarantee that the system will be usable. The reason for this is that the design process itself is very complex and often changes, and therefore matrices that are defined rigidly might divert the development process to ensure that the final system satisfies specific tests, even if overall performances are not satisfactory. Thus, the correct way to view UE is as a toolbox that can be used throughout software development processes, and, by combining the right tools for the appropriate stage of development, it is possible to ensure that the user remains at the centre of the process and the resulting system is usable.

Despite the fact that the usability criteria that were mentioned above (learnability, efficiency, memorability, error rate and user satisfaction, flexibility) cannot be quantified unambiguously, they provide a set of principles that can then be translated into specific measurements and expectations and guide the development process. To further integrate these criteria in the design process, many methods have been developed over the years. These methods cover the whole development process and borrow concepts from many fields of study including Psychology, Anthropology, Ergonomics, and, naturally, the wider field of HCI, turning research outcomes into tools. For example, at the beginning of the software design process, ethnographic techniques can be used to understand the user's context within the process of requirement analysis. At the final stages of development, direct observation studies, where users are asked to carry out tasks with the system, are used to check how successful the system is in terms of learnability or to identify usability problems that have not been found in earlier stages.

In summary, UE as an applied practice is now a maturing discipline with a wide acceptance of its importance and relevance to software development processes. UE principles are now taught as part of the computer science and software engineering curriculum, as they are seen as an integral part of the education of software developers.

1.1.1 Contributing disciplines

HCI is based on multidisciplinary research and draws on lessons learned in Computer Science, Cognitive Psychology, Social and Organizational Psychology, Ergonomics and Human Factors, Linguistics, Artificial Intelligence, Philosophy, Sociology, Anthropology, and Engineering and Design (Preece *et al.*, 1994).

As has already been highlighted, the technological developments and particularly the development of personal computers (PC) lead to the establishment of HCI as a discipline. HCI plays an essential role in the design of the user interfaces and thus in **Computer Science**, by investigating the interaction of users with computerized systems. At the same time Computer Science contributes to HCI, as it provides the technological knowledge, which is necessary to implement such designs. Such developments include, among others, debugging tools, prototyping tools and user interface development environments.

The discipline of **Ergonomics** and **Human Factors** is mainly concerned with the physical characteristics of the interaction between humans and computerized systems. For example, among several other issues, the discipline is concerned with the human anthropometry and its relation to the working environment, human cognition and sensory limits, design for people with disabilities, the physical attributes of displays and design for a variety of different environmental settings. Therefore, the discipline is of particular importance to HCI, because it can help understand the human capabilities and limitations and the human factors which influence the use of a system.

Cognitive Psychology plays an important role in HCI theoretically and instrumentally. Theories of cognitive psychology provide the basis for understanding how humans process information based on elements such as perception, attention, memory, problem solving and learning. These theories provide the basis for predictive models for the evaluation of alternative designs before their actual implementation. In addition, it has led to the introduction of new issues to the research agenda of HCI, such as how the design can enhance learning activities or minimize memorization. Notably, experimental techniques used in psychology for data collection were adopted in HCI (e.g. the Think Aloud protocol).

Sociology and **Social and Organizational Psychology** involve theories and models concerned with human behaviour in social and organizational contexts. Such theories are of particular interest for HCI in order to understand how humans behave and interact when they are engaged in common tasks using computerized systems, as well as how the social environment influences this interaction. Within this context philosophical theories also influence HCI, for example Heidegger's philosophy of familiarity.

Anthropology and **Ethnography** brought to the attention of HCI the elements of situated action and the importance of context. An important contribution of this field to HCI was the method of naturalistic observation, which helped HCI researchers overcome problems associated with laboratory experimentation by observing the user in real environments.

Linguistics, the field that examines the characteristics of natural language, for example its structure (syntax) and meaning (semantics) is of particular importance for HCI, especially for the development of natural language interfaces. From a similar perspective, research in the field of **Artificial Intelligence** is significant for the development of intelligent interfaces, which incorporate knowledge and cognitive structures similar to those developed by humans.

Finally, **Engineering** is a discipline mostly concerned with the development of specific systems, devices and product design. The role of Engineering from a HCI perspective is essential for ensuring that products are developed according to specifications. The role of Design in HCI is also important because it enhances creativity and brings into the field additional usability elements, such as aesthetics.

1.1.2   Areas of research and activity

HCI advances were directed by the rapid technological developments and the progress in each of their contributing disciplines. At the same time, the introduction of new systems has established new modes of interaction, thus creating new needs for HCI research.

For example, HCI was traditionally concerned with the design and evaluation of user interfaces, so that they were of a standard to be used by their intended users. Understanding the user requirements and expectations for the design of usable systems is still the main concern of HCI. However, HCI researchers nowadays have to also consider how sound, hand gestures, touch and speech recognition can all be incorporated into the user interface.

Ubiquitous Computing (ubicomp) is another example of how technological developments influence research activity in HCI. Ubiquitous Computing involves a set of computing devices and infrastructures which can support different tasks in the general environment and covers devices that can be integrated in buildings fabrics or outdoors. This new paradigm creates new types of interactions, which are different to the traditional desktop user interface. Context-awareness, interaction transparency, users' mobility, social acceptance and other user experience elements are different in the context of ubiquitous computing, and therefore demand specific investigation.

The advent of the Internet has been a strong influence on HCI. There is now a plethora of different Web-based interfaces and a need for people to access them for the completion of everyday tasks. E-banking, e-commerce, Web-mapping applications and social networking websites are just a few examples of such Web-based systems. These systems should not only be usable by a specific group of people, but usable and accessible to the majority of Internet users. As a result, accessibility (i.e. designing for people with disabilities), universal design, peoples' privacy and security concerns are all concepts which have entered into the research agenda of HCI as result of the use of the web.

Furthermore, HCI has traditionally been concerned with Computer Supported Collaborative Work (CSCW) and how people use computerized systems for the completion of common tasks (see Chapter 4). However, developments such as the use of Intranet by many organizations and e-learning collaborative interfaces have created new opportunities for HCI research.

User Experience (UXP) is a relatively new but yet very popular and interesting area of research within HCI. UXP, as its name suggests, involves more aspects than simply developing usable interfaces based on Usability Engineering principles. The elements of engagement, satisfaction, aesthetics, emotions and many others are all important for the final UXP that a user will have by interacting with an interface, although usability is still central. This means that a system that is not usable, is less likely to satisfy the end-user. However, designing for emotions and affective interaction involves taking more parameters into consideration, compared to a design where usability is the only concern.

At the same time, there is an increasing need for the taxonomy and evaluation of HCI methods, as well as developing new methods. As a result, HCI researchers have started to consider what techniques are most suitable for the evaluation of specific interfaces, as well as the nature of usability problems that each method reveals. This research activity, which is concerned with the evaluation of HCI methods, is very important, especially from the industrial perspective.

1.2   Geographic Information Systems and science history

While the term Geographic Information Science (GIScience) was suggested only in 1992, the history of the field goes back to the 1950s and the early days of computing. A decade after the first digital computer became operational, geographic application started to emerge – notably in the military, which has a long history of map use.

In 1964, Roger Tomlinson and his colleagues at the Canada Land Inventory had the task of compiling an inventory of the national land resources. While the system was mainly aimed at providing tabular output (the area of the plots and their use), it required the digitization, storage and manipulation of geographic information. Around the same time, the US Bureau of Census developed the tools for the Census of Population of 1970, and as part of it, created a digital map of all streets in the United States as part of the Dual Independent Map Encoding (DIME) programme. Here the system was aimed at producing maps and potentially also producing mapping outputs.

The 1960s was also a period of development of geographic information software in various universities, including work in Harvard University's Laboratory for Computer Graphics and Spatial Analysis, where software called SYMAP and later the ODYSSEY GIS first appeared.

Figure 1.1 shows the computing environment that was used to develop these systems, and Figure 1.2 shows a sample output of SYMAP.

Noteworthy is also the work of Ian McHarg in *Design with Nature* (1969). McHarg advanced ideas about environmentally sensitive planning while using the overlay technique that later became one of the major analysis techniques in GIS. In the early 1970s, a computerized implementation of his methods was developed.

Through the 1970s, commercial applications of GIS started to emerge, with companies such as Environmental Systems Research Institute (ESRI), Intergraph and IBM developing bespoke applications and projects that analyzed geographical information.

By the end of the decade, computer costs dropped and their computational power increased to the level that allowed corporations and central and local government agencies to use them for a range of applications. Thus, the early 1980s saw the emergence of commercial GIS software, with ESRI's ARC/INFO appearing in 1982. The 1980s are also significant because of the advent of Personal Computers (PCs) and the increased use of these affordable computers by more and more users. One of the first products that took advantage of the PC's abilities was Mapinfo, launched in 1986.

Figure 1.1   Computers of the type used to produce early digital maps (Courtesy of Carl Steinitz)

GIS continued to evolve during the 1990s, with other geographic technologies joining in. For example, the GPS system, which was operational in 1981, gave an impetus for the creation of companies such as Garmin (established in 1989) that developed consumer-oriented navigation devices. However, until 2000 these navigation devices had a limited accuracy due to a feature of the GPS signal termed 'selective availability,' which restricted the ability to apply accurate positioning to military applications. On 1 May 2000, the US President, Bill Clinton, announced the removal of selective availability of the GPS signal, and by so doing provided an improved accuracy for simple low cost GPS receivers. In normal conditions this made it possible to acquire the position of the receiver with an accuracy of 6-10m, in contrast to 100m before the 'switch off'. Attempts to develop location-aware devices (in what is known as location-based services) started in the mid 1990s, and were based on information from mobile phone masts or other beacons. However, these methods did not gain much market share due to technical complexity or lack of coverage. The switch off of selective availability changed this and by mid 2001 it was possible to purchase a receiver unit for about $100. These receivers enabled more people than ever before to use information about locations, and led to the creation of products such as PNDs, location aware cameras and mobile phones and many other technologies that are based on location and geographic information. The second part of the 2000s saw a rapid increase in the development and deployment of geospatial technologies.

One of the most active areas of development is the delivery of geographic information over the Internet, especially using the web. From an early start in 1993, the use of the web to deliver geographic information (GI) and maps was burgeoning. However, within this period there has been a step change around 2005 in the number of users and more importantly in the nature of applications that, in their totality, are now termed 'The Geographic World Wide Web' or the GeoWeb. The number of visitors of public web-mapping sites provides an indication of this change. In mid 2005, the market leader in the UK (Multimap) attracted

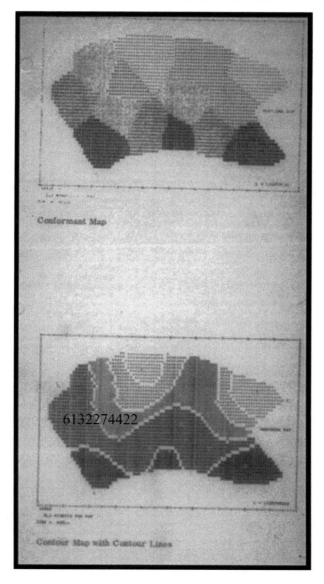

Figure 1.2 Sample of early outputs from a mapping package that Ed Horwood developed (Courtesy of Nick Chrisman)

7.3 million visitors and in the USA, Mapquest was used by 47 million visitors. By the end of 2007, Google Maps was used by 71.5 million and Google Earth by 22.7 million, while Multimap and Mapquest also increased their use. Moreover, by mid 2007 there were over 50,000 new websites that were based on Google Maps, whereas in the previous era of Internet mapping, the number of mapping websites was significantly smaller, due to technical and financial barriers.

An interesting note is that throughout their history, GIScience and geospatial technologies are showing a distinctive lag behind many applications of computers – from the appearance of the first commercial GIS packages in the 1980s – while accounting and database packages appeared in the 1970s, to the rapid growth of web-mapping in the mid 2000s, while e-commerce and many other web applications have been deployed extensively since the mid 1990s. This is due to the complexity of geographic information processing, the need to provide high quality graphic output, as well as the volumes of data that are required in such applications. As we shall see, this has an impact on the attention to, and the development of HCI techniques in GIS.

1.2.1   Contributing disciplines

Despite having been written over a decade ago, the analysis of Traynor and Williams (1995) in their *Why are Geographic Information Systems Hard to Use?* is still accurate. What they noted is that GIS requires its users to have knowledge in multiple domains. GIS builds on knowledge in geography, cartography, statistics, database management and computer programming as core disciplines, with usually an application domain such as logistics or geomorphology where the application is used. The core contributing disciplines are as follows:

Geography contributes to geospatial technologies by providing methods of analysis and ways to consider and solve geographical problems. The use of maps for analysis, or the integration of methods such as overlay analysis, are based on techniques that have been used for many years in geographic research.

Cartography is the discipline that specializes in the production and study of maps and charts. Cartography contributes to GIScience in guidelines to produce maps and techniques such as thematic mapping, where different elements in the map are coloured according to a specific variable (see Chapter 3 for further discussion).

Statistics is significant in GIScience since most of the data are represented as numbers, and many of the queries and analysis rely on statistical techniques. For example, before deciding on the colouring of a thematic map it is advisable to analyze the variable that will be used for the visualization to decide how best to group the variables when displaying the map. There is also a sub-branch of statistics that deals with spatial statistics (or geostatistics) – from understanding the impact of different area units on statistical analysis to development of regression techniques that take location into account.

Databases and data structures are critical to the storage and manipulation of geographic information. First, because the data is voluminous – for example, a modern mapping of building outlines in an area of 1 km sq can contain up to 100 000 pairs of coordinates – and that is before roads and other features are added to the map. Second, databases are required to be designed specifically to be able to deal with spatial queries, spatial indexing and other specific capabilities that are required to manage geographic information. In terms of data structures, GIS require the use of specific geometric data and therefore store the information either as points, lines and area objects (vector data format) or as a grid of values (raster format).

Finally, there is the need to program many GIS in order to perform analysis functionality, since many of them are operating as a toolbox. This means that in many cases the user needs to consider what they are trying to achieve in their analysis task, and then string together a series of actions to achieve the needed outcome. Thus, many GIS include the ability of end-user programming, and the users are expected to be able to use this capability.

Case Study 1.2: What do you mean by 'Field'?

One of the examples of how these disciplines contribute to the complexity of GIS is to consider the different semantic meaning of the term 'field' in the documentation of an agricultural application – it will be included in the discussion on the field (raster) or object (vector) model which are ways of representing the underlying data; to describe fields (columns) in the database that is used in the application; to define fields (areas of data entry) in forms; and to define the attributes of area objects that are captured in the application – the real fields. This is rather obvious to experienced GIS users, but presents major obstacles to non-expert users, since the interface encapsulates a language, world view and concepts that support the system's architecture rather than the user's world view.

1.2.2 Areas of research and activity

The area of research that is associated with GIS and geospatial technologies is GIScience. The term was coined by Mike Goodchild in 1992 and has since received wide acceptance. While this book will focus specifically on HCI aspects, it is worth considering the wider topics that are covered by this discipline. A short definition of GIScience is as the science behind geospatial technologies.

The first aspect of GIScience is linked to the collection of data about objects such as buildings, and activities such as traffic jams, on the face of the Earth. Here, GIScience relies on techniques that come from the field of Geomatic engineering and the use of aerial and satellite imagery for data capture.

The next aspect is representation, since there are many decisions that are being made about the way in which the information is stored and manipulated. The representation can influence aspects such as accuracy, volume of data and the computational effort that is required to manipulate it. Naturally, this is linked to methods of assessing accuracy and uncertainty.

Third, is the development and assessment of a range of analytical methods – from the calculation of the shortest route between different locations, to the advanced geostatistical techniques that were mentioned above.

Fourth, is the need to display geographic information – which can include maps, tables or 3D depictions.

Finally, there are many questions about the impact of geospatial technologies on society and the way they are used by different groups and communities. This is more a concern about the wider implications of using geospatial technologies than the ways in which they operate.

The University Consortium for Geographic Information Science (UCGIS) maintains a GIScience research agenda on their website (www.ucgis.org) and the website is also a good source of information on GIScience research.

1.3 Human-Computer Interaction and GIScience research

The first research publication which deals directly with the HCI aspect of GIS can be traced back to 1963. Malcolm Pivar and his colleagues used a PDP-1 computer at MIT to allow for an interactive analysis of oceanographic data that was collected during the International Geophysical Year of 1958 oceanic research cruises. The vision of an interactive atlas was the motivation behind their experiments:

> [I]n preparing a printed atlas certain irrevocable choices of scale, of map projections, of contour interval, and of type of map (shall we plot temperature at standard depths, or on density surfaces, etc.?) must be made from the vast infinitude of all possible mappings. An atlas-like representation, generated by digital computer and displayed upon a cathode-ray screen, enables the oceanographer to modify these choices at will. Only a high-speed computer has the capacity and speed to follow the quickly shifting demands and questions of a human mind exploring a large field of numbers. The ideal computer-compiled oceanographic atlas will be immediately responsive to any demand of the user, and will provide the precise detailed information requested without any extraneous information. The user will be able to interrogate the display to evoke further information; it will help him track down errors and will offer alternative forms of presentation. Thus, the display on the screen is not a static one; instead, it embodies animation as varying presentations are scanned. In a very real sense, the user 'converses' with the machine about the stored data. (Pivar et al., 1963: 396)

The text goes on to describe a rudimentary system that projected a series of dots onto the screen, with the very basic ability to zoom in and visualize the data. Nevertheless, what Pivar and his colleagues describe above is a vision that would not materialize for average users for three decades!

It should not be surprising that the geographical information was used to demonstrate man-machine interaction (a term that is the precursor of HCI). The information lends itself to graphic displays and interaction for this type of information, as a lot of it is stored and presented in the form of maps or charts, and the graphical representation is a critical aspect of many cartographic and geographic analysis activities. Thus, the maps were required as both input and output of computerized processing, in contrast to the textual and numeral information that was a common input and output of most computer applications at the time.

As noted in the overview of GIScience and GIS, during the 1960s and 1970s research and development focused on the basic technologies. The quality of output from early GIS was basic, or in the words of Ian McHarg in an interview in 1995: 'Absolutely terrible. I mean there wasn't a left-handed, barbarous, mentally deficient technician who couldn't do better than the best computer. Terrifying'.

The need to develop the fundamental data structures and the first analysis techniques led to a focus on the functionality of GIS. In other words, the developers were concentrating on achieving the required task with computers, and were not concerned with the way it should be used by end-users. A correct assumption in that period was that any user of GIS was, by necessity, very familiar with both computers and with low level programming.

Only with the introduction of commercial GIS packages in the 1980s, did the issue of interaction with GIS come to the fore. Yet, for the first half of this decade most users were specialists and had significant GIS and computing knowledge. However, by the end of the decade, with the rapid advent of PCs and the proliferation of UNIX-based engineering workstations, the range of users and applications had increased dramatically and researchers turned their interest to HCI aspects of GIS.

By the late 1980s, cognitive aspects of HCI for GIS were discussed in workshops held at larger conferences about GIS and automated cartography, or in sections of books where HCI issues in GIS were not their primary focus. The early 1990s, however, saw a strong international research interest develop in the topic. Evidence of this can be found in four workshops that were held between 1990 and 1994 in the US and Europe which explicitly discussed HCI aspects in GIS. There were also at least two books – Medyckyj-Scott and Hearnshaw (1993) and Nyerges *et al.* (1995a) – published solely on the topic.

Concerns in the GIS community about increasing processing speed and storage requirements were still the main topics on the research agenda, but they expanded at the beginning of the 1990s to include how GIS were used and how they could accommodate users' needs. This had not been a matter of substantial interest previously, as advances in GIS functionality to satisfy expert user needs were the focus of computer systems' designers and developers and the systems' manufacturers (Medyckyj-Scott and Hearnshaw, 1993). It was realized, however, that GIS were 'more likely to fail on human and organizational grounds [...] than on technical ones' (Medyckyj-Scott, 1992: 106) as the deficiencies of the systems in terms of human factors could compromise their future success.

This early research highlighted the many domains of knowledge that GIS incorporate. Furthermore, it noted that GIS require users to be computer literate and invest enough time to use 'an interface that reflects the system's architecture' (Traynor and Williams, 1997: 288). This is as true today as it was two decades ago, and one of the core issues with GIS is that its users vary in expertise and may use the technology in one of a large number of application areas, as well as demand different functionality and analytical power. Accommodating such a wide spectrum of needs is a challenge in its own right that must take into account a number of factors, such as the components and requirements of the users' work, their capabilities and limitations, the types of support the system can provide, and where it can be provided most effectively.

Several strands of research emerged during this period. This included an interest in usability and design aspects – such as the design of new graphical interfaces to manipulate geographic information; understanding spatial cognition and the way in which humans understand spatial concepts and use them in their daily life; developing task models and understanding task analysis for GIS; use of GIS in collaborative settings where several people use GIS together in a cooperative process of geographic problem solving; and cartographic visualization and the use of computer graphics' ability for novel visualizations of geographic information.

Human-Computer Interaction remained one of the main strands of GIS research. Today, it appears in several areas which are covered in the UCGIS research agenda and in various research groups of international organizations, such as the International Cartographic Association commissions. Cognition of geographic information remains an active area of research with a growing interest in the way different people perceive and use it for various tasks – Chapter 2 is dedicated to this area. The area of visualization of geographic information, now known as geovisualization, deals with the various possible ways to visualize geographic information through the use of computers. This topic is covered in Chapter 3. The interest in collaborative and cooperative applications of GIS has continued and even increased, following the advent of the web, and Chapters 4 and 12 discuss these aspects. With the increased use of mobile devices and the proliferation of applications, GIScience researchers are developing novel interfaces that take into account the size of the device and the context of the user – these devices are covered in Chapter 11. The interest in the general use of GIS and geospatial technologies has gained momentum in recent years and there is a renewed interest in understanding how people utilize geospatial technologies as part of the design process. This aspect is covered throughout the book. Finally, the representation of human semantic understanding of geographic information in the form of computerized ontological representation has received much attention in recent years and it is covered in Chapter 2.

Summary

This chapter has provided an introduction to and an overview of the fields that form the basis for the topics in this book – Human-Computer Interaction and Geographic Information Science. We have seen that both areas are strongly interdisciplinary, that is, they rely on many other areas of science to provide theories, methods and techniques. Both areas are now mature and well established, and therefore for a person who is new to these areas they might seem puzzling and bewildering. At the same time, they offer a very wide range of lessons that can be learned and integrated into a coherent framework according to the needs of a specific project. In the past two decades, special attention has been given to aspects of HCI for GIScience and the lessons from these research projects are the basis for this textbook.

Further reading

As noted in the introduction to this chapter, this book is not intended to be a primer to either HCI or GIS, but to provide information about the intersections between the two areas. To learn more about these two specific areas, the interested reader can use the following sources:

GIScience and GIS

There are now many resources that are available online and offline as an introduction to GIScience and GIS. The most popular textbook which provides an introduction to this area is Longley *et al.*, 2005, *Geographic Information Systems and Science*. The book covers the principles of GIS, its applications, techniques and practical implementation aspects in the operational use of GIS in business.

For a general introduction to GIS, the website http://www.gis.com/, which is maintained by ESRI, is a very useful introduction and provides many examples for GIS use. A more technical explanation of GIS principles is provided on the UK Ordnance Survey website http://www.ordnancesurvey.co.uk/oswebsite/gisfiles/.

Beyond these introductions to the tool itself, there are websites which are dedicated to GIScience. As noted, the UCGIS http://www.ucgis.org/ provides a list of issues that are part of the GIScience research agenda, announcements on conferences and scientific meetings. Further information and links to European activities are available on the Association of Geographic Information Laboratories in Europe (AGILE) site at http://www.agile-online.org/.

There are also many standards that have been set in the area of GIS, and many of these can be accessed through the Open Geospatial Consortium website http://www.opengeospatial.org/ as well as ISO technical committee 211.

Leading academic journal publications dedicated to GIS include the *International Journal of GIScience*, *Transactions in GIScience* and *Computers, Environment and Urban Systems*.

HCI and usability

As a wide and more established area than GIScience, there is a multitude of resources that provide an introduction to the area of HCI and Usability Engineering. Two popular books that provide an extensive introduction to these areas are:

Sharp *et al.*, 2007, *Interaction Design: Beyond Human-Computer Interaction*. The book takes an holistic approach to design, and explains the concept of user-centred design through the different stages of the design process. Another book, by Dix *et al.*, 2004, *Human-Computer Interaction*, takes a more traditional approach to the discussion of the elements of HCI, and covers most topics in a comprehensive manner.

For a general introduction to HCI and Usability Engineering, the Association for Computing Machinery Special Interest Group on HCI (SIGCHI) website provides a lot of information including a suggested curriculum for HCI http://www.sigchi.org/. Another popular site, focusing mostly on web usability, is Jakob Nielsen's http://www.useit.com/. The information on the site provides methods and techniques that can be applied across a wide range of activities and not only Web-based ones.

On the research side of HCI, the online bibliography http://www.hcibib.org/ is useful for finding academic publications in relevant areas. In addition to ACM SIGCHI, the British Computer Society has its own HCI group http://www.bcs-hci.org.uk/.

There are now international standards in the area of HCI, including ISO 20282: Usability of everyday products, ISO/IEC 18021: Information Technology – User interface for mobile tools and many others. Of these, the ISO 9241 is the most significant, covering principles and practical aspects of working with computers. It also covers ergonomic aspects of working with computer screens as well as the design of interfaces and forms.

Leading academic publications in the area of HCI include *ACM SIGCHI Interactions*, *ACM Transactions on Computer-Human Interaction*, *International Journal of Human-Computer Studies* and the proceedings of the CHI conferences which are available on the ACM digital library.

HCI and GIS

There are several books that have been dedicated to HCI and GIS which, although they are from the mid 1990s, still function as useful resources.

The collection of papers edited by Medyckyj-Scott and Hearnshaw, 1993, *Human Factors in Geographical Information Systems* deals with aspects such as the actual work environment of GIS users, and different aspects of designing and implementing GIS. The main advantage of this book is that it includes lessons from human factors and ergonomic studies.

Nyerges *et al.*, 1995a, *Cognitive Aspects of Human-Computer Interaction for Geographic Information Systems*, is based on the outcomes of the NATO workshop that was the first meeting which focused on cognitive aspects of GIS. It provides a lot of grounding in cognitive elements of GIS and many of the observations which are included in this book are still useful and valid.

The Conference on Spatial Information Theory (COSIT) which runs every two years is the premier conference in the area of HCI and GIS, though it mainly focuses on theoretical aspects of spatial information.

Revision questions

1. What are the main milestones in the development of HCI and GIScience as separate fields? Try to set a timeline for the major developments in each area to understand how they evolved.
2. Why did the development of Human-Computer Interaction in GIScience only start in the 1980s?
3. Compare the fields that contribute to HCI and to GIScience. What is the overlap and where do the two fields diverge?
4. Considering the time lag of geospatial technologies and other computer applications, what HCI issues do you predict will emerge in GIScience research in the next decade?

2 Human understanding of space

Clare Davies, Chao (Lili) Li and Jochen Albrecht

2.1 Introduction

By necessity, in this chapter we will take a selective and pragmatic approach to a vast research area: the many aspects of people's understanding of the spaces around and in front of them. This area of research is usually referred to as *spatial cognition*. In particular, since the most common uses of geospatial technologies involve these scales, we will follow researchers as they consider people's cognition of spaces larger than a room but smaller than a country. Given the many interesting and relevant aspects of this topic – see the suggested readings at the end for some longer reviews – we've opted to focus on some areas that appear to have important practical implications for designers, managers and users of geospatial technologies displays and systems.

Noticeably, a lot of research was carried out specifically on GIS, so they will be prominent in the discussion, although many of the issues that we are discussing are relevant to many other geospatial technologies. Some of the implications of research into spatial cognition include the contrast between our understanding (cognition) of the small space of the computer screen, and of the relatively huge space of real-world geography, and when the latter actually matters (or doesn't) in GIS use.

We then move to discuss individual differences among people in their apparent spatial ability, and the problems and pitfalls of psychometric tests and alleged gender differences; or situations where the GIS screen must be matched to the user's actual surroundings, including matching maps to real-world scenes or images, and using wireless mobile GIS within the actual mapped environment.

Other fascinating questions are how people learn and store knowledge of geographic space, how unlike a simple 'mental map' this is, and how our everyday understandings of space and place might affect GIS users' expectations and interaction with the public.

2.2 Spatial cognition: screen versus geography

Obviously, geospatial technologies are intended to represent geographic space on a computer, and so we might expect our skills at understanding, imagining and navigating through such spaces to be important to the way these technologies are used. Our knowledge of those geographic cognitive skills will form a major theme of this chapter.

However, more than one scale of space is involved in geospatial technology use. The external geographic space is being visualized, measured, modelled and reported upon. But as soon as it is visualized on a screen in the form of a map or aerial/satellite image, that visualization in

Interacting with Geospatial Technologies   Mordechai (Muki) Haklay
© 2010 John Wiley & Sons, Ltd

itself is *also* a space. Many of the tasks that users can perform with a GIS, for example, are performed solely with that visualization space, perhaps comparing it to or digitizing it from other visualizations such as paper maps, without having to compare it directly to real-world geographic space at all. Indeed, this ability to simplify geographic problems to the simple manipulation of lines and shapes on a screen, whether 2D or 3D, is arguably part of the power of GIS and other geospatial technologies in the first place.

What happens when a task can be performed solely within the screen space, without having to imagine how the geography looks or works in real life? It may be that human users, in the interest of minimizing effort, will no longer try to visualize Acacia Avenue (or the Grand Canyon, or whatever else is depicted), nor to orientate themselves with respect to it, if all they are trying to do is correct some stored digital mapping details to match the lines on a paper plan.

Indeed, many GIS tasks (particularly in data management and querying) do not require much *geographic* interpretation of the space at all. Examining survey and observation data of professional GIS users in the early 1990s, Davies and Medyckyj-Scott, (1994) and Davies (1995) found that the advanced analysis, visualization and modelling functions of GIS were rarely accessed. On a daily basis the most common tasks of GIS users were the mundane and not especially 'spatial' tasks of maintaining data files, printing out reports and map extracts, and clicking on depicted objects to access database records of their attributes (e.g. customer details).

Therefore, much of the work on geographic-scale spatial cognition, which we will report in this chapter, may have little relevance to users' basic GIS use in some (particularly non-mobile, office-based) GIS usage contexts. This is an important caveat for system designers to bear in mind. Basic knowledge of visual perception and good visual design (as found in general sources on both cartography and human-computer interaction, and in the next chapter) will do the most to improve usability in such situations.

However, GIS usage scenarios are usually not that simple. The inputs to a GIS are often made based on real-world surveys or travel; its outputs are often used in planning or directing real-world operations; checks on the accuracy or completeness of GIS data tend to involve an understanding of 'what's really out there'. Above all, most customized GIS installations are done with at least the *intention* that real-world spatial decision-making and modelling will make use of them, so they should be designed to optimize that potential. Therefore, if we want humans to interpret the spaces we depict on the screen as representing the world 'out there', we have to ask: how do people match those two spaces together?

From the point of view of thinking and language, the two different spaces involved in a GIS imply different spatial **reference frames** (Davies, 2002): different ways of defining or describing spatial relations between things by reference to other objects or directions. Thus we can talk about the 'top' of the screen as well as the 'north' on the map, and we can say that a tree is 'behind' a building in real space but 'below' it on the screen. The physical display of the GIS is part of the immediate space in front of the user, with its own spatial reference frames (e.g. the top or edge of the screen), and the geometry displayed upon it may offer more (e.g. one polygon being 'inside' another). But the map represents a larger and much richer geographic environment that could invoke cardinal directions, fronts and backs of houses, neighbours along (and opposites across) a street or river, the third dimension of terrain and urban multi-storey buildings, and many others.

Taking a cue from earlier research on spatial reference frames in cognition and language (e.g. Carlson-Radvansky and Irwin, 1994), Davies (2002) pointed out that people can switch quite easily (in seconds) between the two spaces and their respective sets of reference frames. However, experiments also showed that someone who has spent time continuously thinking solely in terms of the immediate (screen) geometry (e.g. checking that lines or colours match up to a paper plan, or that polygons are fully closed) will struggle more than people who have been doing more geographically-aware activities with the data (such as checking the relative whereabouts of two named features or streets), if they are then asked to switch to making a geographic interpretation of the map on the screen.

This implies that our knowledge of geographic-scale space can to some extent lie dormant during the more mundane editing or data management tasks with a GIS. Somebody who only does those tasks, all day long, may not be able to remain aware of the geographic implications of the geometric changes they have made.

From the perspective of this chapter, such findings suggest a potential limit of the relevance of geographic spatial cognition to some types of geospatial technologies' use. However, they also imply a potentially important lesson for task analysis and job design. A division of GIS labour that would remove the need for real-world geographic awareness from some users, particularly those who are trying to create or maintain the data, may lead to more editing errors than where the users maintain an awareness of its real-world meaning and the implications of (say) moving or deleting a line. Organizations that either provide or maintain their own GI data are at particular risk of this, wherever the users who perform editing or maintenance tasks are isolated both from data collection such as field surveying and from real-world data use. This is likely to be even more true if data maintenance is outsourced far away from the geographic location in which the data is used in search of cheaper labour.

Meanwhile, we would expect (although this is as yet unproved) that geographic-scale spatial cognition should have far more influence on a user's interaction with a geospatial technology if the user is already familiar with the real-world environment that the map represents, or at least with highly similar environments. Geographic cognition is of course even more critical wherever the GIS visualization is directly compared to actual geographic space by a user immersed within that space, especially in a real-time scenario such as navigation. The following sections discuss our current knowledge of geographic spatial cognition and of people's ability to relate it to map visualizations, especially in a real-time mobile situation.

2.3 Geographic spatial cognition – learning, understanding and recall

2.3.1 The myth of the mental map

In talking of 'spatial cognition', geographers typically subdivide the area into three basic skill areas: the understanding of fundamental spatial relationships, the capacity to visualize these and the ability to orient oneself within the space and hence to navigate through it (Albert and Golledge, 1999; Gilmartin and Patton, 1984; Golledge et al., 1995). These abilities in turn are assumed to build on three types of knowledge of geographic space (Mark, 1993), which in turn are based on a long-standing distinction in cognitive psychology:

1. **declarative** (knowledge of facts such as what is the capital of Kurdistan)
2. **procedural** (how to get from A to B), and
3. **configurational** (awareness of, and ability to visualize, the physical relationships among the objects in the space).

While declarative knowledge is useful for informed decision making, it is the procedural (or route) knowledge and configurational (or survey) knowledge that are often assumed to be essential for learning about a given geographic area, and thence for interpreting its visualization on a GIS by matching it to our existing knowledge. Spatial visualization, whether it is in one's mind or in the form of a (computer) drawing as part of a communication effort, involves the ability to change perspectives and to not become confused when one object changes position (Albert and Golledge, 1999; Gilmartin and Patton, 1984; McGee, 1979). Every cat owner knows how easy it is to confuse a cat by moving just one household item into a new location. To be flexible in one's reaction to spatial changes is a crucial component of human spatial cognition, and a characteristic of our ability to abstract (Velez *et al.*, 2005).

Traditionally, it has been assumed that in order to effectively handle both visualization of and orientation in a space, one would need to develop a 'survey' representation of it within one's mind – in other words, a mental map. Such advanced and complete survey knowledge of a space (Hirtle and Hudson, 1991) has been seen as the highest, and often unreached, level of spatial cognition, as expounded by oft-cited 'stage' theories of spatial learning such as that of Siegel and White (1975). Such theories have proposed that, in the absence of seeing a cartographic map, we can only reach a survey representation through first learning landmarks and routes in the space, and then integrating them together.

It is true that only with complete survey knowledge can we *reliably and accurately* estimate or reproduce all real-world distances and angles, interpret spatial hierarchies of place, identify distributions or patterns across the space, and classify and cluster information into meaningful spatial units such as regions (Golledge *et al.*, 1995). GIS and paper maps usually provide such a geometrically consistent survey representation for us to use. We also know that the hippocampus in the brain is a key area for the storage and recall of routes in the real world, and hence for some kind of mental mapping ability (O'Keefe and Nadel, 1978). However, this does not mean that the knowledge we build in our minds from direct experience of a space, even a highly familiar one, is at all similar to a cartographic map.

Evidence against the 'mental map' idea, and towards a more complex and messy sense of 'naive' geography, has been building since at least the 1980s (e.g. Spencer *et al.*, 1989; Hirtle and Jonides, 1985; Tversky, 1993; Egenhofer and Mark, 1995). Not only do we seem to be able to do many spatial tasks without having built up a complete survey knowledge of a new space (such as in typical psychology experiments in this area), but also even highly familiar spaces seem to be oddly distorted in our minds, and memory for the spaces in between decision points and significant landmarks is very poor.

To some extent, we could expect that we might simplify our memory of space to form an efficient mental representation – after all, this is what cartographers do to make a readable map (Peterson, 1987). However, our mental knowledge in many ways seems to resemble a topological map (such as the famous London Underground one) more than a topographic one, with distance and direction only roughly and inconsistently judged, most detail stripped out even at familiar intersections, and minimal recall of the duller stretches of even well-travelled routes.

Having established what our geographic knowledge is not, researchers have been far less successful in identifying what it is. The evidence has sometimes been simplified to fit with an alternative theory such as non-Euclidean geometries, or systematic distortions predictable from 'hierarchies' of spatial knowledge (e.g. Hirtle and Jonides, 1985), but the reality seems to be messier (Montello, 1992). We seem to hold a hotchpotch of information in our heads about a given space – scenes, routes, signposts, views, half-remembered map fragments and oversimplified assumptions of straight lines and right angles – summed up by Tversky (1993) as a 'cognitive collage'.

This cognitive collage concept is attractive: it appears to match our intuitive reality when we introspect our own fragmentary knowledge of (say) a familiar or half-familiar town or landscape. The problem is that such a loose concept has no predictive power to help us model what people will learn, or how they will behave when trying to recall, navigate or interact with a representation of the space – such as a GIS.

However, careful studies such as those of Tobias Meilinger (2008) have more recently suggested refining the 'collage' concept to consider it more as a 'network of reference frames'. We store key (but not all) visual information from important scenes, such as those at junctions where we have to turn; such a single visible scene was labelled a 'vista space' by Montello (1993). We tend to link these scenes together into a network by storing, for the route segments and perspective shifts that lie in between them, some (at first very basic and approximate) information on distance, direction and limited visual cues. With repeated visits to a space, the information we have on the links between key vista spaces (scenes) is improved, either through experience or salience to us.

Having gained this knowledge, we can plan a route through the space by recalling a series of these scenes, links and perspective shifts. Like many current cognitive theories that draw on our knowledge of the mechanisms of the brain itself, Meilinger proposes that 'spreading activation' (which loosely means heightened accessibility of one memory because it is linked to another that's already been accessed) makes us aware of potential next steps in a route whenever we think of a particular point within it, and so we gradually find a series of activated steps that will take us the whole way from point A to point B. Almost every person will have an experience of getting lost in an unfamiliar space, and the process of finding the right way usually involves tracing one's steps back until the surroundings look familiar again, at which point the memory of the next node in the chain may be recalled.

Meanwhile, for other problems where a survey representation would help, we may or may not know or care enough to recall and arrange our network of frames into a workable 'map', perhaps (although Meilinger did not test this) using the framework of a known shape, outline map or remembered view.

Some of the (as yet untested) predictions of theories like this for GIS use are tentatively suggested below. These are particularly important for situations where lay public participation is sought in a GIS project, such as in local planning, but they can also have implications for any GIS user holding independent knowledge of an area and expecting to reconcile it with the data on display:

- Not every item or route segment shown on the screen, projected on a wall or presented in a printout from a system will necessarily be familiar or recognizable to users, even if they are highly familiar with the represented space. The process of understanding the computerized abstraction and linking it to a personal mental map can be challenging. Thus, in a situation where the display needs to be zoomed in to a large-scale mapping for accurate editing, local knowledge could be difficult to recall and apply, except at key familiar points.
- For a 'local expert' user who has primarily learned the space through direct navigation and experience, those key familiar points will mainly consist of salient landmarks and road/path intersections, not the road sections or areas that lie between them. Some of

these landmarks can be very mundane – the local post box, a specific building or even a sign.

- Such a local expert user may also find it hard to match their internal (simplified, geometrically distorted and semi-topological) knowledge to the geometric layout of a map display, especially if they had limited experience in map reading. For example, they may pan the screen in the wrong direction, and they may need to start by seeing and adjusting to a small-scale overview (which is also more helpful to a user who is completely new to the space, to give them the 'lie of the land').
- Similarly, the GIS's accurate distance and direction calculations may also take some local expert users by surprise (or they may not bother to use the GIS to make them), given the known distortions in people's estimates of these. For some people, it is more important to know what is the internal distance to a location and not the metric one. For example, if you live between two major transport nodes such as underground stations, and you mostly use one of them on your daily commute, the other station will seem further away even if the metric distance might be similar.
- Local expert users may find it useful to rotate the displayed map, rather than keeping it fixed in a north-south direction. However, such rotations could confuse and disorientate a user who was unfamiliar with the space, *if* they needed only to view and find their way around it on the screen rather than applying their knowledge through real-world orientation or navigation (see also later sections of this chapter).
- Aerial views of the space may not do much to help even local expert users to orientate themselves, since this viewpoint is usually unfamiliar. However, linking ground-level images of key vistas to points such as road intersections on the map (perhaps to pop up when the mouse is hovered or clicked) should do more to help them match the display to their existing knowledge. Yet, research has shown that aerial views are more accessible to lay users as a first step to map reading.

2.3.2   Place and 'vernacular' geography

As we saw above, the way that we think about physical space is often not how the GIS sees it – distances and directions are distorted and inconsistent, and we hold a mixture of representations in our minds about it. Another way in which conventional mapping data and GIS concepts fail to match human knowledge is in the concept of place.

Administrative definitions of and attitudes towards place can vary between countries – in the US there are clear single definitions of states, counties, cities and sometimes even neighbourhoods, but this is not so in the UK or elsewhere in Europe. Meanwhile, in any area of the world, locals' place names and place concepts can easily differ from official ones, and vary among the local people themselves. Place boundaries can be disputed or simply vaguely defined: think about where you might decide that your home neighbourhood ends, or a region of natural beauty that you have visited. Locally used names can be multiple for the same area (whether an urban park or a whole region), or the same name can refer to more than one such area in different people's minds.

Naturally, most conventional mapping data does not show all of these multiple names and definitions – indeed, often it is too hard even to collect them. Therefore local users of a GIS may find the presence or absence of name placement or boundaries confusing when viewing data about known places. On the other hand, users unfamiliar with those places may assume too much validity in the displayed names and placements.

This disparity between local 'vernacular' geography (the geography described verbally in everyday language) and the 'professional' geography viewed on a GIS screen can be particularly problematic when GIS users have to communicate with the general public (see case study). Some organizations such as emergency call centres go so far as to develop their own gazetteers of local vernacular place names and place extents, gathered from telephone conversations with local people.

Case Study 2.1: Everyday place geography and emergency services GIS

In May 2007 a media controversy began in England after a boy tragically drowned in a lake in the town of Wigan, apparently without receiving physical help from two nearby auxiliary police officers (who were later exonerated). In the ensuing row over the role of the emergency services in the accident, another issue emerged: the clash between 'official' GI data and everyday geographic knowledge. Assistant Chief Constable Dave Thompson was later quoted by the BBC (BBC, 2007) explaining what had delayed the arrival of help that day:

The initial call to police gave the wrong location. This was no-one's fault, as the lake is known by several different names locally and there are other similar lakes nearby. The Police Community Support Officers (PCSOs – who are support staff without the full powers of police officers) managed to establish the correct location and immediately informed the control room to ensure the emergency services were sent there. One PCSO cycled to the road to alert other emergency services as they headed to the scene, while the other remained at the lake.

This type of problem is commonplace not only in the UK, but in other countries (for example, Pettersson *et al.*, 2004). In everyday 'vernacular' geography a place or geographic feature may have multiple names or means of description; the same name may apply to multiple places; the geographical extent implied by a name may also be unclear (e.g. smaller ponds attached to a main lake, and the vague or disputed extent of an urban neighbourhood). Emergency services waste precious seconds solving these problems, which are not solved by the increasing presence of GPS in mobile devices (since the caller is not always at the incident scene). For professional GIS users to capture these vernacular geographies as expressed by the public, flexible gazetteers and data modelling methods may be needed.

Meanwhile, in the GI science research community, various methods have been proposed for modelling vague and multiple extents within GI data, and these may be worth considering in professional GIS usage scenarios where local place concepts are important to users' tasks (see recent review by Davies *et al.*, 2009). However, such methods have not yet been evaluated with real end-users in such situations, so caution is recommended in their introduction. In particular, any untrained users who struggle with the simplified spatial concepts of conventional GIS might find more sophisticated data models even more confusing. The ability of users to cope with the spatial nature of GI data may, of course, vary among individuals in the first place. The next section will discuss what we know about such differences among human GIS users.

2.3.3   Spatial ability

For all their failures to adequately represent human geographic knowledge GIS are at least completely predictable in their abilities and actions; every manager knows that human users are not. Might some people, and hence some GIS users, cope better with spatial reasoning, navigation or map reading than others? Anecdotally, many of us have heard people comment that they (or someone they know) have 'no sense of direction', are 'hopeless with maps', or 'can't judge distances'. But can we measure or predict spatial abilities reliably, and do they matter for geospatial technologies use?

Unfortunately, after decades of research we still don't have a clear idea of what we actually mean by spatial ability, or which basic cognitive skills it relies on (Lohman *et al.*, 1987; Carroll, 1993). The best we can do is tentatively guess that where people's scores on one so-called ability 'test' correlate well with performance on some other task (whether in an experiment or in real life), the two tasks both measure at least one common cognitive factor. This is not the same as claiming that people's task performance depends on, or is explained by, the test – and it may not even be very strongly predicted by it.

While psychologists continue to debate the nature of 'fundamental' abilities, as far as geographers are concerned spatial ability focuses on the practical skills required to 'do' geography (Golledge, 1992: 5-6):

> 'Spatial abilities include:
> the ability to think geometrically;
> the ability to image complex spatial relations at various scales, from national urban systems to interior room designs or tabletop layouts;
> the ability to recognize spatial patterns in distributions of functions, places and interactions at a variety of difference scales;
> the ability to interpret macrospatial relations such as star patterns;
> the ability to give and comprehend directional and distance estimates as required by navigation, or the path integration and short-cutting procedures used in wayfinding;
> the ability to understand network structures used in planning, design and engineering;
> and the ability to identify key characteristics of location and association of phenomena in space.'

Obviously, since Golledge was talking as a geographer rather than a systems designer, we have to adapt this list with respect to geospatial technology use. The last item above seems to reiterate the first three, but thinking a little further it can help us to understand how these abilities translate to the technological sphere:

- 'The ability to think geometrically' would include understanding and navigating around a visual display such as a map, to calculate distances and use basic geometric 'rules' (e.g. knowing that three points in space form either a straight line or a triangle).
- 'The ability to image complex spatial relations at various scales': as suggested earlier, we may not need to form mental images of a 'real' geographic space in order to perform many tasks with a map of it, but we may need to at least grasp complex relations *within* that map (e.g. notions of neighbourhood and front/back/opposite relations, symbols showing the third dimension such as contours, etc.).
- 'The ability to recognize spatial patterns' – obviously important with maps showing distributions and patterns (e.g. of disease cases or rare plants), but also important in tasks such as picking out roads from waterways, towns from rural areas, or identifying suburbs sharing similar building styles.

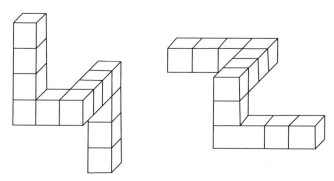

Figure 2.1 A typical mental rotation test item (Source: Jennifer Oneske)

- 'The ability to give and comprehend directional and distance estimates as required by navigation, or the path integration and short-cutting procedures used in wayfinding' – on one level, these could be seen as relevant only to the use of maps explicitly for navigation or wayfinding. On another level, being able to trace a path through a map may be important in route planning or network maintenance. In the case of a digital map there is also another 'navigation' issue: finding your way around a map which extends way beyond the screen boundaries. For example, noting visible 'landmarks' within the map display (such as a particular group of shapes) may help you return to an area you had seen earlier.
- Finally, 'the ability to understand network structures used in planning, design and engineering' is crucial to some GIS applications such as the utilities and highway maintenance, but possibly irrelevant to most others.

In contrast, most spatial ability tests in use in psychometrics have focused entirely on either mental rotation, visualization from a different perspective or spotting specific patterns:

- Mental rotation: for example, the famous 'blocks' test (Shepard and Metzler, 1971); also simpler tests which only depend on 2D rotation such as the cards, flags and figures tests (Thurstone, 1938) – see Figure 2.1.
- Visualizing how things might look from another (non-rotated) perspective or after something is done to them, e.g. after moving in space, or after paper is unfolded; e.g. foam board and paper folding tests (Ekstrom *et al.*, 1976).
- Spotting specific shapes or patterns among larger, more complex figures, e.g. variations on the popular embedded figures test (Witkin *et al.*, 1971).

Note that none of these descriptions seems close to Golledge's broad notions of spatial ability, although all might be relevant to certain tasks within map and GIS use. Of the three, the first is important for identifying a map object such as a building regardless of orientation; the second would be useful in visualizing planned changes to an object or area, and would also be helpful in interpreting contours and other symbols; the third is obviously useful in picking out desired features from a cluttered display and for identifying patterns.

However, most tests fail to cover the ability to notice novel patterns or objects, as opposed to picking out patterns from a different perspective, and they also have little to say about changing reference frame (interpreting the visual display in more than one different, structured, way). There are no well-known standardized tests on describing or identifying routes or network structures, nor on visualizing scenes from maps. Small wonder, then, that spatial ability tests tend to show fairly poor correlations with people's spatial performance in real environments, either with or without maps (e.g. Hegarty *et al.*, 2002). When such studies do manage to find some correlation, they are often surprising: sometimes verbal ability tests seem just as relevant as spatial ones (e.g. Sholl and Egeth, 1982), or else researchers find that a little training and/or the use of more verbal learning strategies can offset the effect of supposedly 'innate' abilities (e.g. Schofield and Kirby, 1994).

The more recently developed Santa Barbara Sense of Direction Scale (SBSOD – Hegarty *et al.*, 2002) does appear to correlate well with wayfinding and orientation tasks in real and virtual environments. However, having been developed and tested on a West Coast population in the US, where grid-layout cities and awareness of cardinal directions (North, South, East, West) are the norm, its cross-cultural stability needs confirming via some ongoing European studies. At the same time, it is also unclear whether such ability on orientation or wayfinding tasks in the outdoor environment is related in turn to the spatial ability needed to manipulate objects in small-scale space such as a GIS screen. Indeed, the same research group that developed the SBSOD has shown evidence that they are separable cognitive skills (Kozhevnikov and Hegarty, 2001).

In other words, we could argue that real-world spatial tasks *include*, as subtasks, the same basic abilities as are tested by the standard spatial ability tests, but since they also include other subtasks as well they may not be usefully predicted by any single-ability measure.

Furthermore, as experts in psychometrics have long pointed out (e.g. Lohman *et al.*, 1987; Carroll, 1993), people can and do use different strategies in solving the same problems whenever they involve more than one single step, even within the artificially simplified tasks of aptitude tests. This makes it difficult to interpret them as pure measures of any single ability anyway, no matter how carefully they are designed. The role of individual strategy variations is even truer for more complex and rich real-world tasks such as those in GIS use.

2.3.4   Gender: women and maps

As well as psychometric tests, people are often classified – and even stereotyped – by more obvious characteristics such as age, gender, ethnicity and education. When it comes to potential differences in spatial cognitive ability, by far the most frequently discussed of these characteristics is gender. Popular psychology literature is littered with books and articles claiming to explain or prove that women, as such authors like to put it, 'can't read maps' (Figure 2.2 clearly shows otherwise).

Since we have already seen that map use is actually complex and not completely dependent on any single cognitive ability, whether innate or gender-specific or otherwise, it should be clear that no such sweeping statements are likely to be true. But what do we know of gender differences in this area?

The literature on gender differences in the use of GIS (as opposed to maps) is sparse. Expanded to the topic of general spatial cognition and map reading skills, however, there has been an enormous effort to settle the issue of gender differences, regardless of whether they are biological or socio-cultural in origin. The latter, of course, being a likely factor in many situations, given the differently socialized early-life experiences of many but not all boys and girls in western society – see Self *et al.*, 1992; Lawton and Kallai, 2002. Yet it appears that both sexes are roughly equal in some tasks, such as developing route and survey

Figure 2.2   Relating the map to space (Source: sanja gjenero, Stock xchng)

knowledge of a specific area of space (Montello *et al.*, 1999). For example, Gilmartin and Patton (1984) found that with regard to map use skills such as routing, symbol identification, visual search, the assessment of spatial properties and right/left orientation, male and female students performed equally well. In a study aimed at determining how well students learn routes, Golledge *et al.* (1995) found that women's assessment of distances and (at least in the case of female geography students) direction were *more* accurate.

One of the most comprehensive studies in this area was conducted by Montello *et al.* (1999), who examined the spatial skills of 79 citizens of the whole adult age range. The participants had to solve a variety of spatial tasks, including psychometric tests, route learning, and memorizing the location of objects with the goal of understanding their spatial relationship from memory. In general men dealt better with mental rotation (as in many previous studies), while women fared better recalling the exact location of objects. There were no significant differences between the genders when learning situations from maps.

In other studies, the appearance or otherwise of gender differences seemed to depend partly on the instructions they were given and the socio-cultural expectations placed upon them (Sharps *et al.*, 1993; Kitchin, 1996). The effects of even subtle changes in such instructions and expectations should not be underestimated. In one educational study, just calling a display a 'map' instead of a 'diagram' seemed to increase the amount that students learned from it (Kealy and Webb, 1995). Caplan and Lipman (1995) found that older women who learned a route using a sketch map were, under certain conditions, significantly worse than younger women at recalling it, although that age difference was not found for men. This could imply that gender differences are stronger among older people. But the same study showed that the older women's performance was actually *just as good* as younger women's, *if* the sketch map was *not* referred to by the experimenters as a 'map' but as a 'diagram'. It is possible that older women may have become intimidated by the *thought* of having to interpret a map, even though they could do it reasonably well.

Similarly, in studies cited by Brosnan (1998), the same aptitude test was administered and described either as a 'perceptual ability' test or as a test of 'empathy'. The test was the Embedded Figures Test, mentioned earlier. Gender differences vanished or even reversed in the 'empathy' condition, and mainly due to changes in women's performance: the women seemed to be performing according to stereotyped expectations rather than stable cognitive traits. Meanwhile Kitchin (1996) studied gender differences in everyday geographical knowledge, and found no important differences. He argued that socio-cultural subtleties in some earlier studies had served to reinforce social stereotypes, which could be especially true if participants were aware that gender was relevant to the goals of the study.

This reminds us that for many laypeople there is a social and emotive aspect to maps, which may be stronger particularly for older women from conventionally gender-stereotyped backgrounds. It has also been noted that women in some studies express greater anxiety than men about navigation and related spatial tasks (Lawton, 1994). Maps may thus be subtly associated for many people with moments of pressure and stress, and with problems more than with solutions (Muehrcke, 1978). This may be worth bearing in mind in geospatial technologies that attempt to engage the public at large. However, note that such studies and remarks were based on the general population: we would expect professionally trained GIS specialists to have no such emotional concerns with mapping, and thus to show few or no gender differences.

Some of the persistent popular belief in female underperformance is based on fact. Across the literature in this area (which is disproportionately vast for such a small effect, due to its endlessly controversial nature), gender differences are not found just as often as they are found (e.g. Gilmartin and Patton, 1994; Vecchi and Girelli, 1998). But when they do appear, it is almost always in the form of women scoring slightly less (not more) than men on certain types of spatial ability (particularly mental rotation), and sometimes also on more complex real-world tasks which draw heavily on the same skills. The actual impact of such differences, however, is hard to assess given that men and women also often choose different strategies to solve wayfinding tasks in the first place (e.g. Lawton and Kallai, 2002).

The generally small size of such differences must also be borne in mind. Equal or larger effects on spatial strategy and performance have also been found or suggested for task expertise (McGuinness, 1994), familiarity with maps (Gilhooly et al., 1988), cultural and even regional (in the US) differences (Lawton, 2001), and many other factors that will always vary among any group of people. More importantly, there is, even in the studies that find a difference, a huge overlap between the two 'bell curves' of spatial ability for men and for women, implying that almost as many men as women will perform worse than any chosen level, and almost as many women as men will perform better. Overall the literature shows that single stereotyping factors such as gender cannot predict any individual's competence at a geospatial technology task.

2.4   GIS in the outside environment: matching maps to geography

2.4.1   Orientation and location with maps

As we summarized earlier, people's knowledge of large real-world spaces, such as towns, tends to include more detailed awareness of particular scenes – nodes in the network, often including key intersections as well as important landmarks – and less of the route segments connecting them together. It follows that for any map or GIS user who has any prior familiarity with the depicted space, matching the map image to that space in the outside world will be easier if the map gives good information at those key points. Matching known or visible landmarks to the map, when trying to orientate oneself (or to match up video or camera images), or to find one's way, is an obvious strategy for reconciling the map to what the users see around them.

The importance of such nodes and landmarks to people's conceptions of urban spaces has been suggested many times since Lynch emphasized them in his classic study (Lynch, 1960). Studies of people performing orienting or wayfinding tasks have also confirmed their importance (Pick *et al.*, 1995; Burnett, 2000; Gunzelmann *et al.*, 2004; Peebles *et al.*, 2007). In addition, maps that are aligned to the direction the user is facing appear to help users to orientate themselves more quickly and accurately (Levine, 1982; Gunzelmann *et al.*, 2004). However, it may be the case that a frequently rotated map makes it harder for users to gain an overview of the local spatial layout, since they do not see a consistent orientation-independent viewpoint.

Beyond the need for real-world landmark information to be present in the map, and for appropriate alignment of it, studies of people locating and orienting themselves with maps have shown that although people's strategies and success rates can vary greatly in solving the task (Peebles *et al.*, 2007, Liben *et al.*, 2008), the characteristics of the user and the visual scene appear to affect people's performance more than the characteristics, shape or orientation of the map itself (assuming the presence of reasonable cartographic design, see Chapter 3).

It is also apparently not possible to predict where orientation will be harder or easier, solely from analyzing the 2D characteristics of the space from the map (Peebles *et al.*, 2007). Rather, performance seems to vary with the distinctiveness of the visual landmarks within the scene that the person is viewing in the real world; a distinctive and landmark-rich environment will be more easily matched than a relatively monotonous one, provided that the most obvious 3D landmarks are unambiguous on the map.

If it is known in advance that a GIS display will often be matched by users to scenes from relatively feature-poor environments, system designers could help users overcome this by including more detailed large-scale mapping within its data, particularly with respect to any available and easily visible 'local landmark' information, such as pub names in the UK, and optimizing the ability to zoom and pan its display.

2.4.2   Challenges for mobile applications and ubiquitous GIS

For almost two decades, a range of information and communication technologies has rapidly developed: the Internet, mobile and wireless handheld communication devices, and location-aware technologies. More importantly than their development, the usage of these technologies in people's daily business and social lives has also rapidly expanded.

This phenomenon has also had its impact on GIS, as Internet maps and wireless GIS are among its growth areas for applications and research. Internet GIS includes services where GIS functionalities are available to users via a browser interface to the Web (Pandey *et al.*, 2000 and see Chapter 12). Currently, the Internet is also being increasingly accessed by users on the move using mobile phones, Smartphones and wireless-enabled PDAs. Therefore, Internet GIS holds the potential for far greater mobility than traditional desk-based systems.

Wireless GIS, resulting from the convergence of wireless mobile technologies and GIS, provide a real-time capability of interacting with external software and accessing and managing data remotely through users' mobile devices (Braun, 2003). Wireless GIS extend the applicability of GIS in mobile situations through its usability, portability and flexibility (Drummond *et al.*, 2007).

Furthermore, GIS are also moving towards more ubiquitous use through applications such as Location-Based Services (LBS). LBS centre on the delivery of data and information services, where their content is tailored to the current or some projected location and context of a mobile user (Brimicombe and Li, 2009). LBS can be viewed, from a technological perspective, as the convergence of GIS, spatial databases, and the Internet.

The above trends in GIS technology have put more emphasis on interaction, usability, mobility and portability. They offer the potential for users to employ GIS in more mobile and diverse situations and for an ever expanding range of applications, in the real geographic environment and in real time. Such developments provide opportunities to enhance and change the ways people access and utilize geographic information, and can lead to 'ubiquitous GIS' through mobile services which focus on the individual in their mobile context (Li and Maguire, 2003). Understanding how users interact with GIS in mobile situations in real-time raises new research questions, particularly regarding the effect on users' learning and knowledge of the spaces they move within.

As we stated earlier, spatial knowledge can be thought of as a mixture of the declarative (knowing facts), procedural (knowing what to do) and configurational (broadly, knowing where things are in relation to one another). Such spatial knowledge can be acquired and developed through various experiences and processes, which may include recognizing and understanding the characteristics of objects and localities, noting the spatial relationships among elements in the environment and internally structuring the knowledge within one's mind. Furthermore, people can acquire and develop their spatial knowledge either through their 'direct' experience, that is the experience gained through activities in a real environment via active learning, or 'indirect' experience, that is experience gained through simplified and symbolized representations such as maps, or both.

There are obvious implications here for how users might access, assimilate and use geographical information through mobile devices. For instance, the device is an 'indirect' source but, unlike the study of a map or GIS screen in an office, it can be immediately matched and verified against the directly experienced space itself. We might expect better integration of people's knowledge in such situations, but only if they have time and mental resources for that to take place. In situations where the primary task is driving a car, for instance, awareness of the surrounding environment is often so limited that the typical Portable Navigation Device (PND) user may not be able to integrate their experiences into a useful configurational whole. Even where time and cognitive load allow greater awareness, the small space of the mobile device screen may limit a user's ability to build a sense of the overall local configuration; their understanding of the space may remain more fragmented than with a traditional map.

Meanwhile, although the advent of built-in digital compasses and GPS tend to remove the problem of self-location and self-orientation discussed earlier, users do still need to match what they see on the screen to their surroundings. Decisions must therefore be taken by system designers as to whether the on-screen map should, or not, move and rotate to match the user's position and facing direction. Such decisions would always need to be based on the context and the task, depending on the extent to which the users' overall goals include building configurational local knowledge, or merely coping with the immediate situation at each location where they find themselves. In the broader design of mobile applications it is also important to consider the dynamic spatial and temporal context, in terms of users and their immediate goals, the device(s) employed and the surrounding environment (Li and Willis, 2006). Therefore the methods described in later chapters, for capturing the contexts and nature of users' tasks, are especially crucial in real-time and mobile scenarios.

The developments of wireless GIS and LBS, discussed above, introduce many new challenges requiring further research into spatial cognition. Such a research agenda has been developed by Brimicombe and Li (2009) in the context of LBS. Spatial cognition is central to gaining insights into how users of ubiquitous geospatial technologies are going to be able to interact with them. There will need to be a better understanding of the reasoning behind the initiation of spatial queries (requests for information), spatial memory and learning, and working with naive and vernacular geographies and natural language. There will need to be more research into human-computer interaction with mobile devices as a means of accessing geographical information, and into user preferences for information delivery.

The means by which people acquire geographical information is inevitably changed when mediated by technologies – mobile devices allow geographical information to be acquired in real time, at any location, but their small size and/or use of verbal output may often limit the amount of spatial information displayed or communicated at any one time. Indeed, the communication mode of geographical information between service providers and users can range from spoken language, written text, gesture, different types of map, to three-dimensional representation, virtual reality and multimedia. Furthermore, different user groups have been shown to have specific preferences for information requirements in real-time scenarios (Li, 2006).

Thus, further research challenges arise in terms of real-time communication of geographical information. These include the semantics of communication between systems and individuals, that is the words and terms that are used and their meaning; understanding the cognitive processes and effectiveness of spatial knowledge acquisition for different modes of communication; the different levels of cognitive load associated with different modes of communication and the scenarios in which they occur; and the representation(s) of geographical information that are most likely to be relevant in given situations.

From a research perspective, a shift from generic to user-centred mapping has been recognized for some time as an important challenge for research and design (Longley et al., 2001). Although there has been some research into user-centred methods for improving information content and cartographic representation for mobile devices (e.g. Reichenbacher, 2001; Nivala, 2005; Chittaro, 2006) and specifically on user needs and preferences for LBS (e.g. May et al., 2007), there has as yet been insufficient systematic research regarding individuals' cognitive abilities, spatial awareness and other related skills when accessing and using GIS applications in mobile situations. Comprehensive design guidelines are thus still unavailable.

Summary

The area of spatial cognition and human understanding of space is the most significant theoretical grounding for geospatial technologies. In this chapter, we have covered several core concepts. First, we have noticed that there are at least two different scales of space and potential ways of talking about the spatial data. These two scales are involved in any interaction with mapping on a GIS screen, but not all GIS use situations strongly to invoke *geographic* spatial cognition. Matching those two spaces together, especially for 'local expert' users with awareness of the actual spatial environment that is represented on the screen, requires some understanding of how we think about large-scale real-world space.

We have also seen that this different scale of thinking about geography is not at all similar to a topographic 'mental map', but is often more like a topological 'network' map of well-visualized nodes and landmarks, linked together by much more sparse and distorted route and survey knowledge. Some suggested implications of this for GIS systems design are listed earlier in this chapter.

People also think more flexibly and vaguely about concepts of 'place' than is reflected in conventional digital mapping data. This can have major consequences for GIS users, especially when using these technologies to communicate with lay public.

Furthermore, although people differ widely in some specific spatial abilities, and some of those have been linked to gender, their performance on most real-world spatial tasks cannot be reliably predicted from tests of those specific skills nor from gender stereotypes, since every task involves multiple steps and skills. The 'node-and-landmark' tendency seen in people's real-world spatial orientation, wayfinding and learning means it is crucial to include landmarks and other scene-distinguishing information on maps that need to be matched to real-world scenes (or to images of them).

Finally, we discussed how research and design for mobile and wireless GIS are still in their infancy, but already task context and goals seem to be crucial considerations in designing or selecting appropriate displays and interaction techniques for their users.

Further reading

Kitchin Freundschuh (2000) *Cognitive Mapping: Past, Present and Future*.
Despite being a collection of multi-authored essays, this book is one of the best primers on geographic spatial cognition. It covers a number of the topics mentioned above in more detail including the distortions of cognitive 'maps', and also gender differences.

Meilinger (2008) The network of reference frames theory: a synthesis of graphs and cognitive maps. The most recently tested theory of local-scale spatial learning and cognition, at the time of writing this chapter, was summarized in this article.

Nyerges *et al.* (1995a) *Cognitive Aspects of Human-Computer Interaction for Geographic Information Systems*. Much of the work and thought relating these spatial cognition topics to GIS use dates back to the early 1990s, and many useful insights can be found in this book.

Cresswell (2004) *Place: A Short Introduction*. A good review of the concept of place from the perspective of human geography.

Davies *et al.* (2009) User needs and implications for modelling vague named places. A recent GIScience-oriented review paper relating issues of vernacular place geography to the needs of real-world GIS users.

Brimicombe and Li (2009) *Location-Based Services and Geo-Information Engineering*. A recent book on mobile GIS and LBS.

Revision questions

1. What are the three types of spatial knowledge, according to David Mark?
2. What are the core considerations for the development of a geospatial technology that will be used by lay users?
3. How would you test the theories that are suggested at the end of section 2.3.1?
4. You are tasked with developing maps that will be put in specific places to help visitors navigate through a university or corporate campus. Based on the principles from this chapter, how would you design such maps?

3 Cartographic theory and principles

Catherine (Kate) Emma Jones

One of the major changes over the past decade is that it seems almost impossible to spend a day without looking at a map of one form or another. We see them used to present daily weather forecasts, help us navigate from A to B with paper maps or Portable Navigation Devices (PNDs), explore holiday destinations using 3D globes, or in news reports locating the latest incident or event. Government reports widely use maps for communicating and providing evidence for policy decision making. Significantly, the wide availability of computer generated maps fuelled this change.

The types of maps produced and their form of representation differs widely depending on why they were created, so finding the most useful map design for a specific purpose can be a challenge because not all maps are useful or usable; poorly designed maps will act as a source of frustration to readers and users. If a map is not created with due care, diligence and attention the results will simply leave the reader unconvinced, confused and questioning how to understand them. In some situations, misinterpretation of maps can lead to costly, and sometimes life threatening actions.

So, whilst maps are common, the process of creating a useful and usable map is not as simple as it may appear to be. To create a well designed interactive or paper map, a lot of initial thought is required. This is achievable if the designer or application developer follows some basic principles and steps following cartographic best practice. Cartography is the art and science of map making, and is based on knowledge that has evolved over many years. In this chapter we will provide a guide to cartographic representation for the general principles of cartography and the theoretical considerations required to make a good map. Later in the book, in Chapter 8, we focus on the practicalities of map design.

3.1 Principles of cartographic representation

3.1.1 What is cartography?

With the rise of computer generated mapping it has been mooted that cartography is a dying practice. As advances in technology brings mapping to the masses 'anyone' can create a mash-up without awareness of how the science and art of cartography can aid spatial understanding. However, although maps are now easier to produce, cartographic best practices are still required. A well composed map can be a very effective form of communication to improve knowledge construction, and cartographic principles support such processes.

We are living in a world bombarded by facts, figures, charts, tables and maps, all trying to communicate information to help us acquire specific knowledge on a subject. This

Interacting with Geospatial Technologies Mordechai (Muki) Haklay
© 2010 John Wiley & Sons, Ltd

bombardment of our cognitive senses means that it is very easy to become overwhelmed and suffer information overload. As map makers and developers of geospatial technologies it is important to remember that if the map is poorly designed or the technology which we use to interact with the map is cumbersome, clunky and/or incoherent, the users may briefly take a look at the map but are unlikely to take the time to seek out the message(s) or will not try to unlock the potential knowledge and insights that could be gained from it.

Maps are simply visual representations of a geographical space and the relations of events and objects depicted in that space. Cartographic representations are developed to encode geographic relationships between real world objects which enable the map reader to build up geographic understanding (MacEachren and Taylor, 1994). They help us to visually represent the complexity of the real world through an approximation of geometric objects such as lines, points and areas. Thus, a map represents the elements of the geographical space with point and line objects (location points marking tube station or car park and lines representing railway tracks, roads, or rivers) alongside area objects (country, administrative boundaries or buildings and parks) and the relations between them. It is the cartographer's job to decide how these objects should be depicted using different types of shapes, colours and other graphic variables to enhance and improve our intuitive interpretation and understanding of how the objects represented in the map relate to the real world.

To create a usable map there is first a process of development that a cartographer needs to undertake. This involves asking questions such as 'What is the purpose of the map?' Within this process there is an artistic side to good map design that requires a thoughtful process of selection, leading to decisions about what data should be included on the map and how it should be displayed to ensure it is interpreted correctly. The design and display decisions undertaken by the cartographer will ultimately influence how the map looks, and will determine the user's conception of the information and how it is then translated into an individual's own knowledge. Therefore cartography requires scientific practice and some artistic flare to produce well balanced compositions.

At the first general assembly of the International Cartographic Association held in Paris during 1960 the organization used the following definition to describe cartography:

> Cartography is the totality of scientific/technical and artistic activities aiming at the production of maps and related presentations on the basis of data (field measurements, aerial photographs, satellite imagery, statistical material, etc.) collected by other disciplines. Furthermore, cartography includes the study of maps as scientific documents as well as their use. In this sense, cartography is limited to 'Cartography proper', i.e., to data presentation up to the reproduction and printing of maps and charts: it will be understood that in the practical application of this definition, the gathering of primary data, field surveying and photogrammetry are excluded as are surveys carried out by other disciplines such as geology, statistics, demography, etc. (Edson, 1979: 165).

From this definition the discipline of cartography can be likened to a process associated with the conception, production, dissemination and study of maps. The discipline of cartography encapsulates the whole process linked to map production. It requires a holistic approach to data gathering, evaluation and processing as well as the artistic design of drawing a map. It is this process that means cartography lends itself to both an artistic and scientific approach, as its nature is truly interdisciplinary.

For contemporary mapping the 1960 definition is somewhat incomplete, and now when referring to map it should mean both the visual printed map and the digital map. Digital mapping is the form of cartography that most developers of geospatial technologies will encounter and so the process of cartography should also be extended to encompass the human-computer interaction between the user and the map because it is this interaction that will ultimately lead to capacity building of a user's spatial understanding.

3.1.2   Importance of cartographic representation

The use of maps in society has a long established history. This is because spatial is special and, 'almost everything that happens, happens somewhere' (Longley *et al.*, 2005: 2) and by knowing where something happens we are better placed to improve our understanding of its complex intricacies. Each map provides an abstracted model of reality which highlights geographical patterns and features that make a significant contribution to the end use of the map and its meaning.

Maps were first adopted by ancient civilizations and have stood the test of time as they evolved independently amongst people living in different parts of the world (Wilford, 2002); they continue to be more and more widely used in contemporary society than ever before with many different application areas and with increasing technological sophistication. Figure 3.1 illustrates the various stages of *reference mapping* which have emerged throughout time; it outlines the periods in history which saw the development of astronomical, thematic,

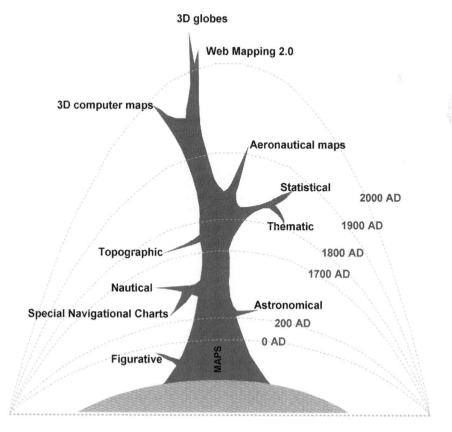

Figure 3.1   Historical timeline of reference maps: adapted from Robinson *et al.*, 1995

statistical, nautical, topographic and contemporary digital maps. We will describe some of these various types of maps later in this chapter.

A key contribution to the field of map mapping can undoubtedly be accredited to Ptolemy. He was a mathematician living during the period between the second century AD and was probably the originator of the coordinate system. He estimated the circumference of the world and used this measurement to divide the world into a grid using notions of *longitude* (lines running north-south) and *latitude* (lines running east-west). Using this coordinate system he documented the locational reference points for thousands of different places all around the world – an incredible achievement for the time. This was the first conceptualization of a reference system based on coordinates to pinpoint individual locations and geographical relationships between places. It is thought that this system paved the way for a contemporary coordinate reference system. The coordinate system that he developed makes most contemporary mapping possible by enabling the projection of a three-dimensional planet onto a flat piece of paper. We will discuss the idea of projections in more detail later in the chapter.

A much more recent influence on the development of mapping techniques was a consequence of the Second World War. As MacEachren (2004) notes, the war experience led map production and cartography away from the previous emphasis on production efficiency and graphic design to a model driven by specific functional requirements. This functional centric model still dominates map making today so emphasis is less about maps as works of art and more about functional scientific tools. The 'interpretation of phenomena geographically depends upon visualization by means of maps' (Philbrick, 1953: 11).

During the 1960s and 1970s, when the emphasis in geographic research was on quantitative methods, cartographers attempted to develop objective rules based on the notion of map functionality. It was during this time that a leading post-war US geographer known as Arthur Robinson identified the importance of recognizing the limitations of human perception with respect to colour, lettering (fonts) and design (MacEachren, 2004). Consequently this led to a series of experiments using *symbols* and *map classification* to identify the most effective types for specific purposes (MacEachren, 2004). This body of research led to the quantitative approach to map classification. This numerical approach was developed using coloured geometric objects in a map and by investigating the creation of data groups based on the underlying data distribution (Jenks and Coulson, 1963; Jenks and Caspall, 1971; Evans, 1977; Brewer *et al.*, 1997; Monmonier, 1996). Much of this chapter will be devoted to the basic rules of this functional type of cartography, exploring its underlying theory and principles.

3.1.3   The role of different maps and spatial cognition

We know maps have had an important role in history and that they have been pivotal to the development of society. This has given rise to many types of maps (see Figure 3.1) which are grouped into two broad categories: reference maps and thematic maps. The objective of a reference map is to provide simple locational information, often at different map scales, with either lots of detail or more generalized maps with less detail (Brewer, 2008). These types of maps are central to our understanding of questions relating to the where? Since they provide answers to everyday questions such as:

- Where is the nearest train station or doctor's surgery?
- Where can I find the local museum?
- How do I get from my house to the beach?

Reference maps provide us with spatial facts which user's link together to create their own mental maps. They can be in the form of a simple street map or more specialized maps such as aero-charts used by airplane pilots, road atlases or hydro-charts for sea navigation. Since the rise in spatial data collection techniques, more sophisticated techniques are employed for data storage and processing using GIS, the role of maps in society is ever changing. More and more often they are being asked to provide sophisticated answers to increasingly complex questions to facilitate geographical enquiry through the use of thematic maps.

Thematic maps use at least one variable for data presentation, communication of exploration and are now very common in the pages of academic journals, newspapers, magazines and government reports. The themes can be derived from social, economic, political, health, cultural, or environmental data variables and are most likely to take the form of static, one-off maps. Examples include land-use maps, maps of cancer clusters and representations of the percentage of retired people within districts in a certain country.

Thematic maps have at least three purposes: they provide information about a specific place or location; they enable the exploration of spatial patterns of themes for a place; and they facilitate the comparison of patterns for different places. Unlike reference maps, thematic maps do not provide a source for navigation, and they depend on reference information to add context for situational understanding. There are many different types of thematic maps and by presenting attribute data relating to spatial units in an intelligible way, thematic maps can be most useful. A more detailed discussion on the methods used to create a thematic map will be included in Chapter 8. Thematic maps can help provide answers to questions such as:

- How will the flu epidemic spread within the population and where will people be infected?
- Which areas are more deprived than other areas?
- What is the temperature in London compared to New York, Tokyo and Dublin?
- Which areas are more prone to flooding?
- Where should a new hospital be sited and what services should it provide?

To answer these questions the maps need to be created using appropriate generalization methods, providing details relevant to the map scale, and displaying suitable information to enable users to engage with and interpret the maps in a sensible manner. All types of maps need to follow a series of key principles and guidelines for colour, shapes and fonts together with map elements that will enhance understanding of a map and improve *spatial cognition*.

When we talk about spatial cognition of maps we are concerned with the way in which maps and geographical data are understood – as we have discussed in the previous chapter; in the context of maps, spatial cognition is concerned with how meaning is derived from them. When cartographers consider spatial cognition they are interested in improving the use and design of maps and other geographic data, information and products which depend upon human interpretation (Montello, 2001). They are looking to improve knowledge building, so for maps to have a purposeful role then it is essential that they can be interpreted meaningfully.

Montello (2001) describes cognition as the 'acquisition, storage and retrieval, manipulation, and use by humans. . . . Broadly construed, cognitive systems include sensation and perception, thinking, imagery, memory, learning, language, reasoning, and problem-solving' (14771). Careful consideration of the cartographic principles will improve the spatial cognition of a map and assist in the visual impact it will have.

3.1.4   Principles of cartographic design for increasing spatial cognition

Cartographic design is unlike other aesthetic design processes. The map maker has a responsibility to produce a representation that portrays the abstraction of reality in a considered and appropriate way and to avoid visualization techniques that exaggerate, overemphasize and mislead the user. If a map is purely about design aesthetics then invariably it will fail to meet its purpose and will most likely not have any viable usefulness. The same can be said of a map that is created purely as an output of technological innovation; if a map is created only for the purpose of exploiting new technology then it is likely to have little practical application. This is because the actual end-use of the map will not be the primary reason for its development. That said there are no ultimate rights or wrongs to map making but there are design guidelines that if followed will assist in the creation of a useful and usable well designed map.

In 1996, the Society of British Cartographers published a list of five principles which should guide good map design; this list is by no means exhaustive but it does enable you to start thinking about what works and what does not when it comes to map design. Before we consider the practicalities of designing maps with colour, map elements and the like it is well worth reviewing these general principles as a starting point:

1. Concept before Compilation
2. Hierarchy with Harmony
3. Simplicity from Sacrifice
4. Maximum Information at Minimum Cost (after Ziff)
5. Engage the Emotion to Engage the Understanding

The first principle is 'Concept before Compilation'. The essence of this principle is to consider the concept of the map before any real design or implementation occurs. This means that every map is thought about holistically at the outset before it is broken down into different parts and elements. This is rather similar to conceptual models of Web pages that are created prior to their design, development and implementation. The visual display of quantitative information, for which maps are just one representation, should have a clearly identifiable message which motivates the design of the map and answers the question of why the map is being created (Tufte, 2001). The function of the map should be clearly identified at the beginning of the design process.

By following this first principle you are guided into thinking about the map and its constituent parts as a complete composition. Here we are interested in the following elemental questions of who, what, where, why and how (see Table 3.1). These questions shift the emphasis towards a user centred perspective for map design. The answers to each of these questions will in turn contribute to the overall conceptualization of the map and how it will be implemented.

The second principle worth following is the notion of 'Hierarchy with Harmony'. What this means is that a map comprised of different data layers should use a harmonious colour

Table 3.1   Questions to be answered during the conceptual design of a map

Why is the map being created?
Who will be using the map?
What will they be doing with the map?
How will they be using the map?
Where will they be using the map?
What is the final medium for the map?

palette. The different components should not compete with each other for attention or detract from the understanding of the map. The identification of colour is immediate and if appropriately used, can accentuate elements of maps and improve spatial cognition and pattern recognition. Thus, there should be a natural integration of the different map layers which will reduce the cognitive load for the map user. The notion of hierarchy is important because with multiple data layers a good design will encourage a natural visual emphasis on the important layers and those that are less so will recede into the background. Less important data should complement the design and not compete with the more important data, so as to provide contextual information that enhances the spatial cognition and knowledge construction of the map user.

Figure 3.2 illustrates this idea of hierarchy with harmony by using OpenStreetMap data and a colour palette proposed by CloudMade and another following the conventions of Google Maps. The colouring of layers can change the emphasis of the map. The map in Figure 3.2a shows how the harmonious pale colours of the different data layers are complementary in such a way as to make all the map layers recessive; roads and parks are not dominant visually but the tube stations are. The data layers presented in this neutral colour palette would be a useful background providing situational data to which further layers of primary importance can be added. In contrast the map in Figure 3.2b is comprised of a set of colours which emphasizes the hierarchy of the road network using two variables, colour and width of the line. Different road types are distinguished both by colour, green, orange and yellow, and width; the width of the road increases in size in accordance with the road importance. The selection of green for the major roads along with the widest road width ensures that for the map user this is the most prominent feature. This map emphasizes the road infrastructure by blending together a hierarchical colour palette and careful use of colour which means the map can also be used by people with colour blindness.

A good method to judge the harmonious and hierarchical components of a map is to stand back from the monitor or printed copy of the map. This will enable the map to be examined as if for the first time, and with an objective eye look at the set of objects or layer that is the most dominant. Is the most prominent feature of the map fitting with the reasons why you are designing the map? Thus a useful way of checking to see if a map you are designing fits with this rule of hierarchy with harmony is the stand back and look principle.

The third principle corresponds to 'Simplicity from Sacrifice'. When we apply this principle to cartographic design we are tending towards the idea that less is more, using a quote attributed to Albert Einstein 'an explanation should be as simple as possible, but no simpler'. We can consider this in the context of cartographic design. A map should be as simple as possible, but should not be reduced to the point that it is too simple, whereby the meaning and detail are lost – if this happens users will not be able to acquire as much potential information and the map could be rendered useless. The map needs to be clear and concise.

In Figure 3.3 two maps are used to show the route for travelling from point A to point B. In Figure 3.3a the map highlights the extreme of taking the simplicity too far. The detail in this map has been stripped out so that only roads on the route from A to B are highlighted, presenting a specific route for a navigator to follow. This map is oversimplified and is far too minimalistic to be useful. Relevant details for effective navigation have been sacrificed. In contrast the map in Figure 3.3b is still simple in its design. It is uncluttered but contains sufficient contextual data to enable the user to follow the route marked and correct any mistakes on route (major roads intersecting the route are labelled – although not extensively as this map is for illustrating a conceptual point). Furthermore the situational context in this map provides the navigator with information to help improve their memory of an area

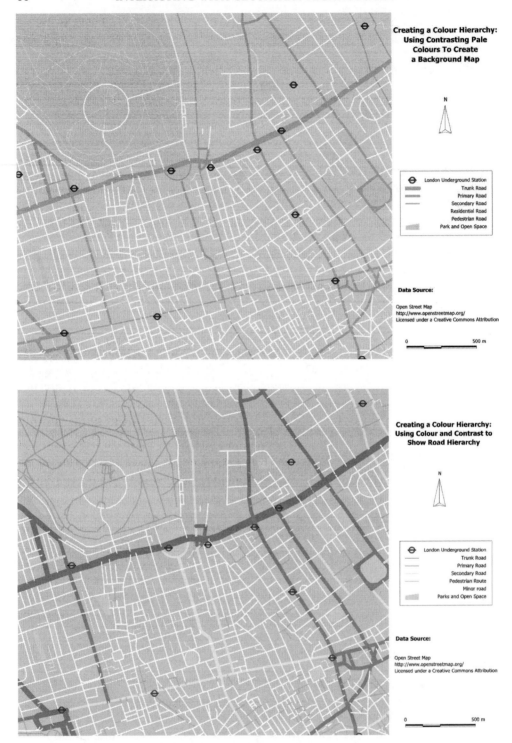

Figure 3.2 Hierarchy with Harmony – same but different (Data source: OpenStreetMap.org) (a full-colour version of this figure appears in the colour plate section of this book)

Figure 3.3 Simplicity from Sacrifice, navigating route in Central London (Data source: OpenStreetMap.org) (a full-colour version of this figure appears in the colour plate section of this book)

(Tufte, 2001); helping users to understand where they are in relation to the surrounding natural and manmade landmarks (parks and tube stations) will aid knowledge construction and will help the memory store and recall the visualizations.

This principle of 'simplicity from sacrifice' can be considered as the rule of clarity. In the same way that a map can be oversimplified there is also susceptibility for a map design to contain too much information. A map that is too busy and contains more details than absolutely necessary will over-stimulate the visual sensors of the user. If the cognitive load required to process any facts and information is too great, it takes a lot of time and effort to make sense of a map and it is likely the user will not make the effort because the benefits are not obvious.

Too much superfluous information, clutter on the map or irrelevant visual elements are often referred to as *map junk*, a term derived from Edward Tufte who discussed the idea of *chart junk* in his book *The Visual Display of Quantitative Information* (2001). Tufte described chart junk as the visual elements included on a chart or graph that do not add to a user's understanding or ability to process the information being presented. Describing it as the interior decoration of graphics that generates ink without telling the reader anything new, he states that these types of adornments are unnecessary and are simply 'non-data-ink' (Tufte, 2001: 108). Chart junk and its manifestation as map junk have proliferated with the growth in technological innovation and the rise in fancy computer graphics. It is good practice to remember that it is not enough to possess a new technology, it must be used to improve designs (Tufte, 2001: 201; Krygier and Wood, 2005). When making an interactive map it is important not to add technical or graphical flourishes that demonstrate the ability of the latest technology but fail to add to the experience of using the map.

Figure 3.4 illustrates two very different examples where the simplicity from sacrifice principle is ignored. In the first image a 3D prism map is projected onto a virtual globe. This type of mapping breaks a number of cartographic rules. The 3D prism map maybe engages the user by creating a Wow! affect and uses the latest virtual globe technology, but it is difficult to compare the results for each area because of perspective and to use the height of each object to represent the information. Once the novelty of the visualization wears off, it is clear that extracting information from this presentation is very challenging. In addition, the 3D prisms on the globe make it difficult to identify different countries and the relative differences between them, and depending on where you centre the globe the visual emphasis of the prisms will be different. Only countries that are viewed together can be distinguished. A simple 2D thematic map would allow the user to complete this task within a few seconds. Thus, as described by Mark Harrower, these types of maps make 'information extraction (1) slower, (2) more difficult, and (3) more prone to reading errors, because of excessive ornamentation and unnecessary design additions'.

Figure 3.4 (bottom) represents a different type of map junk where the simplicity from sacrifice principle has been broken. From the title of this map the user knows the designer is trying to show the 'religious distributions in Belfast'. However, the user experience for this map represents quantity over quality. Too many map layers in clashing colours that have no hierarchical harmony in them are presented, and in addition this the map lacks an appropriate colour classification. The red colour chosen to classify the lowest proportion of Catholics is a poor choice, as in Western culture red implies danger and it is also the most dominant colour on the map, so the map is already culturally insensitive and may be placing emphasis on the wrong interval group. In addition, the contextual data supplied actually reduces the readability of the map. The table in the bottom right is not easy to

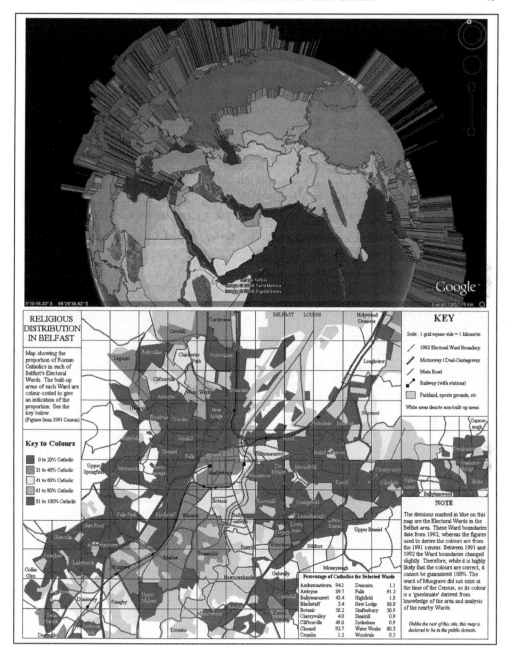

Figure 3.4 Examples of map junk. (Source top: Sandvik, 2009; source bottom: CAIN, 1996) (a full-colour version of this figure appears in the colour plate section of this book)

comprehend and it is difficult to identify the areas on the map it is referring to. The result is an unintelligible map.

The penultimate principle is the notion of maximum information at minimum cost. This principle refers to the balancing between what to include and what to leave out. A balance necessary to ensure the maximum amount of knowledge can be derived from the

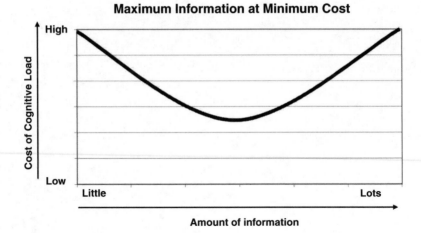

Figure 3.5   Graph illustrating the relationship between information and cognitive load

map without increasing the cognitive load of the reader (see Figure 3.5). This principle is concerned with the functional design of the map – how complex data should be processed and displayed using *generalization, classification, simplification* and *exaggeration* techniques to produce a useful map. These terms will be explained in more detail in Chapter 8, but they refer to graphic manipulations that can be carried out on the map. This principle is useful, especially when creating maps with more than one variable depicted (multivariate thematic maps), since they are prone to become more complex and without careful design.

The fifth and final principle invites the designer to consider the emotional responses to the map: 'Engage the Emotion to Engage the Understanding'. By trying to invoke an emotional response to a map, the designer is tapping into a similar reaction that marketers exploit during advertising campaigns. The parallels between advertising and cartography were drawn by Monmonier (1996) who suggested that the advertising industry creates appealing and clear images in the same way cartographers strive to create usable maps. An extreme form of advertising maps can be illustrated by looking at political propaganda maps.

Designers of political maps create emphasis through the use of different graphic manipulations to portray a political message within the map or how it is interpreted. This elicits a persuasive and emotional response from readers. Maps of these types are designed to influence and engage the audiences' opinions, beliefs and emotions. The propaganda map in Figure 3.6 provides a powerful illustration of this final principle. The map was drawn during the First World War, a social cartoon using various national stereotypes with animals and imagery associated with the different countries to form a powerful graphical map with the ability to provoke an emotional response and stir up nationalistic pride and fervour. The British are presented as a bulldog standing its ground within both France and the UK, sizing up to the German Dachshund (it is biting the German dog on the nose!). The power of the UK is further exaggerated and emphasized to the reader in this map by the careful use of scale – the designer has added to the illusion of grandeur by drawing the United Kingdom so it appears to have a much bigger land mass than is the reality; in the image it is shown to have the equivalent land mass to Norway and Sweden.

From the propaganda map it can be understood that the maps reflect the culture of the designer who produces them as much as they are a representation of an event in time (MacEachren, 2004: 10). Also observed in the propaganda map the UK is symbolized by the

Figure 3.6 First World War propaganda map (Zimmermann, 1914) (a full-colour version of this figure appears in the colour plate section of this book)

colour red – the most visually dominant colour, so by selecting pertinent symbols, imagery relevant to the map users and colours with cultural significance, the *connotative* meaning of the maps can be enhanced. The connotative meaning of a map refers to the way users comprehend what a sign/icon/symbol on a map stands for intuitively. Connotative meanings are implicit to the socio-cultural context of the map and its user (MacEachren, 2004: 331) and assist in improving its interpretive power.

These five principles provide the map designer with some simple ideas of best practice which help make the representations more understandable. The essential thing to remember is that, 'what is mapped is always more important than what is not' (MacEachren, 2004: 334). As a consequence the map designer is responsible for a series of selective decisions which will influence the look and feel of the map and how it should be created. The final map is always a product of the design choices which should be made in order to produce a clear, easy to interpret representation of the complexity of the real world. To develop maps, a good understanding of map scale, projections and generalization must be considered.

3.2 Impact of projections on map design

A major concern for the map maker is associated with the projection system and concerns selecting one which is appropriate for the data. With virtual globes (such as Google Earth and ArcGlobe) it is quite simple to recreate the locations of the Earth's surface, the transformations required to produce relatively undistorted representations. However, these types of representations are not always practical. Positions on virtual globes are measured using

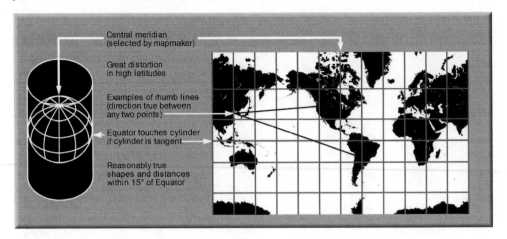

Figure 3.7   Mercator projection (Source: WikiMedia under license of Creative Commons: http://creative commons.org/licences/by-sa/3.0/)

angles in the form of spherical coordinates known as longitude and latitude where latitudes measure the number of degrees from the equator and longitudes the number of degrees from the Greenwich meridian.

However, most maps are produced on flat planes and therefore the issues of transformation arise. This is because all maps are representations of the Earth and most maps are drawn or calculated onto flat surfaces (Figure 3.7). How can a spherical 3D globe be represented as a 2D object on a piece of paper or computer monitor? Figure 3.7 illustrates how one projection system, the Mercator, is created. The transformations are carried out by using a projection system which is a set of mathematical models used to transform spherical global coordinates onto a planar system. Projections also take into account the true shape of the Earth, which is not a perfect sphere.

A *projection system* transforms the curved surface of the earth onto a flat plane enabling the visualization of the Earth or parts of the Earth at a wide variety of scales suitable for different purposes. When this process of projection is carried out, inevitably distortions are introduced, simply due to the stretching and twisting that occurs as the sphere is transformed into a planar surface. Invariably the process changes the geometric relationship of the shapes, areas, distances and directions in the resulting planar maps. Each projection system will lead to different distortions, and the selection of projection system should match the application in which the maps will be used. Figure 3.8 provides an example of the distortions that occur in the Mercator projection, which may be familiar since it is the projection used by Google Maps; the distortions are more apparent as the distance from the equator is increased.

The hundreds of map projections developed all endeavour to minimize distortions of one sort or another, so each come with a set of advantages and disadvantages. Each projection will vary in terms of the nature and pattern of distortions that occur: some are suitable for mapping small areas, some are useful for mapping large areas with an east-west extent whilst others are more appropriate for mapping long transects north-south and some retain the property of direction or distance.

Of the many types of projections that are now available, the most common ones include the *conformal*, *equal area* and *azimuthal projections*. Conformal projections preserve the angles of the original features, thus making them suitable for marine or aviation navigator charts, meteorology and reference maps (Robinson *et al.*, 1995). These types of projections are useful

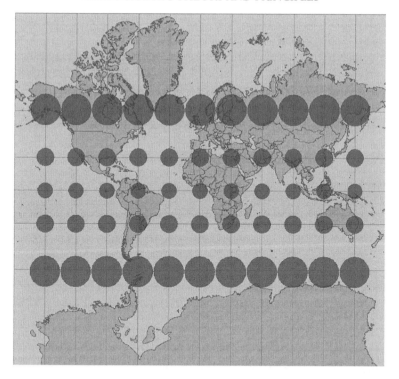

Figure 3.8   Distortions of the Mercator projection (Source: WikiMedia under license of Creative Commons: http://creativecommons.org/licences/by-sa/3.0/) (a full-colour version of this figure appears in the colour plate section of this book)

for small area maps (large-scale) since the area and features are drawn accurately, but for large areas the shapes of countries are not maintained and both areas and shapes are distorted to the extent that they are enlarged beyond their actual size; the distortions are more common at the edges of the maps (see Figure 3.8). It is generally not suitable to use these types of projection systems to create world maps, although it is commonly used in maps produced by the media (see Case Study 3.1).

Figure 3.9 shows the distortions that arise when a conformal projection is used to create a map of the world. In the map, the surface area of Greenland is massive and is almost the same size area as South America. In reality Greenland is approximately 2,166,000 sq km whereas the continent of South America is around 17,840,000 sq km. The distortions at the edges of this map are clear.

Equal area projections preserve area. All regions drawn using an equal area projection will be proportional to the areas on Earth. This type of map projection is not conformal so angles and shapes will be distorted. The projection squashes shapes of countries. It is a useful projection for mapping small scale maps that have large area extents. For example, the United States Geological Survey utilizes equal area projections to make their maps. Azimuthal projections create maps that are projected on a plane that can be centred anywhere with respect to the reference globe (Robinson *et al.*, 1995). These types of projections preserve true directions from one reference point to all other points although they do not always represent the true distances. Azimuthal projections are becoming more and more common since they are effective at mapping airplane routes, satellite paths and other applications where direction is of greater importance than maintaining true area, shape or distance.

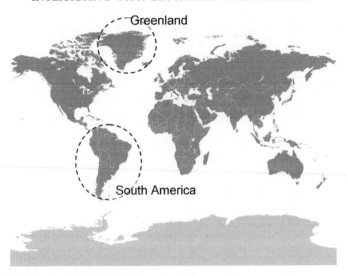

Figure 3.9 Map of the world using the conformal projection (Mercator) showing distortions at either extremes of latitude (Source: Wikimedia Commons)

If projections are used incorrectly the resulting representations will be inappropriate. A case showing how the choice of projection can produce incorrect visualizations and analysis is provided by the international magazine *The Economist*.

Case Study 3.1: *The Economist* map of missiles range

Caution must be applied when choosing a map projection. The wrong projection can have serious consequences. In 2003 *The Economist* ran an article called, 'When bluff turns deadly'. The article discussed the nuclear ambitions of North Korea. To illustrate the article they produced a map indicating the missile threat from North Korea to the rest of the world given the range of the weapons they have amassed.

The map was drawn using the Mercator projection which is a conformal projection type that distorts shapes and areas of countries on the outer edges of the map during the transformation from sphere to a flat plane. In the original map used to illustrate the threat, concentric circles were simply superimposed around a point on the map which represented the location of North Korea. The original map failed to account for the curvature of the Earth and the distortions that arise for the selected map projection as areas become larger at the edges of the map (look back at Figure 3.8).

The result is a map which wrongly assesses the threat of the missile reach from North Korea. The range of the missiles as presented on the map do not conform to a concentric circle and the areas of the world at threat of North Korean missiles should be much bigger than that portrayed in the original map.

The maps below reproduce the error that was made by the *Economist* graphic designers and clearly illustrate the differences that arise by using buffers as concentric circles irrespective of the projection (Figure 3.10) or a geographic circle that accounts for the area distortions in the projection (Figure 3.11).

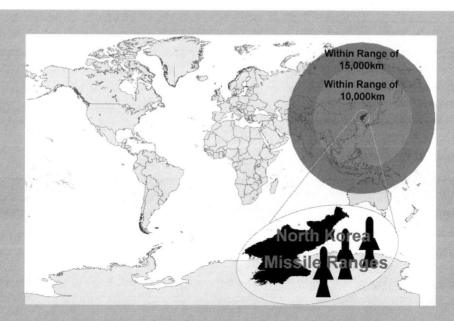

Figure 3.10 Missile range without due consideration to projection

A corrected version of the map was included in the next edition of the journal and highlights the North Korean threat to encompass many more countries than the original map.

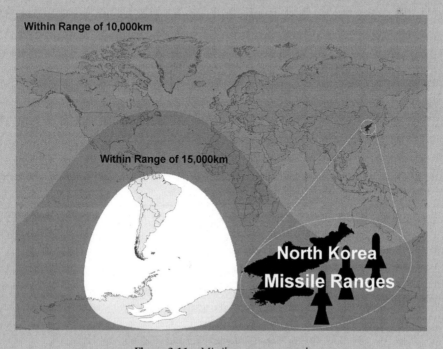

Figure 3.11 Missile ranges corrected

Projections are very important in geospatial technologies. They impact distance calculations, map visualization and many other aspects of the application. It is therefore necessary to understand their impact on the application and ensure that the correct decision about the type of projection and the manipulation of coordinate information is integrated into the design process.

3.3   Impact of cartographic scale on map design

After projections, the next fundamental concern of all map makers is the issue of scale. Scale refers to the levels of representation, experience, and organization of geographical events and processes (Johnston *et al.*, 2000). In this section we will discuss notions of cartographic and analytical scale. An understanding of scale is essential for good map design because the world is a very large place and a paper or digital map by comparison is very much smaller, and without the understanding of how much of the Earth we want to show and the size at which we want to show it, it is difficult to make the appropriate design choices (Monmonier, 1996; Krygier and Wood, 2005).The scale at which a map is drawn will significantly impact the final representation. Without this understanding of scale, creating useful maps will be an arduous task. This is because the scale of a map influences the amount of detail captured and represented. Without knowing the scale the user cannot interpret the map usefully, for instance it would be very difficult to understand the relative distances between different objects.

The cartographic scale determines the amount of details, context and annotation on the map and the level of generalization and classification used to draw the objects within map layers. Generalization methods reduce the complexity of map objects (lines, points, areas) so they are appropriate for the scale of representation. A large scale map will have much more detail than one drawn at the small scale, because the larger scale map is more representative of the features in the real world as it presents a small area of the Earth (Tomlinson, 2007). Notice that maps showing a small area of the Earth, but with lots of details, are usually large scale. This is because the ratio in which they are depicted is large – such as 1:1250 (read 1 to 1250), which means that every centimetre on the map corresponds to 1250 centimetres in the real world. On the other hand, maps that show large areas of the world will use scales such as 1:50,000 or even 1:1,000,000 and are termed small scale maps. As the scale changes from a large scale to a small scale the data represented will require more generalization and classification to improve the readability of the map because small scale maps represent larger areas of the Earth's surface.

The cartographic scale on the map is conveyed to the reader in one of three ways: a short textual description, a graphic object or as a numerical ratio, and it constitutes a fundamental *map element*. Map elements will be introduced in detail in Chapter 8. In essence they should be included on all maps. The scale bar is used to convey to the reader the relationship between the distance on a map and the distance in reality on the Earth's surface that is represented. The textual description is useful only on a paper map and even then if it represents a whole number and not just a random ratio. The map scale conveyed to the reader using a ratio is also known as the representative fraction (RF) and is unit free. By using a cartographic scale the map reader or user can interpret the relationship between distance on the map and distances on the ground because a unit of measurement on the map represents one unit of measurement on the ground. For example, a map drawn at a scale ratio of 1:100 using centimetres would be interpreted verbally as 1 cm on the map approximates 100 cm in reality (1 cm: 1 m) or a map with a scale of 1:100,000 would correspond verbally to 1 cm on the map represents 1 km on the ground. Sometimes a fraction is used instead of a ratio but the meaning does not change.

There are no clear rules about what scale constitutes a small, medium or large map, but as a rule of thumb a map with a scale of less than or equal to 1:50,000 would be a large scale map whereas map scales greater than 1:500,000 are synonymous with small scale maps (Robinson *et al.*, 1995). It would be appropriate to call those in between medium scale maps. Using these ratios or fractions it is easier for the map reader to determine which scale the map has been drawn at, simply because a larger fraction denotes a larger scale; a larger scale map has a small denominator (the number on the bottom of the fraction or to the right-hand side of the ratio).

The graphical scale bar function is similar to a conventional ruler, the distance is presented as linear units. This is a graphical line symbol subdivided into units which highlight the distances on a map and how they convert to distances in the real world. Figure 3.12 summarizes the key points required for understanding scale and it is useful to remember the scale represents the resolution of a map. The actual ratio numbers and the scale may in the first instance be counterintuitive, but really the lower the number in the ratio the larger the scale of the map and the higher the resolution (Tomlinson, 2007). Confusion may arise in the interpretation of scale because in everyday language we describe something being small scale when referring to the granularity of detail, so a small scale area may equate to the neighbourhood and a large scale area a city.

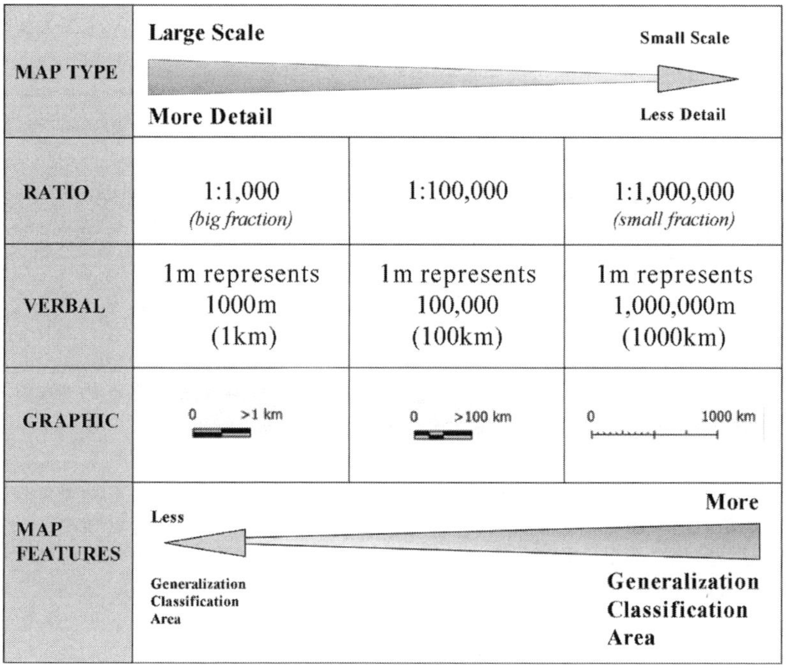

	Large Scale		Small Scale
MAP TYPE	More Detail		Less Detail
RATIO	1:1,000 *(big fraction)*	1:100,000	1:1,000,000 *(small fraction)*
VERBAL	1m represents 1000m (1km)	1m represents 100,000 (100km)	1m represents 1,000,000m (1000km)
GRAPHIC	0 >1 km	0 >100 km	0 1000 km
MAP FEATURES	Less		More
	Generalization Classification Area		Generalization Classification Area

Figure 3.12 Components of Cartographic scale

3.3.1   When to use the different representations of scale

There are three possible methods for representing scale on a map. It is advisable not to use them interchangeably. In instances where it is appropriate to use a graphical scale bar to communicate the map scale it will not always be suitable to assume that the same map can adopt the representative fraction to present the scale to users, this is particularly true of digital maps used in geospatial technologies.

The representative fraction is a versatile description of scale as it can be applied to almost any unit large or small and allows for the amount of reduction in the map scale to be easily understood. There are a number of complications which result from using this type of scale representation. For small scale maps the ratio can be difficult to interpret since the units are difficult for the reader to comprehend and visualize – they require distances to be multiplied or divided by the representative fraction, for example try visualizing how long a road is that is drawn at the scale of 1 map unit to 10,000,000. Furthermore, it is complicated to convert the ratio into one that has understandable numbers – if the fraction is a whole number readers are better placed to make sense of it. You should not produce a map that uses un-rounded representative fraction such as 1:64,239,021. The representative fraction is invalid if the map is reduced or enlarged – as can happen if the map is projected on different screens. If a map originally drawn at 1:100,000 is enlarged, the representative fraction would change, but this would not show up on the enlarged map and will mislead the user. Finally, the representative fraction is not useful for video or web-mapping maps since the size of the screens will vary significantly and the scale behaves unpredictably (Monmonier, 1996). Likewise if a screen map is then projected onto a wall or reduced for use on a mobile phone application the representative fraction will change but this will not be reflected in the wording of the original fraction.

For the reasons outlined above, it is perhaps most useful to use the ratio or representative fraction on static products. This is the case with paper maps and indeed the representative fraction is most commonly associated with topographic maps created by national mapping agencies. For example the UK Ordnance Survey Landranger map series is known by both the representative fraction and the verbal scale at which they are produced e.g. 1:50,000 (2 cm to 1 km, $1\frac{1}{4}$ inches to 1 mile).

The verbal scale is helpful since a simple sentence communicates the relationship of distance on the map in a more intuitive way than the ratio, since it is more descriptive so users can relate to the verbal scale more easily, although there are a number of limitations associated with its use. It is difficult for the user to convert easily between different units of distance. The verbal scale is not easily converted by the map user from one scale to another. If the verbal scale uses centimetres to describe the map scale, e.g. 2 cm to 1 km, it will be complicated for the user to convert to inches if that is the unit they require or is most familiar to them. Like the representative fraction, the verbal scale will not translate if the map is enlarged and it should not be used for screens.

The graphical representation of scale using a scale bar is the most useful and versatile. The graphical representation is similar to the ruler, a tool with which almost all map users will be familiar. Unlike the representative fraction or the verbal description of map scale, by using a scale bar it is possible to enlarge or reduce the final map since the scale bar will resize accordingly with the map proportions (Monmonier, 1996). The scale bar is also beneficial as it does not depend upon the user's sense of distance and skill in mental arithmetic to convert units. With a scale bar it is possible to show both types of units (metric and imperial) in one graphical symbol. It is the most useful form to represent scale on a digital map or Web-based mapping system.

Below is a case study showing how change of scale impacts representation and detail for an online Web-mapping application from the Environment Agency (2009):

Case Study 3.2: The UK Environment Agency 'What's in Your Backyard?'

In 2008 the Environment Agency redesigned a feature of their 'What's in Your Backyard?' website. The online Web-mapping application provides an interactive map system to enable members of the public to find out about the local environment in which they live. Users can type in their postcode or place name and the application returns information about different environmental themes such as flooding, industrial pollution or presence of landfill sites.

This Web application provides an example for the impact of cartographic scale on usability. A screen shot of the website is shown below. In this example the user requested information on river quality in London. Look at the image and ask yourself: What scale is this map drawn at? How long did it take you to find the scale information? Can you intuitively understand the scale of this map?

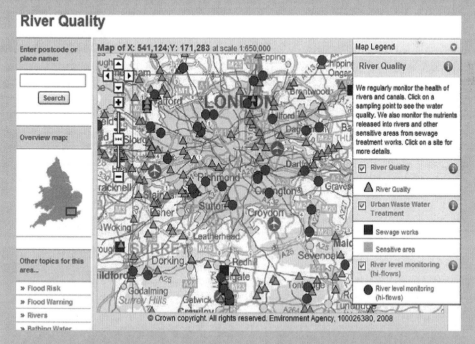

Figure 3.13 The Environment Agency 'What's in your backyard?' map (Source: Environment Agency)

The scale is represented by a ratio of 1:650,000 which is included in the interface above the map, next to the X,Y coordinates of the location searched for (do you know the X,Y coordinates of your town?). This is a small scale map where perhaps 1 cm on the map represents 650,000 cm on the ground. Can you intuitively visualize this distance? It corresponds to 6.5 km. On the computer screen 1 cm cannot relate to 1 pixel. The unit of measurement of a computer monitor is the pixel. Does it make any sense to say 1 pixel represents 650,000 pixels on the ground? A much more appropriate measure of scale for this application would have been to use a scale bar with both kilometres and miles.

3.3.2   Matching scale to the detail of the map

When considering aspects of scale it is important to match the scale and detail required of the map to the appropriate scale of the source data. The scale of the source data often indicates the level of accuracy of the data (Tomlinson, 2007). If your map requires a very different scale of data to the source data you have, it may not be appropriate to use it. If data layers of different scales are mixed any subsequent analysis may yield erroneous conclusions. We have already mentioned that scale affects the detail that is contained on a map. A map is often comprised of many different data layers, and it is good practice to compile the map in such a way to match it to the scale of the smallest scale of data, the data available with the least detail.

The design of a map is influenced by the scale you choose to visualize the data and the scale at which you draw a map will also impact all of the other features included on the map. In turn this is also influenced by the medium for which the map is being produced and the size of the paper or size of the on screen window for which it will finally be used.

Below is a case study showing how change of scale impacts representation and detail for an online Web-mapping application by Multimap (2009).

Case Study 3.3: Use of multiscale cartography in Multimap Web-mapping site

Launched in 1996 and with a recent overhaul in 2007 improving its design, usability and functionality, Multimap's mission statement is to deliver online maps, point-to-point driving directions and geographical ('where's my nearest?') searches to businesses and consumers.

Multimap use a range of cartographic styles appropriate to different map scales, providing scale appropriate content. As the maps become larger scale and the reader requires more detail in order to use the map, the cartographic style changes. For example minor roads, parks, landmarks, building names etc are included in the representation and are identifiable by their labels. Within London the larger scale maps use a different data source and cartographic styling for generalizing and classifying the roads and other features.

The detailed street maps use the Harpers and Collins styling which Londoners are more familiar with since it is a common mapping product. This is different to the smaller scale maps which source the data from the same supplier as Google and therefore have the now more familiar styling that is synonymous with Google Maps.

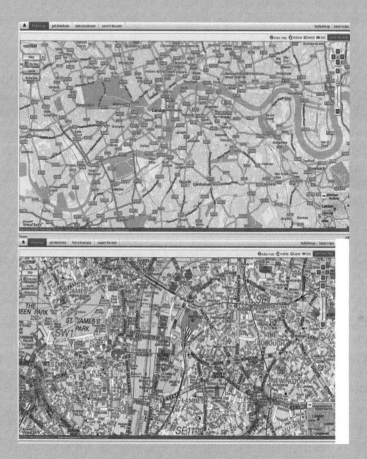

Figure 3.14   Multimap maps at two different scales (Data source: Multimap.com) (a full-colour version of this figure appears in the colour plate section of this book)

3.4   Generalization

The first choices any map maker must take are associated with the selection of the projection and scale of the map. Once they have been correctly identified, the data layers used to create the map must be presented suitably with these choices. The method of map generalization is a useful technique for transforming data by reducing the scale and complexity of existing map objects (point, line or areas) to produce new sets of objects more fitting to the scale and projection of the map. In effect map generalization is a process of simplification where the most important elements and characteristics of the object and its location are maintained.

Generalization is an integral feature of map design and the various techniques utilized allow us to effectively condense geographic information and re-represent it and, in so doing, emphasize issues of particular interest (Richardson and Mackaness, 1999). For example, if the purpose of the map is to present local data on health services for different neighbourhoods in one district, the data layers would need to be detailed enough to provide an easy understanding of where data relate to: individual streets (major and minor – classified) with individual buildings and medical facilities labelled. By comparison, if the task is to produce a map of health services for an entire region, it is more suitable for a medium scale map, so only major roads which could be classified by type and medical services as point objects are shown, since showing individual buildings may make the map too busy.

To create the medium scale map it is advisable to generalize the data to avoid visual clutter, improve legibility and harmonize content. As we progress from a large scale to a small scale map we are tasked with the challenge of providing large areal extents often for the same physical area on the map, there will not be enough map space to contain lots of detail. It is necessary to compromise on how data are displayed, even if this means that some objects or features will be removed or the distortion of features will be unavoidable. The essence of the original input data, however, should be preserved so compromises are necessary. These types of problems are particularly true for computer applications where the map window will most likely be a consistent size regardless of the scale of map. Changes of scales would be best handled using different methods of object generalization for the different scales.

A case study showing how changes of scale are articulated using different levels of generalization is given below.

Case Study 3.4: Generalization in public Web-mapping site

Selecting the correct information appropriate for the scale of geographical enquiry will ensure the map is more user friendly and contains contextual information that aids spatial cognition. In this case study we look at an interactive Web-mapping system developed for Vantaa City, Finland. The purpose of the application is to provide information and interactive maps about the city by providing users with information on addresses and corresponding land uses. Much of the data is in the form of points stored in layers which the user can switch on and off but for our discussion here the interest in this application is the use of generalization.

For the large scale city maps of local neighbourhoods the mapping background of the city plan has a considerable amount of detail with individual buildings, plots labelled with numbers and street names and even pavements visible. In contrast, the smaller scale, city-wide map highlights the hierarchy of the roads using different colours to show the motorways, primary and trunk roads but road names are omitted ensuring an uncluttered map. Generalization methods have been applied to the small scale data to group together blocks of contiguous open spaces and built up areas to form a generalized view that is scale specific. The developer has used generalization methods to develop scale dependent representations of the data.

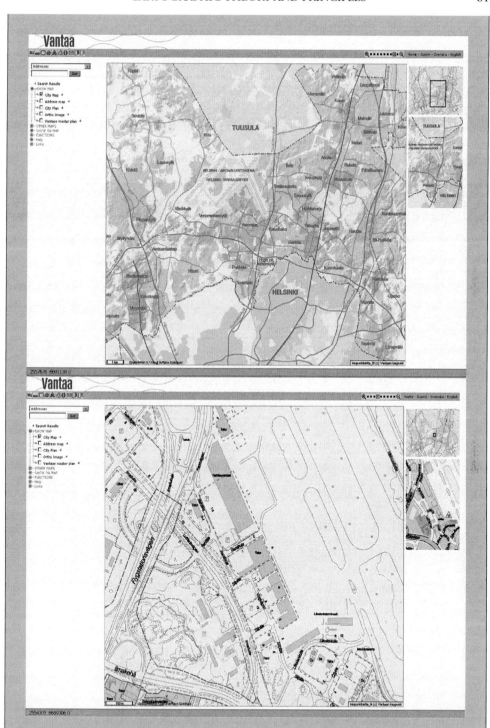

Figure 3.15 Vantaa Maps on two scales. Notice the generalization in the upper image (Source: http://kartta.vantaa.fi/Default.asp)

Table 3.2   Methods and uses of generalization techniques

Action	Terminology	Graphical changes to object geometry	Conceptual changes using object attributes
Choose important features to map and eliminate unimportant ones	Selection	✓	✓
Simplify the graphical complexity of objects	Simplification	✓	×
Certain objects are made bigger	Enlargement	✓	×
Group small objects together as one larger more simple object	Merging	✓	✓
Move the location of map objects to reduce overlaps and visual clutter	Displacement	✓	×
The symbols of map objects are changed according to the theme and purpose of the map	Symbolization	×	✓
Certain map features are emphasized in order to highlight them	Enhancement	×	✓

There are six common methods for generalizing spatial data which can de divided further into two distinct types of methods: graphic versus conceptual generalization (Shea and McMaster, 1989). Graphic generalization is concerned with the graphical look of the objects and this type of generalization is focused on the changing of the geometry of the point, line or area objects and the relationships amongst them. In this type of generalization the object symbol does not change at all, so a point remains a point, a line a line and an area an area (Kraak and Ormeling, 2003). The five types of graphic generalization include simplification, selection, enlargement, displacement or merging of objects, see Table 3.2 for more information.

In contrast, conceptual generalization techniques take the attribute data associated with the various objects and by carrying out techniques that lead to conceptual changes, the symbology of the original object changes – so the newly generalized object is seen and interpreted by the user differently. Symbology is the cartographical term used to describe the colour, shape, pattern, font and other graphic attributes of a map feature. Processes that can be categorized as conceptual changes include: selection, merging, symbolization and enhancement.

During the process of selection the map maker needs to identify what are the most important characteristics of the data that must be retained to ensure the map is fit for purpose. A clear understanding about the end use of the map will make the choice about what to keep and what to remove. Simplification is used to reduce the complexity of objects by modifying the geometric properties and by altering the shape of objects so that they have more simplified features. In merging, the manipulation is such that several objects are combined together to eliminate clutter. Object enlargement is necessary when mapping at larger scales, where sometimes the original symbol size would be too small and illegible. Map objects are displaced when it is necessary to reduce the visual clutter and overcrowding of maps;

this process will maintain the relationship between objects (Krygier and Wood, 2005) and it preserves the shape of the object but will change its location to improve the clarity of the map.

Conceptual generalization often involves enhancement of objects, usually by changing the symbology so that it is more emphasized. When we think about generalizing the symbology of an object, it can mean changing the actual graphical object to an icon that can be interpreted by the map reader which taps into the natural connotative meaning. The icon's meaning is intuitive. A discussion on the types of symbology that can be used in practical cartography is outlined in Chapter 8. Weather maps provide an example of the use of symbology to generalize geographical data whilst simultaneously improving the user's immediate understanding of the map.

Case Study 3.5: The use of symbology in weather maps

The key topic of small talk in the UK is the weather, filling the gap in all sorts of awkward silences. The public is therefore consuming weather information which, as Monmonier (1997) notes, is presented via television, radio, Internet or the daily newspaper. With the exception of the radio, a map is invariably used to communicate where and when it will be raining. The weather map presents a cartographic snapshot of the atmosphere most commonly in the form of a forecast map (Monmonier, 1997). There are many different ways of presenting the weather but commonly the audience of weather forecasts are members of the general public who are not habitual map readers. The purpose of a weather forecast map is to present, to a mass audience, a summary of the most likely weather for the day. Therefore, it is necessary for the maps to convey the information clearly, simply and with representations appropriate to the audience.

In 2005 the BBC (British Broadcasting Corporation) underwent a radical re-design of the maps they used to present the weather. The new maps were said to be 'Virtual reality weather forecast maps with a realistic 3-D landscape set to revolutionize the BBC's forecasts. . . . presenting the biggest change to the way the weather is presented in 20 years' (BBC, 2005). Examples of these new weather maps can be seen by accessing the BBC weather website and selecting to play the forecast video (http://news.bbc.co.uk/weather).

In the forecast an animation moves the map so that the camera flies over the UK. In the new 3D graphics, the UK landmass is represented as a muddy-brown with a relief shading to highlight mountainous regions. Areas that are forecasted cloudy weather are then represented by using light and shade to make the landmass darker to depict clouds moving over the land. Areas for which rain is forecast are coloured in light blue with animated rainfall falling from the sky. Temperatures are portrayed on the map using square symbols shaded from blue to dark orange using a sequential colour palette to highlight cold to warm temperatures; they are labelled with the temperature in degrees centigrade. The underlying premise of the redesign was to improve clarity over detail. When the forecast maps were first released some of the public complaints included that the landmass was brown and not 'green and pleasant' as the British like to think about their country, the fly through animations were too fast and the angle at which the map was originally produced distorted the view and size of Scotland due to the animated flying path.

The new map signifies the loss of familiar symbology of weather phenomena. The previous incarnations of the BBC weather forecasts were not as technology driven, providing a much more simple representation. Familiar icons symbolic of various weather elements such as the sun and grey, black and white clouds communicated with ease to the audience the expected weather for the day. An example of the old style weather maps at the time of writing could be seen on the BBC website (http://news.bbc.co.uk/1/hi/in_pictures/4551822.stm), but an example of the style of icons they use is illustrated in Figure 3.16.

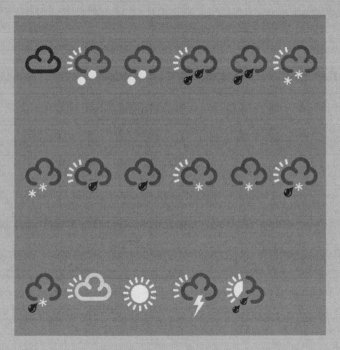

Figure 3.16 Traditional weather map icons (Source: WikiMedia under license of Creative Commons: http://creativecommons.org/licences/by-sa/3.0/)

The symbols used in these traditional maps had a connotative meaning for the reader and harnessed their intuition, so they could be interpreted quickly and easily without the need for lengthy explanations. By using this type of context specific icons the map designer has considered the importance of how, why and when generalization is necessary. The more simple static representations using icon symbology for weather forecasting maintained clarity appropriate to the end use with content specific for a given scale (McMaster and Shea, 1988: 242). With the move to computer graphics the BBC weather maps now take the users more time to interpret and have resulted in a loss of clarity that was specific to the intended audience (Shea and McMaster, 1989).

There are six useful conditions for which the generalization methods discussed above should be implemented: congestion, coalescence (merging), conflict, complication, inconsistency and imperceptibility. Data layers will require generalization if one of the following situations occur: (1) too many objects are located close to each other so that the map will appear congested; (2) the objects are touching because either the map scale or the type of symbolization selected results in them merging together; (3) an object is in conflict with its background layer; (4) ambiguities arise from the change of scale; (5) data is non uniformly distributed; (6) the scale of the map means that certain objects or features cannot be seen (Shea and McMaster, 1989). If the data you are working with conforms to any of these six conditions then generalization methods can be used to improve the clarity of the map and its design, reducing the amount of map junk, its complexity and clutter.

Summary

In this chapter we have covered the basic principles of cartography that are important to any map design. We have covered the five composition principles for maps, which should be considered before creating the map. This was followed by a discussion of projections, scale and generalization, which are all critical in the process of preparing data for visualization and presenting useful maps. These aspects are relevant for paper and digital maps. Naturally, this chapter is all but a brief introduction to the area of cartography.

Further reading

There is an extensive body of literature that encompasses research about the study of functional maps and cognitive theory, the research is nicely summarized by MacEacheren (2004) in *How Maps Work*. For a comprehensive and detailed introduction to cartography, the book by Robinson *et al.* entitled *Elements of Cartography* (1995) is still highly relevant and informative. For a more entertaining read, Mark Monmonier's books, and especially (1996) *How to Lie with Maps* is valuable reading, as well as John Krygier and Dennis Wood (2005) *Making Maps* which is highly accessible, useful and informative. For more detailed information on map projections see the book *Datums and Map Projections for Remote Sensing, GIS and Surveying* (Iliffe and Lott, 2008).

Revision questions

1. Before creating a map what questions do you need to answer?
2. How can scale influence the design of a map and how should it be represented in a web GIS?
3. How do map projections influence the look and feel of the map? Consider a map that you might want to create and consider how it will be impacted by different projections.
4. Using the principles that were described in this chapter, analyze some existing maps and critique their composition – how well do they respond to the criteria? By examining existing designs, you can learn lessons that you can integrate into your work.

4 Computer-mediated communication, collaboration and groupware

Mordechai (Muki) Haklay

In the previous chapters we have looked at the personal foundations of interaction with geospatial technologies – human spatial cognition and cartographic representations. In this chapter we move from the personal to the social realm, where a group of people use geospatial technologies collaboratively to solve a geographical problem. This method of using geospatial technologies is known as collaborative GIS, and includes several types of systems. In order of complexity, these are: multi-user GIS, collaborative Spatial Decision Support Systems and Participatory GIS (see MacEachren, 2000 for discussion of the last two).

In a multi-user GIS many users need to access and manipulate geographic information as part of the task that they are trying to accomplish – such as designing a water mains network or maintaining a national database of land ownership. In these situations, many of the issues that relate to collaborative working can be automated and managed by the GIS software. There are, however, some requirements to ensure that such automation is successful.

In collaborative Spatial Decision Support Systems (SDSS) settings, users are trying to solve spatial problems which are more complex and where multiple alternatives exist. The people who are involved in the decision making process need to negotiate and discuss possible solutions in order to reach a decision. Many of the problems that are dealt with through the use of collaborative SDSS are planning problems, such as locating facilities (e.g. schools or ambulances). In these situations there is a need to use a significant amount of data to provide the underlying information. In addition, special algorithms are used to support decision making, such as multi-criteria evaluation, where the users are requested to weight a set of criteria and the system provides feedback on the implication of this selection. Most importantly, due to the nature of the process collaborative SDSS requires a higher level of communication among participants than is the case in multi-user GIS.

Finally, participatory GIS, or PGIS, is a term used to describe the use of GIS with the widest possible group of users in order to discuss spatial problems. This can happen in public consultation regarding a new development or reorganization of a citywide transport system. In such situations there is a need to reach many participants and the use of public communication means such as the web is necessary. While the sophistication of the computation is somewhat lower than in collaborative SDSS, the social interaction and communication element is heightened, and there is a need to take into account the skills and abilities of the potential users. A deficient design can lead to misunderstandings and delays in the implementation of the proposed activity. Thus, this type of collaborative GIS requires special considerations and is the most challenging in terms of design requirements.

At the core of all these types of collaborative GIS is the use of computers and computer networks for communication and therefore Computer-Mediated Communication (CMC) holds a special place in their development. The communication between the users will frequently determine if the whole application is considered successful or not. Therefore, this chapter starts with a discussion about CMC.

Another aspect that is crucial for the success of collaborative GIS is the social setting in which it is embedded. The interaction among group members and the way in which the social dynamics evolve during the use of the application can influence the outcomes of the application use. The second section of this chapter is dedicated to social dynamics and group decision making.

Once the foundations of collaborative use of computer systems are covered, we turn to a short overview of the field of study within HCI that is dedicated to the use of computers to support collaborate work (Computer Supported Collaborative Work or CSCW) and the development of software that is specifically aimed at group work (Groupware). The background to this area of research is provided in the third section of the chapter.

Finally, the chapter turns to the principles of collaborative GIS and the issues that need to be taken into account in the design of such systems. More detailed discussion on specific issues of collaborative GIS design is provided in Chapter 12.

The chapter ends with a case study of a collaborative GIS that relies on Computer-Mediated Communication – OpenStreetMap, a large scale collaborative geographic information project.

4.1   Computer-mediated communication

At the core of any collaborative work is the need to facilitate communication between the participants. Therefore, it is not surprising that the most common and successful applications of computers are about communication – from email through to the web to the huge increase in the use of mobile phones across the globe.

A good way to understand the relevance and impact of Computer-Mediated Communication (CMC) is to consider its spatial and temporal aspect. In their discussion of groupware, Ellis *et al.* (1991) suggest a taxonomy of interaction that is based on these elements (Table 4.1).

The four modes of interactions that are suggested by Ellis and his colleagues are:

- Face-to-face interaction – when the participants in a collaborative project are at the same physical location at the same time, the most effective interaction is face-to-face discussion. There might be a use of technology during this meeting, but most of the communication is carried out without the mediation of computers. This is the richest mode of communication between humans (Brown and Duguid, 2000).

Table 4.1   Time and space aspect of collaborations (after Ellis *et al.*, 1991)

	Same Time	Different Times
Same place	Face-to-face interaction	Asynchronous interaction
Different places	Synchronous distributed interaction	Asynchronous distributed interaction

- Asynchronous interaction – this can happen when participants in the activity are at the same physical location, but at different times. For example, when different people are working on digitizing a map, but due to work times, they are using the same computer at different times of the day. In such situations, communication can be handled through the use of post-it notes, or by leaving digital notes.
- Synchronous distributed interaction – when the participants are in different locations, but know about the availability of each other for collaborating on a task, we can have a synchronous distributed interaction. A telephone call is an example of such interaction.
- Asynchronous distributed interaction – in this case, which is the most complex in terms of communication, the participants cannot be at the same place or coordinate a suitable time for phone conversation. Here, there is a need to facilitate communication despite the temporal and spatial differences. An answering machine provides such a service – the two sides in the communication are in different locations, and because the recipient cannot answer the phone, the answering machine records and preserves the message. However, we are all aware of the irritating nature of broken conversation that can result from continuous exchange of answering machine messages.

Geospatial technologies and CMC can have a role to play in every type of interaction, though more commonly in asynchronous and distributed interactions where they have become essential. In the following sections, we look at each of the communication media – face-to-face, video and audio communication and text-based communication, and see how they work with each type of interaction.

4.1.1   Face-to-face interactions

To understand the advantages and limitations of CMC, it is worth paying special attention to the nature of face-to-face communication. When observing face-to-face conversations, it is easy to notice that a very significant element of the information that is transferred among participants is not conveyed within the words that are spoken during the conversation. Many other elements are influential in setting the tone of the conversation, its progression and outcomes (Dix *et al.*, 2004). The physical setting of the conversation is significant – a meeting in a coffee shop has an informal atmosphere, which is difficult to maintain in a board room with a formal meeting table. The clothing and the posture of the participants also have an effect on the conversation. In a community meeting, if one participant appears in a three-piece suit while everyone else is in their daily clothing, this person will receive more attention during the discussion. The position that the different participants take around the table or around the discussion space is also influential.

Another set of elements relate to participants' bodily behaviour. Eye contact and gaze have an influence on the way people perceive the conversation as truthful or how they are engaged in the discussion with the person who is talking. In addition to gestures, body language, the range of back-channels such as the small 'um' or 'er' sounds that are frequently used in conversations, also influence the tone of the conversation and its meaning. Furthermore, accents, pitch and other characteristics of the spoken words and sentences are also influential.

There is significant coordination that occurs in face-to-face conversation – gazes, gestures and back-channels are frequently used to convey the message to the person who is speaking that another participant would like to say something. In group settings, an experienced facilitator will use these cues to manage the discussion and to ensure that all participants have an opportunity to participate in the conversation. All these elements make face-to-face conversation a very rich mode of communication. It is also the reason that in complex decision-making situations, when the stakes are high, this mode is preferred despite the range of electronic communication tools that are available today.

When the issues that are discussed in the meeting relate to geographic information or have a strong spatial element the use of language is not enough. Even when the topic is relatively simple, the use of graphic aids such as a sketch is necessary. Imagine a conversation where you try to explain to a group the route from your home to your childhood school and the position of landmarks along the route without using any graphic annotation. Although it is possible to describe this through text, this is rather challenging. Therefore, even in face-to-face situations we can find an extensive use of geospatial technologies. The use of a paper map or sketching on a paper can be a very effective way to represent information about the geographic element. Participants can gather around a map and use it to discuss the topic at hand, while pointing to a location or annotating the paper. Figure 4.1 provides an example for such a discussion. Notice how the use of the map causes the participants to turn their attention to the map and not to look at each other, although in reality people are alternating their gaze between the other participants and the map. This is also the case if the maps are projected on the wall using a data projector, or, if input from participants is important, on an interactive whiteboard. In such a case, the participants have to stand around the projected map or information to have a shared discussion. Usually a single person is controlling the map by using a computer

Figure 4.1   Face-to-face discussion using map

(because this person is 'driving' the application, the term 'chauffeur' is used to describe her) or a suitable device to write on the interactive whiteboard. The board offers a more afford-able technology in terms of enabling untrained users to participate in the discussion and annotate the information. With projected maps, there is usually a requirement for a knowl-edgeable operator to drive the system. On the other hand, a development of applications that allow high affordability requires more investment in time and resources, and it is not immediately available in most GIS packages, while projecting information from a GIS by using a data projector does not require specialized software.

There are also geospatial technologies that allow participants to see the information on the top of the table while the participants are sitting around it. An example of such an interactive environment was developed by Eran-Ben Joseph and his colleagues at MIT and is called 'Luminous Planning Table' (see Figure 4.2) (Ben Joseph *et al.*, 2001). This computerized workbench allows the participants to use physical models that are being positioned on the table and to see various computer simulations which are projected in coordination with the physical models. The participants are engaged in a discussion, and can take turns in changing things on the table and seeing the consequential impact. The configuration of participants

Figure 4.2   Face-to-face discussion using Luminous Planning Table (Image © Eran Ben-Joseph) (a full-colour version of this figure appears in the colour plate section of this book)

around the table allows for eye contact between them and for a discussion which is more similar to a usual face-to-face interaction.

In a situation where the information is presented on the table, the participants are all able to see each other and the participation is open to all the people around the table. In the whiteboard and data projector situation, the information is presented at a certain position in the room, and the participants can either gather standing around the projected image, or sit and look towards the screen, with frequent gazes to the speaker.

In short, the need to introduce geospatial technology in a meeting will, by necessity, change the dynamics and the development of the discussion. Different configurations will impact on the level of interaction between participants. Thus, there is a need to evaluate carefully what are the goals of the meeting and to choose the technological elements that will facilitate the required outcome in the best way.

4.1.2   Audio and visual based CMC

Because computers are generally networked and with integrated audio and video capabilities, they are now widely used as communication devices, with applications such as Internet telephony and online video conversation being used daily by millions of users.

The use of video or audio CMC is mostly as part of synchronous distributed interaction. While they can be used for asynchronous interactions – as in the case of leaving voice messages, or tagging a common plan with video annotation, they are less effective than text-based communication for these settings. The conversation cannot take place unless all the participants are available at the same time in places where they can access computers that are linked to high speed networks. Between the two modes, audio requires limited bandwidth, while high quality video requires significant bandwidth. The wide availability of affordable web cameras and microphones has made computers an effective communication tool for many, with most conversations involving two participants. Yet, even today there are severe limitations to running large group discussions over the Internet. Running a video conference with over 10 participants who are in different locations requires specialized computer hardware and software, or service from a company that can provide dedicated communication links to allow for multi-participant conversation.

Compared to face-to-face conversations, video conversations provide some of the non-verbal cues, but not all. For example, the participants don't share the same physical experience of the meeting space, so some participants might be at their home while others might be at the office. Or because of the use of cameras the participants observe only part of the body and the viewing position is static unlike the face-to-face case where the point of view changes constantly. However, the ability to see the participants' face and body means that gestures, facial expressions and other information about the participants can be gleaned.

Naturally, audio conversation provides even fewer cues to the participants. In such a situation there is a need to introduce protocols into the conversation – such as the 'hallo' response to a phone call, as a way to identify one caller to the other. The lack of cues can be advantageous in some situations – for example, when you want to participate in a business meeting but you are currently at home and do not want to dress up.

Significantly, Audio and Video CMC works fairly well when the conversation is only between two parties. The information is rich enough to provide an interaction level that is close to the face-to-face experience. However, as the number of participants increases, the need to provide more rigid protocols to the discussion become greater. For example, control over the process of turn taking during a conversation is needed and therefore in radio communication the messages such as 'over' and call signs were introduced. Even when the communication is handled by computers there is a need to have a clear role of chairing the meeting and managing the conversation. There is also a need for an alternative mechanism for a shared discussion – such as an online notepad area where the participants can write questions and comments.

As with the face-to-face case, when using Audio and Video CMC to discuss a geographical problem, there is the need to share additional information in a graphic form. The reduction in the richness of the communication channel makes the additional information more important. The use of geospatial technology during such group conversations is by providing an add-on to the communication software that allows the users to see a shared application. At the most basic level, the participants can see an application that is run by one of them. A more sophisticated application allows participants to take control in turns to discuss an issue, point to a specific location or change the visualization.

4.1.3   Text-based CMC

Based on the analysis above of the richness of face-to-face communication, the use of text-based communication seems limited in its expressive ability. Yet, of all forms of CMC, this is the most common. It is also the most flexible as demonstrated in the success of the Short Messaging Service (SMS) over voice messaging as an application of mobile telephony.

The most common synchronous CMC is based on text communication, where the participants 'chat' by exchanging short textual messages. The success of instant messaging Internet applications such as ICQ or Microsoft Instant Messenger is based on this mode of communication. It is also one of the oldest forms of CMC, with Internet Relay Chat (IRC) dating from the 1980s. This mode of communication creates a whole plethora of unique and new psychological and social issues (Turkle, 1984, 1995). For example, because of the lack of almost any social cues, text-based communication allows participants to pretend that they are someone else, or assume fictitious characteristics which they don't have in real life. Another aspect is mistakes in interpretation of the meaning or intentions of the writer, which led to the introduction of emoticons (such as the now ubiquitous 'smile' :-) or 'wink' ;-)). On the positive side, text-based communication has been shown to be empowering for people who otherwise would avoid speaking in public – for example, for a person with speech impediment. Another advantage is that the history of the conversation is recorded and can be used as a reference at a later stage.

As in the case of audio and video CMC, synchronous CMC requires that both participants will be near their computers to participate in the conversation. The extension of instant messaging software to mobile phones increases the ability of participants to be on the move while being part of the conversation. The discrete and short messaging nature of the discussion allows participants to engage in several conversations simultaneously, though clearly with an impact on the attention of the participants.

Because of the reduced need for bandwidth and the ability of modern systems to indicate if one of the participants is currently writing a message, text-based CMC is very suitable for communication amongst a large group of participants.

A case study of using text-based CMC with geospatial technology is provided below from Thomas Friedman's *The World is Flat* (2006), 38–39.

Case Study 4.1: CMC for Intelligence interpretation

'In the fall of 2004, I accompanied the chairman of the Joint Chiefs of Staff, General Richard Myers, on a tour of hot spots in Iraq. We visited Baghdad, the U.S. military headquarters in Fallujah, and the 24th Marine Expeditionary Unit (MEU) encampment outside Babil, in the heart of Iraq's so-called Sunni Triangle. The makeshift 24th MEU base is a sort of Fort Apache, in the middle of a pretty hostile Iraqi Sunni Muslim population. While General Myers was meeting with officers and enlisted men there, I was free to walk around the base, and eventually I wandered into the command center, where my eye was immediately caught by a large flat-screen TV. On the screen was a live TV feed that looked to be coming from some kind of overhead camera. It showed some people moving around behind a house. Also on the screen, along the right side, was an active instant-messaging chat room, which seemed to be discussing the scene on the TV.

"What is that?" I asked the soldier who was carefully monitoring all the images from a laptop. He explained that a U.S. Predator drone – a small pilotless aircraft with a high-power television camera – was flying over an Iraqi village, in the 24th MEU's area of operation, and feeding real-time intelligence images back to his laptop and this flat screen. This drone was actually being "flown" and manipulated by an expert who was sitting back at Nellis Air Force Base in Las Vegas, Nevada. That's right, the drone over Iraq was actually being remotely directed from Las Vegas. Meanwhile, the video images it was beaming back were being watched simultaneously by the 24th MEU, United States Central Command headquarters in Tampa, CentCom regional headquarters in Qatar, in the Pentagon, and probably also at the CIA. The different analysts around the world were conducting an online chat about how to interpret what was going on and what to do about it. It was their conversation that was scrolling down the right side of the screen.'

The advantage of text-based CMC in the context of geospatial technology is that it does not consume a large space on the screen, and, as we shall see, one of the more important aspects of designing geospatial technology is to provide as many as possible 'screen assets' to the geographic information that is presented – in the case above, the image that is captured by the drone. In addition, the history of the messages provides the participants with the context of the discussion, while allowing them to take turns in suggesting ideas or questioning the information that they are viewing.

Text-based CMC is the most effective tool for the asynchronous, distributed case, which is the only mode in Table 4.1 which has not been covered so far. In this case, there is no expectation that the participants will be at the same place or interact at the same time. It allows participants to be involved in conversations that are progressing at a slower pace and to consider their contributions at length. In many ways, in asynchronous distributed interaction, text imitates 'off-line' media such as use of mail, memos and notes, but the temporal difference makes them distinct and different – for example, the speed at which electronic mail travels across the world compared with regular mail. Thus, it is an accelerated modality of communication.

The two common media for this type of communication are either a personal tool where the participants are dealing with all messages according to their own priorities, as in the case of email, or in a shared interface where the focus is around the discussion, as in the case of discussion forums.

In both cases, the conversation progresses through the exchanges of messages. The messages can be broadcast to many readers – either by displaying them in a public space that all participants can visit, or by sending them to their email inbox by using mailing lists. The responses to the messages can be made by any member of the list or the discussion who wants to respond to the issues. Some issues develop as threads, with responses and development of conversations, and it is common that in this mode of communication several topics are discussed simultaneously.

The main issue that arises in the use of asynchronous text-based CMC with geospatial technology is with the need to link the conversation to a geographic location. This is because the participants in the conversation need to know which geographical area or object they are talking about. This requires adding a marker on the map interface that shows that there is a comment that relates to the object, and in the discussion board area there is a need to create a link or a button that allows the reader to return to the map. Another way to represent the discussion is by allowing the discussion to be integrated with the map by making it pop-up when a user clicks on a marker that provides a cue that a discussion is associated with the area.

4.1.4   Advantages and disadvantages of CMC for geospatial technology

On the basis of Ellis *et al.*'s (1991) typology, Mitchell (1999) and Shiffer (1999) developed their analysis of the role of geospatial technologies in public discourse. This analysis is also useful in summarizing the points that have been raised in this section. In this analysis, 'same place' is exchanged for 'presence', and 'different places' with 'telepresence' to highlight the communicative element. The temporal labels are also adjusted with 'synchronous' and 'asynchronous'. Table 4.2 summarizes this analysis and provides examples for 'traditional media' and ICT. Each cell is divided into three areas. The top-most are the advantages and disadvantages, with the advantages marked with a plus sign, and disadvantages with a minus sign. This is followed by examples of traditional uses for each medium. The last cell provides CMC examples.

The advantages and disadvantages are assessed by taking into account the monetary and non-monetary aspects of organizing and running a collaborative session. Some of the direct costs are the need to hire a space or bring all the participants to a location and thus there are transportation costs. There are also organizational costs such as coordination and the ability to get feedback from the participants and reaching a conclusion to the discussion.

Table 4.2 shows that none of the communication modes is ideal. Each mode is suitable for a specific context and some of the issues that are noted as advantages and disadvantages may be the other way around for a specific application or context. It is therefore important to take into account these aspects when considering the most suitable CMC for a specific application, taking into account the spatial, temporal and channel (face-to-face, video, audio or text) elements. The decision should not be solely based on technical limitations, but on balancing the technical aspects with the level of the discussion that is needed to deal with the specific context of the application.

Table 4.2 Advantages, disadvantages and examples for modes of communication (After Mitchell, 1999 and Shiffer, 1999)

	Synchronous	Asynchronous
Presence	+ Intense, multimodal – speech, visual . . . + Immediate feedback + Few technical obstacles – High transportation costs – High space costs – Need for coordination between participants, space . . . – Requires full attention	+ No need for coordination + May allow some division in attention + Requires some mechanisms for communication between participants – Limited by storage and playback capabilities – No immediate feedback – High transportation costs – High space costs
	Public meeting	Public exhibitions of plans
	Shared use of information systems	Community information kiosks
Telepresence	– Need for coordination between participants + May allow some division of attention + Immediate feedback – Limited by bandwidth and interface capabilities + Reduced transportation costs + Reduced space costs	+ Reduced transportation costs + Reduced space costs + No need for coordination between participants + Allows multiple activities and transactions in parallel – Limited by storage, bandwidth and interface capabilities – No immediate feedback – High reliance on technology
	Telephone/conference calls	Letters, Email communication
	Video conferencing, chat, IRC Interactive TV Phone-in radio or TV discussions	Email lists Internet discussion groups Web systems

4.2   Social dynamics and group decision-making issues

Once we move from the specific aspects of the communication channel, there are considerations that are linked to the social interactions that occur in such situations. In collaborative systems, there is a need to consider the roles of the different people who are going to use it and the interactions among the participants who form the group.

If the group that will use the system or will participate in the decision-making process already exists, the participants will bring their existing relationships and hierarchies to the discussion. In the case that the group does not already exist, there will be necessary stages of introductions and forming social relationships among members of the group. There are various models for understanding the process that group members will go through when setting up a team. Such a model was suggested by Bruce Tuckman in 1965 and is still widely popular. The Tuckman Model suggests that a group will go through four stages of development – forming, storming, norming and performing with a final stage of adjourning when the group dissolves. Forming is the first stage in the process, where participants explore the interpersonal boundaries and set up the behaviour and relationships within the group – including identification of a leader. The second stage, storming, includes some conflict and possible polarization among group members. This stage is useful for each group member to identify their position in the group and to establish working relationships inside the group.

In the norming stage the group develops cohesiveness and roles for the task are established. Finally, the group ends the process of establishing the social dynamics and turns to deal with the task itself – therefore focusing on performing the task.

The development of various roles is another aspect of group dynamics that needs to be taken into account. Without a direct facilitation and design, some member of the group might assume the role of leadership and will set the tone for the discussion and the relationships within the group. This can occur even over asynchronous distributed means, for example, when one person is starting to post multiple messages in a discussion board or respond to all other comments. Therefore, if the aim of the collaborative process is to ensure inclusion of many participants, it is important to introduce a role of facilitator to ensure that the process itself is managed carefully.

In short processes, in which the use of geospatial technology is occurring over a period of a day or two, the creation of a cohesive group with defined roles is possible. However, since many of the processes are more complex and require a longer time frame to evolve, there is a need to consider the way in which participants will join or drop out from the process, as well as assuming different roles. This is an especially significant issue for the design of collaborative GIS, as in some cases programmers prefer to set rigid roles to different members of the group as this simplifies the application development. However, this can create problems when participants change roles.

The composition of the group and the differences and social roles that participants are bringing with them to the process are another significant factor. For example, gender issues can play a significant role. A group which is composed of a single gender can behave in a different way than a mixed group. A mixture between elderly and young people can influence the involvement of different participants due to cultural convention. People from different ethnic backgrounds might also feel that they cannot participate in the discussion on an equal footing.

General experiments in group decision making using computerized decision support systems in a laboratory setting have led to some conclusions. For example, they have shown that group performance depends on the familiarity of users with the technology that is being used. The more experienced they are, the faster the process. This can be compensated with an experienced facilitator. Other studies have shown that in simple decision-making processes, technology does not speed up the decision making, whereas in complex issues it can assist by providing the needed information and algorithms. It was also found that the presence of a facilitator improves participants' satisfaction with the technology and the process. This means that the social dynamics that are occurring during the meeting have a significant influence on the perception of the technology and the way it is used (Jankowski and Nyerges, 2001).

Unfortunately, there is a lack of studies that evaluate the social dynamics that occur during the use of geospatial technologies in collaborative settings. In one of the few studies that evaluate social dynamics in group decision making, Jankowski and Nyerges (2001) have shown that users use general purpose maps to visualize the location of alternatives and to evaluate tradeoffs among alternatives. They have also shown that in different stages of the problem-solving process, there are different levels of conflict among group members, with the exploratory stage, with which the problem exploration starts, characterized by low levels of conflicts, while the analytical phase, when the different alternatives are evaluated, is characterized by high levels of conflict.

A final aspect that needs to be taken into account is the goal of the process – is it mainly aimed at reaching a decision that is accepted by a majority of participants, or by core stakeholders, or is the aim to establish a decision through consensus? If the aim is to reach a decision, regardless of opposing views, a system can be configured to allow people to vote for a preferred solution and this solution will be the one that will be put forward. However, if the aim is to reach a consensus among participants, there is a need to facilitate a discussion, increase trust in the system and allow opinions to be discussed. In the latter case, the importance of efficient and trustworthy CMC is especially significant. For example, Reitsma and her colleagues (1996), in another rare study of group decision making while using geospatial technologies, have shown that the availability of geographical models to participants helped in identifying some alternatives that meet suitable constraints. This can be useful in developing consensus – as long as the participants trust the model itself.

4.3   Computer Supported Collaborative Work and Groupware (CSCW)

Although communication is both critical and fundamental to any collaborative work, the area of Computer Supported Collaborative Work (CSCW) and the development of software that supports group work (Groupware) go beyond communication systems. Figure 4.3 provides a simplified model of the interaction between participants and the objects that they are working on. The participants who are involved in the work (the circles marked by 'P') need to develop a shared understanding of the task that they are trying to accomplish. To achieve this shared understanding of the task, the participants need to communicate in different modes (face-to-face, video, audio and text) and pass messages that build shared understanding, confirm this shared understanding, and build on it with new ideas. In addition to the interaction between the participants, there are also the artefacts that are part of the task. Sometimes, the artefacts are tangible – for example, a map that the participants are using. However, the artefacts can be digital or even conceptual. The existence of the artefacts necessitates additional interaction between the participants and the artefacts, providing control over the object while changing it, and feedback about changes that occur in the artefacts through other participants. While the simple model shows the situation with two participants, in most cases the requirements are to allow collaboration among a large group of participants, sometimes with tens of users that working on the same task.

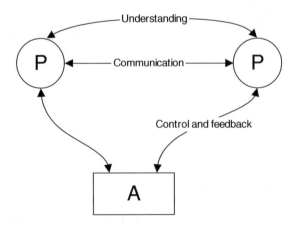

Figure 4.3   Cooperative Work Framework (after Dix *et al.*, 2004)

The design of CSCW is therefore a combination between an effective communication system that allows the development and maintenance of shared understanding, combined with a technical framework that provides the control and feedback of the artefacts to participants.

When we looked at the integration of geospatial technologies with different modes of communications in the previous section, some devices that were developed by CSCW researchers were introduced, such as the integration of interactive whiteboards and the sharing of a computer application across all participants in the distributed synchronous case. GIScience researchers have contributed to this area in a significant way, as the example of the Luminous Planning Table demonstrates. This is the result of the need to include geographic artefacts (plans, maps or charts) as the element that is being discussed by the participants.

In today's workplace, much CSCW software – or groupware – is being used without the users even noticing. Word processors that provide the ability to co-author documents by allowing users to track changes or shared calendars that allow users to identify when a colleague is free and set a meeting are now ubiquitous, yet their origins are in CSCW research.

When considering CSCW, it is worth refining the spatial/temporal matrix that was introduced in Table 4.1, and considering the different levels of coordination required regarding access to the artefacts and to the communication amongst participants. The range of options, according to Dix *et al.* (2004) include concurrent synchronized, mixed, serial and unsynchronized forms of interaction. In a concurrent synchronized system, the participants need to agree on the control of the artefact and usually only one user has full control at a given moment. For example, in a case of using a shared computer to manipulate geographic information that is presented on the screen in the meeting room, only one user at a time can actively control the mouse and keyboard and manipulate the information.

Mixed interaction is relevant when users want to works in both modes – synchronized and unsynchronized. This can happen when a design team works on the same sets of maps, with each member dealing with a specific issue. They might want to work from home, or while mobile, but also to continuously update the shared database when they are connected to the network again. This type of system requires testing for conflicts that can arise from different users updating the same area of the map, unaware of the actions of other users. The process of merging different versions of the same information can be very complex and time consuming. Even for non geographic information – such as editing a document – there is no groupware system that solves the problems of mixed interaction perfectly.

The serial mode of interaction is useful in cases where users can work on a shared document or information, but it is possible to force the process of turn-taking automatically. In the example that was just discussed, it is possible to decide on a spatial or thematic lock. In such a case, while one of the participants in the discussion is changing the data in a specific area or a specific layer, no other participant can change it until the person who works on it has completed their activity. Argumentation systems such as Argumap (Rinner, 1999), also force the participants to be explicit and structured in the way that they are engaged in the discussion. At the same time, there is no need for close coordination and different users can engage in different parts of the discussion at the same time.

The unsynchronized interaction is relevant in cases where participants are working together and the conditions of the task are such that there is no need to work on the same information at the same time. In cases like these, the system can allow many users to annotate information on a shared working space, with each one of them doing it in their own time. This is a useful mode in many participatory GIS settings.

Connected to these different variations of synchronization is the way in which the artefacts are involved in the communication itself. In many cases, this communication is implicit – for example, when one user edits an object in a layer in collaborative GIS. The other users can see that a change has been made and do not require further information. In other cases, there is a need for the participants to explain what change they have made to the system. In such cases, the application might demand an explanation of the change that was made before committing it to the shared database.

4.4   Principles of collaborative GIS

Having looked at the foundations of collaborative GIS, this section turns to review the principles of the different types of collaborative GIS.

As noted in Chapter 1, the development of GIS in general is frequently hampered by limits in computing. In one of the first discussions of the requirements for collaborative GIS, Armstrong (1994) identified three main obstacles to the implementation of such systems. First, there was a need for further development of hardware to facilitate CSCW – as in cases where the participants were requesting a simulation of a scenario which might be computationally complex, or the need to display complex visualization when the screen resolution is limited; second, specialized software was required to facilitate group activities in GIS which at the time was mainly aimed at single users; and third, conceptual frameworks for collaborative GIS implementation had to be formulated to facilitate the development of coherent and consistent systems.

More than a decade later, significant progress has been made in all these areas. On the hardware side, most obstacles have been removed. Modern personal computers (PCs) are equipped with fast processors, large storage and memories, and strong graphics capabilities. The ability to display detailed colourful maps, 3D models, or animated maps is integral to the computer systems that are available today. Furthermore, the networking capabilities in a wired or wireless environment are adequate to enable sharing of geographical information resources within organizations, as well as with other organizations and individuals. Noteworthy is the development of mobile computing, and the increased use of mobile telephones, which are, in effect, information devices capable of Internet connection.

On the software side, a great deal of improvement has occurred in GIS in general, and in collaborative GIS in particular. With advances in programming tools, knowledge in geographical visualization, and spatial decision support functionality, today's software tools offer sophisticated interfaces that enable users to explore many dimensions of the problem at hand. Other tools for collaboration, such as shared discussion boards are also available on the web, and are relatively easy to implement.

As for conceptual frameworks, Jankowski and Nyerges (2001) offer a comprehensive framework for the development of collaborative GIS, known as Enhanced Adaptive Structuration Theory (EAST2). This is a macro-micro strategy drawing on multiple disciplines. It is based on sociology, philosophy, management science, operational research, planning theory, GIS researchers, and their critics. EAST2 provides theoretical grounding for many facets of collaborative decision-making processes. Other frameworks are offered in the writings of Carver and his colleagues (Carver, 2001; Carver et al., 2001), although less explicitly, as well in Balram and Dragicevic (2006).

Despite all these significant developments, research into collaborative GIS is far from complete. Collaborative GIS seems to be a 'wicked problem' all by itself. Wicked problems are problems that change their characteristics as a result of the attempt to solve them and frequently mutate as a result of external factors. In the case of collaborative GIS, the problem arises from the fact that as GIS is becoming a common tool for planning activities at various organizational and spatial scales, the variety of users increases, as does the range of computer skills that the system must accommodate. These skills include map reading, computer operation, and data analysis, to name but a few. The growing reliance on the Internet as the medium for facilitating collaborations has introduced a new set of issues, from the speed of access to the use of the Web-browser as the main interface. The current transition to ubiquitous computing, and the need to consider situations where the user is accessing the system from a wide variety of devices, such as computers, digital television sets, and mobile phones, raise further issues.

In the following sections, we take a second look at the three types of collaborative GIS – multi-user GIS, collaborative Spatial Decision Support Systems and participatory GIS. This time, we will review them in light of the issues that were covered in previous sections.

4.4.1   Multi-user GIS

In the case of multi-user GIS, most collaboration can be managed and policed without the need to create a dedicated CSCW application. Many of the issues that relate to the co-ordination among users can be done by implementing rules and functionality that are available within the GIS or the database that stores the geographic information.

Of importance is the nature of transactions among the participants. In many non-geographic multi-user systems, the handling of transactions – the process of updating shared information in a way that will guarantee that the shared copy is the authoritative one – is usually a short process. Therefore, the mechanism to ensure that only one user at a time updates the information is carried out by locking the information to all other users; updating the shared copy with the new information; and finally releasing the lock. Even if many users are working on the same information, the delays that they experience due to locks are tolerable.

In contrast, in a multi-user GIS it is common to have 'long' transactions. For example, when one of the users updates the base map from an aerial photograph, this process can take several days, and while the work is going on, it is important to ensure that the underlying information is not changed by other users. Therefore, many GIS packages and spatial database systems provide the functionality to lock some objects or an area in the application for a long period, and to allow a process of 'check out' and 'check in' of updates. Note, that in this case, the artefact (the GIS database) needs to convey information among users about who is working on what area, so other users can resolve conflicts in terms of editing and changing the information. This communication can be implicit, so the system registers only the details of the user who checked out the information. However, from a design perspective it is more useful to ensure that users are communicating in a more explicit way – for example, by filling in a form every time they check out information which explains the operation that they are carrying out, or when they expect to complete it and release the information for further editing.

In this type of collaborative GIS, the work can be carried out in a mixed mode (synchronized and unsynchronized), with suitable mechanisms to lock information, allow users to extract it for editing and merge it back to the main database. The level of interaction with the end user can be limited to providing information about changes that were carried out.

While the potential for combining such a system with more comprehensive CSCW applications is apparent, so far this has not been done by any of the major vendors of GIS.

4.4.2 Collaborative Spatial Decision Support Systems

Spatial Decision Support Systems (SDSS) differ from multi-user GIS by the type of geographic problems that they are used for. Multi-user GIS are used in situations where the users' tasks are clear and don't require sophisticated algorithms and data sets. For example, when the system is used as a repository of geographic information and the main function of it is to coordinate updates to the shared database. In Spatial Decision Support System, the aim is to help the user to solve a more complex geographic problem, where the information that is required to solve it belongs to several domains and is usually extensive. In addition, specialized computer algorithms are used to calculate possible alternatives and to evaluate the range of solutions. SDSS are being used for a range of tasks, from routine tasks such as route planning to specialized and complex tasks, such as decisions about setting school catchment areas or location decisions for new retail outlets.

Collaborative SDSS are used in cases where the decision-making process involves more than one user. Their most common use is in planning problems, where the solution to the problem is based on multiple areas of knowledge or involves many stakeholders. Thus, the problems that collaborative SDSS are dealing with are ill structured, 'wicked' and it is usually impossible to define a simple optimal solution. While in some SDSS, the user trusts the computer to provide a solution – for example, when an SDSS is used for taxi dispatching decisions – in collaborative SDSS the decision is based on social, economic and environmental tradeoffs and therefore on group decision making. Thus, while in SDSS there is a need to communicate with the user to explain and justify the recommendation that the computer has made, in collaborative SDSS the communication among group members is paramount. In such situations, it is expected that the participants will have wide variability in knowledge of the problem domain and of computer use. Therefore, it is important to make all the communication lines that are presented in Figure 4.3 explicit – not just among the participants, but also with respect to computer models, with explanations of the actions that were made by the computer and allowing users to experiment with different scenarios and configurations of the problem situation.

While the literature offers examples of asynchronous and distributed interaction applications of collaborative SDSS, they will usually be used in conjunction with face-to-face meetings due to the political and economic implications of the decisions. Therefore, in many cases where a collaborative SDSS is used, geospatial technologies are used during group discussion and the social dynamics of introducing computerized tools in the discussion room should be considered carefully, before opting for a 'chauffeur' based solution or providing all the participants with computers so they can view the information for themselves. In the latter case, there is a risk that some users will not participate fully in the discussion, because they are doing a task with the computer in front of them. Recent research has shown that in group settings, if a user spends 10 seconds on a specific task on their laptops, they will disengage from the discussion for a significant period of time (Newman and Smith, 2006).

In the asynchronous and distributed case, there is a strong need to make sure that the communication does not introduce misunderstanding and confusion to the discussion. Systems like Argumap, which allows participants to explicitly state their positions regarding specific spatial locations and issues, have been designed to allow a structured discussion. While this can be helpful, it does remove some useful ambiguity and the ability of changes in positions by the participants, with the risk that positions will become entrenched since a change of views will be logged and be visible to all.

4.4.3   Participatory GIS

The final type of collaborative GIS is participatory GIS. This area of research emerged in the 1990s, due to the increased use of the web to deliver geographic information to the public, and increase in multimedia capabilities of personal computers as well as a growing interest of GIScience researchers in using the technology to engage communities in decision-making processes that influence their lives. Two terms are being used to describe this area – Participatory GIS (PGIS) and Public Participation GIS (PPGIS). The extensive PPGIS literature covers both the technical and societal aspects of setting up and running such systems (see Dunn, 2007 and Sieber, 2006).

While there are examples of participatory GIS use to support face-to-face public discussions, only in very small groups is it possible to use methods where participants use computers directly. This is because many public meetings in which participatory GIS is used are carried out with large groups of participants, many of them with limited computer knowledge. Moreover, unlike the case study with group SDSS, the knowledge of the problem domain is also limited.

In the cases where geospatial technologies are being used in very large face-to-face public meetings, they need to be simplified to ensure that participants can use the system successfully. One of the good examples for facilitation of large group discussion was demonstrated in the 'Listening to the city' event in New York in August 2002, when 4500 participants discussed the development plans for the World Trade Centre area. During this meeting, a 'chauffeur' was assigned to each group of participants to show the information on development alternatives. Simple voting devices (of the type that are used in many TV shows) were used to allow participants to vote on their preferred choice. In smaller consultations, where the number of participants is several dozens, projecting information from a computer with a 'chauffeur' is also common.

Yet, the literature also includes cases where the participation was facilitated by using more basic geospatial technologies such as paper maps or papier mâché models. These information sources are then transferred to digital form and integrated with other layers of information.

One of the common applications of Participatory GIS is to use it for community consultation about proposed plans such as changes to the transport system or the built environment. The need to include many participants who are dispersed over a large area and with difficulties in attending a meeting at a specific time means that asynchronous and distributed modes of interaction are the most suitable. Furthermore, because the information needs to be provided over public telecommunication networks, the use of the web is very common in this area (Carver *et al.*, 2001; Kingston, 2002).

Participatory GIS represents several design challenges: first, most participants will use the applications only a few times, and without support, so the interface should allow users to learn the application quickly. Second, CMC should support communication among groups of participants who are not familiar with one another, and thus the risk of misunderstandings is heightened. Third, the knowledge of using and manipulating geospatial technologies – even at the level of zoom in and out – should not be taken as given and some users might have difficulties with spatial literacy and map reading. Finally, because of the difficulties in communication, the role of the digital artefacts within the process increases. It is therefore important to ensure that the way in which objects are controlled and how feedback is provided to users is clear and informative. In summary, participatory GIS represents a major design challenge (Haklay and Tobón, 2003).

Case Study 4.2: OpenStreetMap – large scale distributed collaboration

By the mid 2000s, the wide availability of broadband Internet connections, computer speed and web design enabled collaboration among large groups of users. Through this collaborative effort, users can work on shared tasks, such as creating an online encyclopaedia (e.g. Wikipedia). The web has made it possible for a distributed group of users who are loosely coordinated to work towards a shared goal.

The project OpenStreetMap (OSM) is the best example of this type of collaboration in the area of geographic information. OSM is a project to create a set of map data which are free to use, editable and licenced under new copyright schemes (see Figure 4.4). A key motivation for this project is to enable free access to current GI where, in European countries, digital geographic information is considered to be expensive. In the US, where basic road data is available through the U.S. Census Bureau TIGER/Line programme, the details that are provided are limited (streets and

Figure 4.4 High resolution map from OSM of the area near University College London (Data source: OpenStreetMap.org) (a full-colour version of this figure appears in the colour plate section of this book)

roads only) and do not include green space, landmarks and the like and is not updated frequently.

OSM includes many thousands of people who participate in data collection and a core group of about 50 people who develop the core technology and maintain the servers. Significantly, even this highly active group meets face-to-face very rarely and therefore relies on CMC for communication. Therefore, OSM is an example of a collaborative GIS and the way in which it operates can provide many lessons on how to coordinate the work of a large group of participants.

OSM data can be edited online through a Wiki-like interface, where once a user has created an account, the underlying map data can be viewed and edited. A number of sources have been used to create these maps, including uploaded Global Positioning System (GPS) tracks, out of copyright maps and aerial photographs through collaboration with Yahoo!

Because data collection usually requires a survey on the ground to collect the data, the OSM community also organizes a series of local workshops (called 'mapping parties') which aim to create and annotate content for localized geographical areas. These events are designed to introduce new contributors to the community with hands-on experience of collecting data, while positively contributing to the project overall by generating new data and street labelling as part of the exercise. The OSM data are stored in servers at University College London, which contributes the bandwidth for this project.

The project includes editing tools developed by participants with a lightweight editing software package that works within the browser and another standalone version, more akin to a GIS editing package.

Involvement in the project requires the participants to be knowledgeable about computers and GPS technology, in order to know how to collect GPS tracks, upload them to their computers and then edit them and upload them to the OSM server. The use of the data also requires knowledge on how to extract the information from a database and convert it into a usable format. This means that technically, the users are fairly computer literate, though the technical hurdle is not as significant as using a fully fledged GIS package.

OSM provides a good example for the social and technical aspects that were highlighted in this chapter. The project provides synchronous communication among users in the form of Internet Relay Chats (IRCs) as well as using mixed interaction like mailing lists and twitter channels. More information is provided on the main Wiki, which contains instructions on how to join the project, what information to record as well as participant diaries. The importance of the Wiki is that it allows any participant to update the information without the need to define a specific role of document owner, editor etc.

Within the set of tools that are available to participants, OSM Mapper demonstrates how a distributed group of users can coordinate their work with little communication. The tool was developed by ITO World, a UK company that specializes in providing Web-based services for transport professionals and transport users. OSM Mapper allows participants to see many details of the information that exist in their area and to consider what information is missing in the database. It also allows participants to see the user-names of other participants who are collecting data in their area (Figure 4.5).

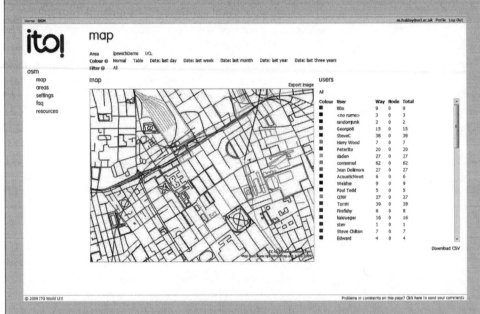

Figure 4.5 OSM Mapper, showing the list of users who collected the data (Image courtesy of ITO World Ltd)

In addition to OSM Mapper, another application allows users of the maps to point to errors in the underlying data (http://openstreetbugs.appspot.com/) and several other tools.

OSM demonstrates how general CSCW can be adjusted to serve a geospatial technology project. Because of the collaborative nature of the project, participants regularly suggest ideas for tools that can improve data collection and quality.

Summary

In this chapter, we have looked at the collaborative GIS and geospatial technologies. The three types of applications that are common in collaborative GIS – multi-user GIS, collaborative SDSS and participatory GIS – share common characteristics which need to be taken into account when designing them. The most important element in any collaborative geospatial technology is the communicative element. Therefore, it is important to understand the role of CMC and to ensure an appropriate theoretical framework – such as the one proposed by Jankowski and Nyerges (2001) – is selected before starting the development process.

CSCW and groupware research clearly influence the development of collaborative geospatial technologies. However, there is a need to take into account the unique characteristics of geospatial technologies – the use of geographic information, or reliance and use of map-based representations.

Further reading

For an introduction to the HCI aspects of CMC, CSCW and Groupware, see Sharp *et al.* (2007) Chapter 4 or Schneiderman and Plaisant (2005) Chapter 10. The latter provides more academic context, while the former includes many examples and is very accessible and comprehensive. Dix *et al.* (2004) provide a comprehensive discussion of Groupware in Chapter 19.

In the area of geospatial technologies, there are several books dedicated to group work. The most detailed discussion of collaborative GIS is Jankowski and Nyerges (2001) *Geographic Information Systems for Group Decision Making*. The book discusses the EAST 2 framework developed by its authors, and provides examples for collaborative SDSS. Another recent edited collection by Balram and Dragicevic (2006) *Collaborative Geographic Information Systems* covers many aspects – from usability to community aspects of collaborative GIS. Finally, Laurini (2001) *Information Systems for Urban Planning* provides a discussion of the use of groupware in urban applications of GIS.

Revision questions

1. What are the advantages and disadvantages of the four possible channels (face-to-face, video, audio and text) when used in conjunction with geospatial technologies?
2. Web GIS, Standalone desktop GIS, Luminous Planning Table – how can these geospatial technologies be used in different spatial and temporal settings?
3. What is the role of CMC in each of the three types of collaborative GIS?
4. What will be your topmost consideration for a successful public decision support system for locating recycling centres?
5. What are the implications of group dynamics for the design of collaborative applications such as OSM bugs?

Section II

Framework

5   User-centred design

Mordechai (Muki) Haklay and Annu-Maaria Nivala

User-Centred Design (UCD) is a framework for hardware and software development that emerged in the mid 1980s and from then became one of the guiding principles for designing usable technologies. As we shall see, UCD is both a philosophy of design which can apply to digital as well as non-digital products, and also a framework for application development. To understand the concepts of UCD, the chapter starts with a background to UCD, followed by the main principles of this framework. The chapter also discusses participatory design, which can be considered as a method to implement UCD.

But first, in order to understand why UCD is so important, we turn to a case of non-UCD development, which provides a sobering tale of the results of ignoring the needs of users in a development process of geospatial technologies.

Case Study 5.1: San Jose Police Department dispatching system

Aaron Marcus, *President, Aaron Marcus and Associates, Inc.*

In June of 2004, the San Jose, California, Police Department (SJPD) rolled out a new mobile, in-vehicle communication system for police officers. This rollout completed the entire replacement of a dispatch and mobile system that had been in place since 1990 and that the SJPD and the developers had spent many years perfecting. Through this long development process, SJPD had learned much about what made a very usable system for their officers, sergeants, lieutenants (managers of teams of officers), dispatchers, and dispatch managers. They considered themselves among the more knowledgeable police departments in the USA, having had many years of experience debugging and using their system. However, it gradually became apparent that a system based on a long-obsolete OS (Windows 95) was in need of overhaul or replacement. Among other factors, connecting with new communications technologies such as wireless was not possible with the current system.

A new system was commissioned by a committee of 10 members comprising SJPD managers, information technology (IT) representatives, and managers, but no officers or dispatchers. At the end of the tender process, they concluded in their estimates that upgrading the closed, customized system would be more costly than using 'off-the-shelf' components from a commercial vendor. This was a significant shift in development philosophy. After further consideration, they selected

---

Figure 5.1   Policemen with the new information system installed (Image courtesy of Aaron Marcus) (a full-colour version of this figure appears in the colour plate section of this book)

off-the-shelf modules from a major GIS vendor, which were additional elements of a dispatching system that the SJPD had also recently installed.

Customization and debugging of the system was an ongoing process since initial rollout, involving teams from the vendor, the SJPD's IT department, and other parts of the SJPD. Early experience with the new system prompted many concerns, including safety concerns, on the part of many SJPD officers and supervisors because this system for police officers had crashed upon first rollout.

The officers were given only a fifth of the hours of training that were recommended by the software vendor: although the software vendor recommended 16 hours of training, the officers were given only three. Furthermore, the training had to be done on desktop PC systems in a training classroom instead of the actual touch-screen LCDs in the vehicle, or at least on actual units in a classroom setting.

In addition, map data appeared to show incorrect or missing data. Officers were finding the system so difficult to use that some were turning it off, reverting to only audio mobile communication, which was limited and placed additional burdens on the dispatch staff. Some officers became confused by the system and were performing poorly. Finally, versions were constantly being rolled out among partial vehicle fleets, which meant extra complexity in keeping track of which users were using which system in terms of bug reporting, evaluations, etc.

In short, concern was mounting over the way in which the new system was performing.

Unfortunately, a conflux of circumstances led to less-than-best practices. Because the SJPD, officers' managers, and the IT people were not necessarily familiar with user-centred design, not being UI professionals, and because the software vendor did not supply that guidance, the SJPD was able to proceed only with good intentions, and immediately ran into problems.

At this point, more than four million dollars had been spent on the software, and it was performing poorly, with many usability problems that even novice usability analysts could not help but notice. Reports of problems were leaking to the press. The City Council of San Jose and the union of police officers, the San Jose Police Officers Association (SJPOA), were becoming involved as tempers rose and delays escalated. At this point, a usability consulting firm (AM+A) was called in to evaluate the system.

Upon examining the acquisition and set up of the system, some of the consequent problems seemed not too surprising. Among other factors that seemed to contribute to the low level of initial success, the following were identified by the usability consultants:

- System acquisition seemed not to make usability a priority criterion in the evaluation of proposals from vendors (as opposed to functionality or cost). This oversight is typical for technical people, business managers, and marketers. The focus is on the issue, 'does the system do function X?' rather than, 'if the system does function X, what is the general usability of that function by itself but, more importantly, in relation to typical task scenarios A, B, and C for users L, M, and N under conditions E, F, G?' Of course, this query is more complex and expensive to determine. If vendors were to provide data for these, presumably typical functions, users, and tasks, it would be possible for customers to evaluate software more easily, but, of course, they don't, for their obvious cost and marketing implications.
- The system acquisition team did not consist of any user representative (officer or dispatcher), but rather *managers* of those users, who might be presumed to make reasonable decisions about the users they were representing, but might also have management bias. For unknown reasons, perhaps communication or bureaucratic slip-up, the police officers had been invited to send representatives, but none had been sent.
- The system acquisition team did not query users in evaluating the finalists or in setting up initial conditions for customization for the initial rollout. In other words, user focus groups were not established for the purpose of elucidating user preferences, use tasks, and use scenarios.
- In fact, the entire development of detailed user-interface development, including customization of functions, layouts, colours, typography, textual content, symbols/icons, etc., was carried out with little or no significant input from actual users, especially police officers.
- No one from the development team constructed user profiles and use scenarios, even though these would have been relatively easy to construct given the detailed knowledge of all available subject-matter experts and the somewhat limited,

predictable, and closed world of most-typical circumstances. Best-practice knowledge might have been collected, organized and archived, and made available to all involved in user-interface development.

- The SJPD decided to use maps from the city's department of public works (DPW) as the basis for geo-location functions. While this choice may have proved to be sound in the long run, the short-term consequences were quite negative. Technical differences between the software vendor's system for handling geographic information and the city DPW's system for handling the same information resulted in dramatic display and accuracy problems.
- For cost reasons, the SJPD software rollout team decided to reduce training to a minimum amount and did not establish good feedback loops from initial users in order to tune the system quickly and effectively.
- The onsite team from the vendor initially assigned to the SJPD project had either little incentive or little knowledge of client-relationship best practices. They were not as cooperative as they might have been under the circumstances in offering and implementing improvements.

The evaluation revealed serious errors in design. Some of the more serious ones include: -

Reliance on indirect rather than direct controls. Officers were accustomed to command-line entry in the previous system. Although that system required the memorization of arcane codes, once learned, the system allowed one-handed performance of many powerful functions without looking at the keyboard at all. Although the new system sometimes gave officers many options to perform the same function, e.g., touchpad, touch screen, and keyboard combinations, these techniques usually required the officer to be aware of more than one part of the user interface at the same time, resulting in serious cognitive distraction.

Numerous issues with mapping and routing. Symbols, colours, and text appeared in a confusing display. Maps were not set to show the most useful level of data. For example, rather than showing the location of a new 'event' or the officer's car in relation to that event, the system first showed the location of the officer's car, presumably something already known! Frequent encounters with erroneous data undermined officers' confidence in the system, and in at least one case led to a dangerous misunderstanding.

One of the most serious problems in the design was the lack of understanding of the users' context. At times when the officers required intense focus on driving or on a scene in front of them, they were essentially operating the mobile device 'as if blind'. The officers were able to take their eyes off the scene in front of them for only a second to check a key piece of data. Yet, this was not taken into account in the system's design. For example, they might need to keep an eye on a suspect in a vehicle stopped in front of them while they ran a license-plate check to determine if the vehicle had been reported stolen or whether the vehicle or its occupants matched any characteristics of persons with warrants for arrest or current important events. This would determine in advance the caution an officer would take when approaching occupants of a vehicle.

They would also have to keep their eyes on the occupants to ascertain if they made any suspicious movements, e.g., reaching for weapons or disposing of any objects.

The story of this system reveals the terrible consequences of not undertaking a user-centred design process. The toll is extensive: possible dangers to officers, including life-threatening risks; annoyed and angry users; involvement and time of other agencies and organizations, like the union, the city council, and the press; to say nothing of the problems for dispatchers, who had equally serious concerns (which were not addressed at all during this project), and the time/expense required by the software vendor to address and solve as many of the usability issues as possible.

[Adapted from Marcus and Gasperini, 2006]

The case of SJPD is, unfortunately, too common in GIS and geospatial technologies applications that are purchased and developed without intensive engagement of the users. In this case, the fact that the errors in the design are actually risking the users was severe enough to provoke them into action, including involving the union and using the local press. In all likelihood, many such projects occur but because the end result is mainly frustration of end-users and loss of productivity, these failures remain hidden.

The case study provides some hints to what could be done differently. Most of the issues that are mentioned in it could be avoided if a UCD approach was used. Therefore, we now turn to discuss the origin of this design philosophy.

5.1   Background

During the 1970s, as computers started to spread in the workplace they moved out of the specialized data processing centres and into the general office environment (see Table 5.1).

Table 5.1   Development of computers and usability issues (after Shackel, 1997)

Period and type of computers	Users	Issues
Research computers – 1950s	Mathematicians, scientists	Reliability, user is expected to understand the machine and programme it
Mainframes – 1960s & 1970s	Data processing professionals	Users of the output (business people) grow disenchanted with delays, costs and rigidity
Minicomputers – 1970s	Engineers and other non-computer professionals	Users must do programming and learn the technology, usability influences productivity and use
Microcomputers – 1980s	Almost anyone	Usability becomes a major problem
Laptops, PDAs – 1990s	Anyone and often in mobile situations	Complexity in providing usability, especially with new input/output modalities
Mobility – 2000s	Multiple devices, ubiquitous wireless network accessibility	Usability of multiple devices and in a connected environment

This increase in use means that the people who have using these systems are no longer specialists who dedicate all their time to operating computing machinery, but people with different tasks within the organization that can be supported and enhanced through the introduction of computers. Some of these first 'general' users include the people who dealt with numerical data such as accounting clerks, warehouse managers or engineers. For these users, the computer was not the centre of the tasks that they were trying to accomplish. They were interested in entering ledger data, auditing the stock or designing a bridge, not in learning how the computer works or what is the appropriate way to represent the ledger as a computer file. Thus, for these users the computer is acting as a support tool that assists in carrying out their job. Therefore, the ability to accomplish the task effectively and quickly is important.

As a result of this increased range of users, the importance of designing computer applications that take into account the skills, knowledge and abilities of the users gained importance. Therefore, this period is marked with a growing interest in human-computer interaction and the establishment of some of the first research centres that are dedicated to HCI research (Shackel, 1997).

Early research on HCI focused on developing techniques for the design of better applications and metrics to compare different design options – for example, the Keystroke-Level Model that allows scientific measurement and analysis of the performance of a task with a computer system (this and other methods will be discussed in Chapter 6).

By the mid 1980s, a more comprehensive theoretical framework for HCI design was starting to take shape. Of special importance is the work that Don Norman has carried out in his laboratory at the University of California in San Diego and which led to the publication of two books that are seminal in the development of UCD. The first is an edited collection of papers titled *User-Centred System Design: New Perspectives on Human-Computer Interaction* (Norman and Draper, 1986), as well as the highly popular *The Psychology of Everyday Things*, which in later editions was renamed *The Design of Everyday Things* (1990). In the following years, the concepts of UCD were promoted through academic papers, conference presentations and books. An example for such analysis is provided by Thomas Landauer in his *The Trouble with Computers* (1995). Landauer justifies the need for UCD on the basis of productivity. He associates the productivity paradox with problems in HCI. The productivity paradox is emerging from the fact that despite the heavy investment in computing in the 1960s and the 1970s, the productivity of office workers in the US did not increase substantially. Considering the immense productivity gains of industrial and agricultural workers from the introduction of machinery, it would be expected that the automation of office work will lead to significant gains, too. Yet, this is not the case. Landauer's explanation of the paradox is to note that because of usability problems, many software products are not providing the needed support to their users and therefore the potential gains are wasted on the interaction with the computer. For example, he argues that the average software contains about 40 design flaws. As a result, the loss in productivity can be up to 720% compared to properly designed software. Significantly, Landauer's argument justifies the investment in UCD on the basis of delivering productivity improvements which without it can harm the organization that uses the software.

As result of the efforts in the mid 1980s, UCD gained popularity, until it became the central methodology and philosophy. Furthermore, UCD was also enshrined in international standards. The first standard that used UCD concepts was a German standard that considered Visual Display Units (Shackel, 1997). Other standards and regulations that integrate UCD continue to emerge, culminating with the establishment of ISO 13407 'Human-Centred

Design Process for Interactive Systems'. The standard is aimed at achieving quality through the integration of a UCD process and defines it as a 'multi-disciplinary activity incorporating human factors knowledge and techniques with the objective of enhancing effectiveness and productivity, improving human working conditions, and counteracting the possible adverse effects of use on human health, safety and performance' (UsabilityNet, 2006).

As with other aspects of computing and geospatial technologies, it took about a decade for UCD to travel from computing to GIScience. Only in the early 1990s was the concept of UCD noted by researchers in GIScience (Lanter and Essinger, 1991). Lanter and Essinger note that part of the reason for the lack of attention on user issues is the need to dedicate much more software code and development effort to the user interface. While in other software packages about 66% of the code was dedicated to the user interface, only about 35% was dedicated to the user interface in GIS. However, the realization that UCD is important for GIScience became clearer through the 1990s, and especially in the early 2000s (Haklay and Tobón, 2003). The acceptance of UCD by GIScience researchers increased, especially in the area of geovisualization (Robinson *et al.*, 2005), as well as for the design of an interactive national atlas (Kramers, 2008) and mobile mapping applications (Nivala *et al.*, 2008). A notable increase in UCD during the last decade came with the new focus on Web-based GIS, where a lot of software development is dedicated to the user interface. Unfortunately, while the knowledge that UCD exists is now common among geospatial technologies developers, quite often it is understood in a very generic form, without taking into account the full implications of turning the development process to a user-centred one, as was exemplified in the case study of SJPD. In the next section we turn to the concept and principles of UCD.

5.2   Principles

The basic philosophy of UCD is to put the user at the centre of the development process. This means that when developing a new software product or an application, the designers and developers need to focus on what the real people who will use their product are going to do with it. Instead of considering the functionality of the application or an eye-catching gimmick in the way the software presents maps, the developers must take into account the real scenario of where the software is going to be deployed and for what purpose. A properly developed system will take into account the skills and judgment ability of the users, and will directly support their work.

During the development of UCD, several sets of principles were offered. The most widely adopted one was described by Gould and Lewis (1985). Yet, the set of principles which Don Norman described in *The Design of Everyday Things* (1990) and Ben Shneiderman's eight golden rules are helpful in understanding the underlying design philosophy of UCD. Therefore, before turning to Gould and Lewis, these two sets of principles are presented.

Norman's seven rules are presented here in the way that he describes them (1990: 188-189) with additional explanations and examples:

1. **Use both knowledge in the world and knowledge in the head.** Knowledge in the world is provided through clear instructions and a manual, while knowledge in the head is the conceptual model that the user uses to understand the system. A single sheet of 'read this first' with a clear graphic explanation of assembling a device can be an effective

way to ensure that the user is starting to use the device properly and build a suitable mental model of its operation.

2. **Simplify the structure of tasks.** Users have limited ability in both short and long term memories. Thus, it is important to simplify the set of actions that they need to perform and not expect them to learn or remember a long chain of actions.

3. **Make things visible: bridge the gulfs of execution and evaluation.** Users should be able to figure out, without additional information, what different parts of the interface are doing. A door with a plate on it invites the user to push, whereas a handle invites the user to grab it and pull the door towards them.

4. **Get the mapping right.** The use of graphics is valuable, but the position of different elements should match the conceptual model of the user. In physical devices an oft cited example is the location of gas burners on a stove and the set of knobs that operate them – if they are organized in the same order as on the stove, the use of the stove becomes easier.

5. **Exploit the power of constraints, both natural and artificial.** Constraints can help in navigating the user through a series of actions and provide an indication of what it is possible or not possible to do at a certain point. A greyed-out toolbar is an example of such constraints, notifying the user that using this part of the interface is not possible at this point.

6. **Design for error.** Plan for the possible errors that can be made by the user and include them in the design process to allow the user to recover from an error. The ubiquitous undo option that is common in computer systems is implementing this principle by allowing the user to revert to a previous state.

7. **When all else fails, standardize.** Sometimes, because of the need to manipulate abstract objects, there is a need to make an arbitrary decision. In these cases, it is important to set standards that either apply throughout the product or, better, follow industry or international standards.

Shneiderman suggested his eight golden rules in the original 1987 edition of *Designing the User Interface*. The version that is presented here is based on the fourth edition (2005: 74-75). The rules are:

1. **Strive for consistency.** Consistency in the interface and the behaviour of the interface is tricky to implement, but yet very important. Consistency is needed in naming conventions, layout, colours and commands as it all assists the user throughout the process of learning and using an application.

2. **Cater for universal usability.** Shneiderman's concept of universal usability is to recognize and cater for the need of diverse users. Understand the differences between expert and novice, people from different age ranges, users with disabilities, or how the different genders will respond to the application.

3. **Offer informative feedback.** The system should provide feedback for users' actions with feedback that is proportional to the action – minor or frequent actions should be responded to with modest feedback, and more substantial actions should receive significant feedback.

4. **Design dialogs to yield closure.** A set of actions should have a clear beginning, middle and end. Informative feedback at the end of a set of operations can provide a satisfying feeling of accomplishment to the user.

5. **Prevent errors.** As much as possible there is a need to design the system in such a way that prevents the users from making serious errors – for example, deactivate parts of the interface that should not be used during certain operations. In the case where errors are made, a simple, constructive and clear route for recovery is offered.

6. **Permit easy reversal of actions.** Actions that are reversible are important for relieving anxiety because it is clear to the user that errors can be undone and they should not feel that they will 'break' the system by one mistaken action.

7. **Support internal locus of control.** For experienced users, it is important that they feel that they are in charge of the interface and the interface responds to their actions. Inconsistent and surprising responses, lengthy and tedious data entry actions, lengthy information retrieval or very long sequences for common operations all add to dissatisfaction in users.

8. **Reduce short-term memory load.** Take into account the limitation of short-term memory ability to store 'seven, plus/minus two' chunks of information. Keep displays simple, and provide sufficient training for code, design sequences of actions. Provide access to command-syntax forms, codes, short-cuts and similar information where possible.

It is not surprising that both lists share some principles, such as the need to take into account the human memory ability, or the way in which errors are handled. Notice how both lists do not talk about the application's functionality, complexity or the number of features that are included in the product. They are about the user, and how the system is perceived by her. Because the interface *is the system*, the guidelines are about the way in which it is working and interacting with the users.

While the sets of principles are useful as shorthand for the main aspects of UCD, a more comprehensive framework was offered by Gould and Lewis (1985) (presented in Sharp *et al.*, 2007). They provide three core principles that are the basis for the implementation of UCD in design projects. The principles are: early focus on users and tasks, empirical measurements and iterative design.

Early focus on users and tasks. For a successful application design there is a need to understand who the users are and then engage them in the design process. The focus on the users means that the design process should include observing what the current tasks are that the users are doing in their daily jobs, understanding the full context of these tasks and then involving the users in the development of the product. The observations and understanding of users' tasks should include their conceptual models, actions, demographics and knowledge.

This element of the design can be expanded further in five ways: first, the users' tasks and goals should be the driving force behind the development process. Instead of considering how a new technology can be deployed and what needs it can answer, the focus should be on what the user needs to do and how the technology can support this. Second, the users' behaviour and context of use should be studied in detail when designing the system. Therefore, the focus should go beyond the specific task in order to understand the full context of use. Third, users' characteristics should be captured and taken into account during the design, for example, consider the visual ability of the user when designing a map for them. Fourth, users need to be consulted throughout the design process and their input should be seriously taken into account. Finally, all design decisions must consider the users' context, work and environment.

Empirical measurement. The design process should be based on measurement of the different elements of the design – from the recording of users' reaction and performance with a prototype to the evaluation of the final version.

Iterative design. The design process should not be linear, that is progressing from the early concept of the product to the released version by completion of each stage and moving forward relentlessly. When problems are identified in a user testing, the designer should go back and redesign elements of the product. These cycles of design are at the core of the method, and the development should progress through the 'design, test, measure, and redesign' cycles.

The main difference between this framework and Norman's and Shneiderman's principles is that it provides a clear operational guidance to a development process. By so doing, the framework is easier to implement and communicate. In addition, the framework was developed through engagement with the development process of products, such as the telephone messaging system that was developed by IBM for the 1984 Summer Olympics.

Gould and Lewis's framework was translated in ISO 13407 into a set of instructions that help achieve user needs by utilizing a UCD approach throughout the whole life cycle of a system. The study starts with planning the project, at which point a decision must be taken on what kind of information is needed to ensure the usability of the end product: information about the usability of an existing product, ideas for developing a new product, or information for comparing products already on the market. The factors affecting the project planning process are strongly related to the amount of resources: money, time, people, etc. In addition, it is preferable to decide during the early stages of the project by whom, how, and when the usability evaluation will be carried out, i.e. usability experts or users, with usability tests or questionnaires, and at which stage of the project.

The three-step design is an iterative process, as described in Figure 5.2. The first step is to ascertain the user requirements. These can be analyzed by studying potential users and the context in which the application will be used. A decision must also be taken on which usability criteria are to be emphasized in the study: effectiveness, efficiency, satisfaction, memorability, and/or minimal errors (these criteria and others are discussed in Chapter 6).

The first design prototypes and preliminary mock-ups can be designed on the basis of the user requirements. The following stage must analyze whether the defined user requirements

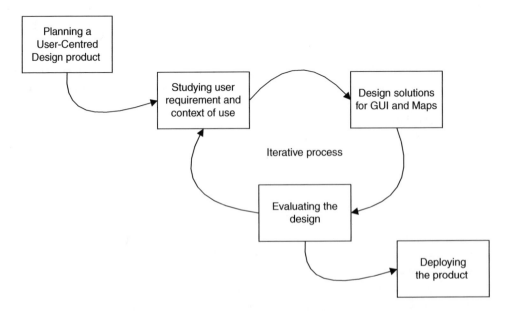

Figure 5.2   A user-centred design (UCD) cycle for geospatial technologies

have been met. Evaluation can be carried out using various usability engineering methods, and if the results indicate that the user requirements have not been achieved, the iterative process goes back to redefining the user requirements. If the requirements are met, we can consider the design outcomes as ready for deployment.

User studies should not end there either, since the market situation may change quickly. Therefore, evaluations on existing products should be carried out accordingly, to obtain information e.g. for future versions, and for making sure the product still satisfies user needs.

5.3   Applying user-centred design in geospatial technologies

While there are examples of the use of the UCD framework in the development of geospatial technologies in the literature, it is not the main framework for the development of geospatial technologies to date. As Case Study 5.1 demonstrated, even recent projects by major GIS vendors are not implemented in a UCD approach, to such a degree that they offer a case study of non-UCD. Part of the reason for this is the complexity associated with the core technology itself. We have noted before the computational challenge that GIS represent to developers. The implication of this technological challenge is that developers of GIS and their clients are focusing on functionality, data standards and many other technical aspects of the application, but not the users. The unfortunate outcome is that the story of SJPD can be told on many GIS projects, and should not be taken as a special case of failure.

To balance the picture of Case Study 5.1, Case Study 5.2 provides a comprehensive description of the application of UCD within a geospatial technology project

Case Study 5.2: UCD in the GiMoDig project

Geospatial Info-Mobility Service by Real-Time Data-Integration and Generalization (GiMoDig) was a three-year project (2001-4), with the objective of the project being to develop methods for delivering geographical information in real-time and with generalization to mobile users. The project created a seamless data service infrastructure to provide access, through a common interface, to topographic geo-databases maintained by the national mapping agencies. The need to consider usability issues during the product design process was one of the main concerns. Evaluations started at the beginning of the project and continued throughout the whole lifespan of the project.

The practical outcome of GiMoDig was a working prototype (but not a complete end-user application) for a mobile map service. During the project some inconsistency was encountered with regards to developing a research prototype and its evaluation by users. This was due to some scientific and research goals that were integrated into the project and could not be reconciled with user needs. Here, we focus on the user's point of view. The implementation and success of other research objectives are not discussed here.

The UCD process starts by identifying all the primary and secondary users and classifying them in a meaningful way according to the project's objectives. The preliminary requirements for the system can be decided by identifying the user requirements for these groups and the real context of use. This is an important stage of the study, since the first mock-ups were based on these results.

The first usability goals can be created from the user requirements, against which future designs can be evaluated and tested. The acquired information can be structured e.g. by compiling use scenarios and use cases, which specify how users perform their tasks in specified contexts. They should include information about which activities should be performed by the user and which by the computer when the user is performing a certain task with the system. After compiling the scenarios they can be validated according to their relevance and importance. Usability goals can be established from the results.

For the context of GiMoDig, the study started with analysis of user requirements and market analysis of mobile map services. The potential usage areas of the GiMoDig service included: information services, safety, emergency, restrictions on use or movement, guidance or navigation, logistics and military services. The potential users of the GiMoDig service were identified and classified into two user groups: end-users, and technology users.

Endusers are the potential users of the service and they can be further divided into professional users and consumers. The preliminary expectation of the GiMoDig project was that it would result in a working prototype for the endusers.

The analysis of users' goals (Jakobsson, 2002) suggested the following typical tasks: locating your position, and being located, locating other people, locating objects, obtaining guidance, obtaining information and obtaining help. Twelve different use scenarios in which the user could benefit from using the service were compiled using the information on user goals and usage areas.

Technology users are those organizations which provide value-added services for LBSs. These include, for instance, yellow page services etc., as well as tourism and leisure activity organizations. Technology users consider a critical success factor for the project as being the ability to exploit the results from GiMoDig and provide their own datasets above the topographic datasets for users with mobile devices.

Following the analysis of requirement, the focus turned to the context of use, which was studied by arranging a field test in a national park. The purpose of the tests was to obtain basic information on user requirements related to the use context, and also to find out how usable existing topographic maps in mobile devices were. This information was used in creating preliminary design principles. At this point the project did not have any real prototypes to test. Therefore, it was decided that existing maps with existing hardware and software should be tested. The aim was not to test the software or the hardware, but to use them only as a means of utilizing mobile maps.

These two studies provided the first step in the UCD cycle. On the basis of them, a list of requirements was established for each group.

For endusers, the requirements that were established were Easy-to-use User Interface (UI), based on an observation that users had problems during the test tasks with the commercial navigation software; suitable cartographic presentation, based on difficulties in map reading that were observed during the early study; seamless integration of different data sets; and creating context-aware maps. Specific goals were established for the technology user too. The next stage in the UCD process was to move forward to the design phase.

To evaluate the GiMoDig UI from the endusers' point of view, the evaluation started soon after the first prototype became available. The mock-ups were evaluated and new user requirements were established, followed by new design implementations. The iterative process continued until the prototype met the user needs, while also taking into account the limits of the project's research goals.

The usability evaluations in the GiMoDig project were carried out using four different methods: heuristic evaluations, expert evaluations, usability tests and intuitivity tests (these are described in Chapter 10).

(Adapted from Nivala *et al.*, 2005)

The GiMoDig case study provides the evidence that even within the limits of a research project, which runs on a tight budget and with clear scientific objectives, it is possible to integrate UCD successfully. Furthermore, it provides guidelines for creating a UCD map application. Cartographic systems are specific in a way that the usability of the system is composed of different system elements. First, mobile map usage may vary a lot (outdoors/indoors, PDA/PC, navigation in the forest/tourist navigation in the urban area). The context of use should be studied beforehand to compile realistic user requirements, and during the design process to ensure that it is suitable for the use context. Second, the mobile map has to be user-friendly and usable in the context. Third, the graphical UI must also fulfil the needs of the user. Therefore the usual model of the context of use should include a specific element that focuses on the cartographic design.

Because of the significance of the use of maps in navigation, field tests are extremely important. The experience also shows that involving cartographic experts is the best way to evaluate the cartographic design itself. They possess knowledge about the visualization possibilities and restrictions, as well as the spatial cognition of map users. Thus, these experts have a realistic insight into the design. When an expert cartographic evaluation is carried out, the context of the use should be taken into account as well, e.g. maps in mobile devices have to be evaluated outdoors in varying light conditions. Involving users is necessary for finding answers to questions such as the aesthetic quality or intuitivity of the map design. But it must be kept in mind that asking users about the design's aesthetic appearance is always such a subjective matter that valid results will not be gained unless there is a sufficient number of users involved in the evaluation. It should also be borne in mind when involving users in the map evaluation that in many cases the use of a map is strongly related to getting used to something: if the user expects a traditional map, the differing visualization may be a distracting factor.

5.4   Participatory design

One of the big advantages of UCD is that it can be integrated, fairly easily, into software development projects. While UCD takes users' concerns into account, it is not mandatory to include the users in the design team itself but to be in constant communication with them throughout the development cycle. The argument to support this approach is that 'users are not designers,' that is, they might have difficulties imagining alternative futures and different ways to perform their tasks unlike designers, who are familiar with the latest developments. It is therefore more suitable to carry the UCD process through the interaction of experienced designers with the users and with the software and hardware developers.

Participatory design follows a different path. In participatory design the aim is to work collaboratively with the users throughout the design, and involve the users in every stage of the design process. The differentiation between users and designers remains, but the introduction of innovations or new ideas to the final product is done through deliberation with the users.

Participatory design emerged in Scandinavia in the late 1960s and is based on labour unions having a level of democracy in the workplace, and therefore the workers can have a say about changes to their work environment (Sharp *et al.*, 2005). This requirement of involvement by the end-users is changing the UCD process from one that can be largely conducted as one way communication, in which the designers extract information from users, to one where two way communication is required.

The challenges of a successful participatory design project are quite significant. As noted above, there are significant cultural differences between users and designers. As in any activity that involves people who are coming from different knowledge domains – in this case the users and the designers – a significant amount of time is needed for translation of concepts between the two sides and continuous examination of the terminology to ensure that a shared understanding emerges. This, by necessity, means that more meetings and discussions are required.

Dix *et al.* (2004) identify three main characteristics of participatory design. First, is focus on improvement of the work environment, so the evaluation is inherently context oriented and not system oriented. Second, the collaboration with the users is at the centre of the process and they are involved throughout the design process. Finally, as in UCD, the process is iterative, and in each stage the proposed design is negotiated with the users.

There is a range of methods that are especially suitable for participatory design, as they are geared toward the knowledge sharing aspect which is critical to the success of the project. These include (Dix *et al.*, 2004):

Brainstorming – a process in which the participants in a meeting suggest ideas to solve a specific problem. This should be done in an open meeting where all participants are involved and encouraged to suggest ideas without fear of criticism from other participants.

Storyboarding – the storyboard approach is a way for participants to describe their work practices and communicate them to the designer. This can be done by using images and text to describe the various activities.

Paper and pencil exercises – these are exercises in which the participants are asked to describe their work environment and workplace using paper and pencils. Since there is no technical barrier, the participants can use the paper to draw and write and discuss the visualization of the process with other participants. This can be done either during a design meeting, or as preparation for it.

Workshops – workshops are critical to the process of knowledge construction. They can involve presentations from users who will explain to designers concepts about the users' work context, or from the designers to educate users about the development process and possible technologies. The workshops should include significant time for questions and answers from both sides, to ensure mutual understanding.

In terms of geospatial technologies, participatory design can be used in several ways. First, it can be used in the design and development of software that will be used in a workplace, or by a community. Second, it can be used for the development of applications that customize off-the-shelf software for a specific task or use. Third, it can be used for the design of the information that will be held in the system, so while the way in which the system operates is not being negotiated, the meaning of the information that is held in it is discussed collaboratively. Fourth, participatory design can be used to discuss the output of a system, to ensure that the visualization is suitable to the needs of the users. Finally, the system itself can be used within a participatory design process in which it serves as a tool.

The use of participatory design of geospatial technologies that are being introduced in the workplace is still a rarity, and apart from recommendations in the research literature (Harvey, 1997) it is not widely implemented.

The use of participatory design for the customization of systems, setting up the organization of information and visualization is common in participatory GIS (Sieber, 2006). In this type of application, the GIS are being used as a tool to empower communities and to allow them to describe their needs through the use of maps and other geographical visualizations. For example, Al-Kodmani carried out a participatory process in which the community of the Pilsen/Little Village area in Chicago discussed future plans for the area. The discussions with the community included visualization of future plans which were drawn by an artist who participated in the meeting. This method helped in overcoming problems in understanding planning maps and diagrams, which tend to be rather abstract. The use of the maps, artistic visualizations and structured workshop with robust ground rules to ensure focused discussion led to a successful use of geospatial technologies (in this case GIS) within the design process.

Another example of participatory design process was carried out at University College London as part of the Mapping Change for Sustainable Communities project, implemented by Haklay and others (Ellul *et al.*, 2008). In this project, a structured process was carried out with each community in which an online community map was developed to establish which themes are of most interest to the community and what type of information should be included in each data point. The results of this discussion led to the development of specific maps, layers and icons.

More generally, the final application of geospatial technologies within participatory design process is by far the most common. Geospatial technologies can be used in design processes such as Participatory Rural Appraisal, Participatory GIS and Public Participation GIS (IIED, 2006). All these processes use geographical information extensively to discuss a specific issue which concerns the local community and develop possible solutions to it. In many of these processes there is no attempt to change or adapt the technology, and it is used within a participatory design context. However, as Haklay and Tobón (2003) noted, because of the difficulties in operating GIS and other geospatial technologies, they should be deployed carefully, as they can harm the participatory process instead of assisting it.

Summary

In this chapter, we have covered user-centred design and participatory design, which can be viewed as a 'democratic version' of UCD, or as UCD with higher levels of user engagement. We have seen that UCD provides a framework for the whole design process, from initiation to implementation and emphasizes the requirements of the users and not the systems. This is an especially important point with respect to geospatial technologies, since in many of them there are significant technical challenges that attract the attention of their developers to sophisticated algorithms, data structures and other such issues, so the importance of the user is sometimes forgotten.

UCD requires early attention to users' needs, iterative design and empirical evaluation. These principles are further elaborated through Norman's and Shneiderman's principles.

Participatory Design requires a higher level of engagement from users, and the education of both designers and users to achieve a common understanding of the required product and to ensure its success.

The case studies and examples from the area of geospatial technologies showed that consideration of UCD issues must include analysis of the data used, its representation and visualization in addition to considerations about the software and the hardware in general. Participatory design is especially suited for cases of Participatory GIS, but a decision must be made about the degree of involvement of the users with the technology to ensure that all the appropriate aspects have been taken into account in the design process.

Further reading

Don Norman's book, the *Design of Everyday Things* (1990) is an excellent starting point for understanding the principles of UCD, and why it is needed. In addition, the details of the ISO 13407 standard are a very useful starting point for authoritative and common implementation of UCD into a software development process.

In geospatial technologies, the NCGIA report from 1991 (Lanter and Essinger, 1991) provides a review of the early attempt to integrate UCD into GIS in the late 1980s, with a focus on graphical user interfaces.

For participatory design, Schuler and Namioka (1993) *Participatory Design: Principles and Practices* is recognized as an important, and still relevant source of information. Specifically for geospatial technologies, the existing literature on participatory GIS and Public Participation GIS (PPGIS) provides a wide range of techniques and lessons that can be applied in various projects. Sieber's review (2006) provides a good entry point to this area, as well as the website PPGIS.net.

Revision questions

1. What are the core principles of User-Centred Design? Can you describe them as a set of words?
2. Try to merge the three sets of principles for User-Centred Design that have been described in this chapter. Where do Norman's and Shneiderman's principles fit within the Gould and Lewis model?
3. Imagine that you've been hired by SJPD as an adviser at the beginning of the new dispatching system development, and you are part of the design team. What steps would you take to ensure that your project follows UCD?
4. How would the GiMoDig UCD process change if the designers decided to use a participatory design approach? Is participatory design suitable for this project at all?

6 Usability engineering

Mordechai (Muki) Haklay, Artemis Skarlatidou and Carolina Tobón

In Chapter 5 we covered the design philosophy of User-Centred Design (UCD). UCD provides a framework for the design process as well as guidance through principles for the design process. While this framework is very useful, there is an inherent difficulty in translating it into action. This is because in any given project not all the people who are involved will be familiar with the principles, and even if they are, there is always a level of ambiguity in the way that they are interpreted.

To solve this problem, the area of Usability Engineering (UE) emerged. Usability Engineering is an approach aimed at integrating central concepts and lessons that were learned through HCI research into software design processes. The integration of UCD principles in the software development process is done through the creation of frameworks, techniques, and matrices that can be deployed systematically and rigorously. By developing such methods and tools, UE aims to ensure that the concept of usability is translated into measurable criteria and a set of actions that the developer can carry out through the life cycle of the software.

Of course, since UE is reliant on cognitive models of tasks and abstract manipulation of information and since the final product will be used by a range of users with differing culture, age and educational attainment, UE is not an engineering discipline where criteria and methods are rigidly defined and where predictions will work deterministically in every case. Furthermore, in the software development processes, it is unlikely that presubscribed matrices that were set at the early stages of the design process guarantee that the system will be usable. The reason for this is that the design process itself is very complex and often changes, and therefore matrices that are defined rigidly can divert the development process to ensure that the final system satisfies specific tests, even if overall performances are not satisfactory (Dix *et al.*, 2004). Thus, the correct way to view UE is as a toolbox that can be used throughout software development processes, and, by combining the right tools for the appropriate stage of development, it is possible to ensure that the user remains at the centre of the process and the resulting system is usable.

In this chapter, we provide an introduction to Usability Engineering, explaining its background and the criteria that are used to evaluate the usability of a product, and show how these translate for geospatial technologies.

6.1 Background

During the 1980s, despite the rapid technological advances and the increasing awareness about the HCI methods and techniques, only a few companies paid special attention to

Interacting with Geospatial Technologies Mordechai (Muki) Haklay
© 2010 John Wiley & Sons, Ltd

usability and tried to develop methods that can help in the integration of usability principles into the development process. The early work on usability engineering was carried out by Whiteside and colleagues at IBM, where one of the first usability laboratories was established in 1981 (Whiteside *et al.*, 1988) and by Nielsen at Bellcore (Nielsen, 1994a). Apple had a very good usability programme during the 1980s, 'running frequent user tests on designs for the Apple II, the Lisa and the Mac', while at the same time companies like Microsoft began to organize their usability groups after 1988, therefore 'the initial versions of Word and Excel and even Windows 1 and 2 were designed without usability input' (Yank, 2002).

Usability Engineering suggests that UCD goals can be achieved by the application of engineering frameworks to the design process. Like other areas of engineering this can be achieved by establishing a background of meaning, agreeing on useful criteria and measurement goals, and then evaluating products and designs against these criteria and goals. As a result, the effort within the research area of usability engineering is to develop methods and test criteria that can be used in the design process.

It is therefore necessary to define what usability is, as without a clear definition of this core concept, the other criteria cannot be developed. One of the early definitions of usability was offered by Shackel (1986, cited in Faulkner, 2000) and highlighted the aspects of effectiveness, learnability, flexibility and attitude. Effectiveness is defined for a particular range of users and their ability to accomplish a task within a given time limit. Learnability is defined as the time frame that the users will require in order to become familiar and proficient with the system, flexibility as the degree of adaption and variation in which the task can be accomplished, and attitude as the aspects that are linked to human costs such as tiredness, or discomfort.

Nielsen (2005) defines usability as 'a quality attribute that assesses how easy user interfaces are to use. The word "usability" also refers to methods for improving ease-of-use during the design process'. A more detailed definition is available in ISO standard 9241-11 which defines usability as 'the extent to which a product can be used by specified users to achieve specified goals with effectiveness, efficiency and satisfaction in a specified context of user'.

One of the outcomes of UE studies is a body of knowledge which defines how to develop usable computer systems. For example, over the years, the definition of a usable computer system evolved to include the following criteria:

- Effective – allowing the user to achieve the specific goal accurately and completely.
- Efficient – achieving the specific goal accurately with as little work and time as possible.
- Error-tolerant – in any system, it is expected that the user might make mistakes. The system must recognize these mistakes, and allow the user to recover from mistakes that she has made (as in the case of undo).
- Learnable – the system should help the user to learn its functionality, as well as learning more powerful options as they develop their knowledge about it. It should also be easy enough for infrequent users to work with the system easily without extensive retraining.
- Satisfying – the work with the system should ideally be enjoyable, or at least pleasant and satisfying to use.

Despite the fact that all these criteria cannot be quantified unambiguously, they provide a set of principles that can then be translated to specific measurements and expectations and guide the development process. To further integrate these criteria in the design process, many methods have been developed over the years. These methods cover the whole development process and borrow concepts from many fields of study including psychology, anthropology, ergonomics, and, naturally, the wider field of HCI, turning research outcomes into tools. For example, at the beginning of the software design process, ethnographic techniques can be used to understand the user's context within the process of requirement analysis. At the final stages of development, direct observation studies, where users are asked to carry out tasks with the system, are used to check how successful the system is in terms of learnability or to identify usability problems that have not been found in earlier stages.

In summary, UE as an applied practice is now a maturing discipline with a wide acceptance of its importance and relevance to software development processes. UE principles are now taught as part of the computer science and software engineering curriculum, as they are seen as an integral part of the education of software developers. In this chapter we examine the process of product development under the usability engineering paradigm, giving an overview of the main stages in product development and how usability engineering methods are integrated within them. Examples that are relevant to geospatial technologies are presented at each stage.

6.2   Usability engineering and product development process

Developing a technology that is based on digital computers is a tricky process. Because of the flexibility of modern computer chips, the abundance of memory and storage, and the wide range of hardware options that can be used, designers have an immense flexibility when they are setting out to develop new products and applications. The complexity that is part and parcel of digital product development is increased due to the range of applications – from a simple device that will record locations at a press of a button to national scale GIS. Sometimes a product is designed, developed and deployed by a single person, but in most cases the development will include a larger group of designers, programmers and developers who work in concert. To help plan these processes, many development methodologies have evolved over the years. These methodologies can be highly structured and hierarchical. An example of this is the waterfall model for development – in which the product is progressing in an orderly manner through the requirements, design, implementation, testing and deployment. The process drives decisively forward, and at the end of each stage there is no possibility of returning to it, thus the 'waterfall' metaphor. At the other extreme, there are methodologies such as Rapid Application Development in which the development is done with little requirement analysis and through iterative development in which prototypes are quickly developed and tested. These many methodologies are all part of the field of software engineering, and there are probably several hundred of them. Each of them is suitable for specific types of product, organizational settings, structure of the development team and budget.

Because these methodologies are being developed for specific domains, it is not surprising the specific methodologies emerge to deal with the large scale end of geospatial technologies – the development of large scale GIS projects. One of the main reasons to develop a specific methodology for geospatial technologies is the special features of these products – which we have covered in previous chapters. Of specific importance is the need to include special considerations for geographical data capture and manipulation, and the need to include in the system some reference maps that will put the information that is displayed in the product in the appropriate context. Roger Tomlinson, who is credited with the development of the first large scale GIS, has also developed specific development methodology for GIS. The overall methodology is described in Tomlinson (2007) and covers the process from a strategic analysis of the organization and how the system fits within it, through the evaluation of available technologies and the design of information products, to the implementation. There are several other books and academic papers that discuss GIS implementation through the utilization of existing development methodologies and adapting them to the special properties of geographical information. Probably because of the major technical challenges of developing a successful GIS application, the vast majority of these specific GIS methodologies pay little or no attention to the issue of usability and user interaction.

Unlike the GIS specific methodologies, some of the general methodologies explicitly integrate user-centred design principles into their structure, to ensure that the final product is answering the criteria that were mentioned above. For example, some implementation of Rapid Application Development calls for a high level of engagement from the users throughout the development process, and the testing of the prototypes during the development process is aimed at allowing the users to comment on the usability of the application. Others, for example Mayhew (1999), provide, a detailed methodology that allows developers to integrate usability engineering into their existing software engineering methodologies in a structured way.

Regardless of the methodology that is being used in a specific project, we can identify several stages in the development that appear in all of them. These stages will be explored here, to provide an overview of how usability engineering is integrated in each stage. The aim in this chapter is to provide a high level overview of the way in which user-centred design consideration can be taken into account at each stage. In the following chapters, we discuss each of the stages in more detail with a fuller explanation of the techniques and approaches that we have described here.

The common stages in any geospatial technologies development are the gathering of requirements and needs, followed by application development, then, when the application is in the advance stages of completion (or in some cases a suitable prototype is available) evaluation and testing with typical users and finally, deployment. In the following paragraphs, we provide a high level overview of these stages from a UCD perspective.

The first stage in the development of any product is to develop the requirements. Sometimes the application is being developed without an explicit request from the users, as is the case with most commercial products that are being brought to market by their developers in the hope that they are answering a need. An example of these is GPS receivers designed for recreation activities. Another example is if the users or their organization are feeling the need for an application and they approach the developers with a request for a solution to an emerging or existing problem. Most organizational GIS are developed in this way, starting with request for help in organizing their geographical information. In both cases, the stage is dedicated to establishing the purpose of the application. This means answering questions such as: What is the functionality that is needed? How will the product fit within the overall set of activities that the user is carrying out? What are the users' needs? In a UCD framework, these activities always include end-users – the people who will use the application or the product. The aim is to understand the context of the users fully, and not to assume that high level IT managers in the organization or a group of developers can understand the user without consultation. In most development methodologies the outcome of this stage is a document that defines the purpose of the application, the functionality and an indication of the way in which the application will be used once developed.

The second stage of development is the development of the application of the product. The development of the product can either follow a linear progression in which different elements are developed and signed off, and then the next element is being developed, or in an iterative way in which different parts of the application are being developed and tested, and the whole process is progressing towards the final product. The iterative development works better from a UCD perspective, as it allows for ongoing engagement with end-users. This can take the form of involving them in the testing of prototypes, consulting with them about the layout and structure of the interface or, in the cases of collaborative design, continuously consulting with representatives of the users about every significant design decision.

The third stage in product development is the testing. This is aimed at ensuring that the final application answers the requirements that were set for it and allows the users to use it effectively and efficiently. The testing stage usually includes some examination of the interface and the way in which it is used, so in most methodologies an element of usability evaluation is included, even if only at a rudimentary level. Unfortunately, many developers still think that only at this stage should usability considerations and testing be introduced and problems rectified. However, many issues with usability are not at the interface level, but in the way that the software is conceptualized and data structures organized – and discovering problems in these areas at the testing stage means that it is too late to change some of the element of the application without incurring significant costs.

The final stage is when the application is being rolled out and used by its users. Even at this stage there is a need to continue to monitor how it is being used and what kind of changes will be required while it is in use. Often the introduction of a new device or application can lead to changes in use practices. For example, the wider availability of GPS receivers at an affordable price led to the innovation of geocaching – an outdoor activity in which participants hide and seek 'treasures'. In order to navigate to the places where the treasure is hidden the participants use GPS coordinates. This was not an application that was envisaged by the designers of the devices. However, by monitoring the way in which the receivers are used, a vendor will notice that geocaching is a growing activity and therefore there is an opportunity to develop a device that answers the specific requirements of those who participate in this activity. Sometimes the changes are less dramatic and require a small adjustment to the application.

In the rest of the chapter we take a more detailed look at each of these stages, and explain what the role of usability engineering techniques is at each stage.

6.3 Understanding user requirements and needs

As noted, the first step in application development is the process of establishing the requirements from the product or the application. In a UCD framework, this will mean that the functionality will be derived from the needs of the people who are going to use the application in a way that will ensure it fits smoothly into the work or recreation activity that they are carrying out. To achieve this, the first stage is to understand who the users are, and what is the full context in which they are going to use the application. The most basic aspect of such investigation should include an understanding of the users in terms of who they are. Falkner (2000) suggests differentiating among direct, indirect, remote and support users of the system. Direct users are the people who most application designers will think about – they are the people who use the computerized device to operate the application. In many cases, they are the only users. Indirect users are people who ask someone with access to the system to perform an operation with it on their behalf. An example of these can be the residents of an area who ask the planning officer to run a query with the GIS to provide them with a site plan. The reason for understanding these users is that their questions need to be answered by the direct users of the system. Finally, there are support users – people such as system administrators, helpdesk personnel etc., who will need to provide support for the system. We should also differentiate between mandatory and discretionary use of the system – in many cases geospatial technologies are used in a work context where the users must use it to accomplish their tasks. For example, in many cases where a GIS is central to land registration, the user doesn't have a choice and must use the system regardless of how awkward or complex it is to complete the task. In other work situations some users might find certain computerized tools to help them complete their task, or in most cases where a product is sold through retail outlets, the use of such devices is discretionary. Off the shelf PNDs (SatNavs) are discretionary systems.

In addition we should differentiate between the levels of expertise of different users. Novice users, who don't use computer systems frequently, might find using a complex set of instructions difficult and therefore will be reluctant to use the application. Expert users, on the other hand, will look for ways to automate their work through keyboard short cuts, writing up short scripts and using powerful query languages.

In addition to Falkner classification, there is also a need to understand the demography of the user population – an application that is aimed at children should be designed differently from one for elderly people. There is also a need to consider users with differing ability. In many cases, taking into consideration users with limited mobility or vision can improve the overall usability of the application. For example, as was discussed in Chapter 8, making sure that the colour range of your maps is suitable for users who are colour blind can make your maps more vivid to users who are not limited in their colour vision.

To avoid the situation where the application designer is imagining the users without any factual basis, usability engineering offers a range of techniques and methodologies that allow factual data about the users to be collected. The techniques are common in social sciences and include questionnaires that are structured to provide the demographic makeup of the users and which are distributed to potential users or the people who will be using the information; analysis of existing statistics – this is especially relevant in cases of improving an existing application; interviews of potential users and possibly observation of them while they perform the types of tasks that the proposed system will support.

It is important to consider the appropriate research tool that will be used to collect the data carefully, as well as the research questions that will be explored. Things to consider are: what type of information about the users do we need to gather, and at what level of detail? Which questions should be answered through quantitative data sources such as statistics, and which through qualitative sources?

Secondly, and in addition to the generic understanding of the users themselves, it is very important to understand the context of use. As we have seen in Case Study 5.1, when discussing the San Jose Police Department, the application developers did not consider fully the context of use. For example, that it is a common thing for a police person to ask for an ownership search on a vehicle while they are driving and therefore while they have a very limited ability to enter input to the system without risking an accident. The best way to understand the user context is through qualitative research methods, many of them taken from ethnographic studies. There are many ways to get such information, from asking users to take a photo of their work environment to detailed observation studies where the researchers 'shadow' the users for a period of time to see how they carry out their work.

Another element of analysis that is frequently carried out during this phase is task analysis, which is covered in Chapter 7. Task analysis allows the designer to understand the tasks that users are carrying out in a structured way which helps the design team to develop the applications around this task.

Finally, once the analysis has been completed, and to make the users more 'accessible' to the development team, there is the possibility of establishing personas. Persona is an imaginary user, who is given a history, name, sometimes an image (taken from an image bank) and description of his or her background and interest in using the application. By 'materializing' the type of user who will use the application, the design team can think of concrete features that will fit the needs of each persona.

Case Study 6.1: Learning how GIS users utilize their monitors through screenshot study

(A. Zafiri and M. Haklay)

One possible method of gathering information about the way that existing users of geospatial technologies, and in particular GIS, are using their screens, is to run a screenshot study. The aim of the screenshot study is to understand the effectiveness of the interface of the existing systems that are used by the users. The study is carried out by asking users to provide a screenshot of their entire screen while they are in the middle of their work with GIS. A short questionnaire accompanies the screenshot to provide additional information about the users and their tasks.

The screenshots constitute the primary means of collecting the essential information and achieving the objective of such a study. Most importantly, they provide a more objective measure for assessing the usability of the GIS interface and they can be combined and correlated to the subjective measures derived from the questionnaire.

In a Graphical User Interface (GUI) environment, the user is capable of changing the size of windows in a given application, setting toolbars, and organizing certain aspects of the interface. The use of the GUI creates tension between the integration of User Interface (UI) elements, which are used to drive the application and perform tasks, and the available area for the actual application. The size of the map area as a proportion of the total interface can have significant implications on the effectiveness of the application. Yet, the ability to customize the interface means that the user is capable of reducing the effective area of the map. It is, therefore, important to understand how GIS users organize their interface before designing a new application.

For a detailed discussion of such a study, which was carried out more generically to explore the typical screen allocation practices of GIS users, see Haklay and Zafiri, 2008.

6.4   Application development

Many developers think that once the requirement stage is finished, they can detach themselves from the users, shut themselves up in a room and develop the application uninterrupted. When they finish, they will test if their creation satisfies the users. Some may even consider that dealing with usability issues is only relevant when designing the interface, so it can be left to the last minute. Nothing can be further from the truth.

Not considering user issues throughout the development process is likely to lead to less useful applications than one where considerations about the user and aspects of usability are integrated at this crucial stage of development. With geospatial technologies this is especially important, as the integration of reference mapping, the visualization of the specific data for the application, the functionality that the application provides and the interface are inseparable from the point of view of the user. For example, a decision on using pre-rendered reference mapping, the way that most Web-mapping applications work, means that the user will have parts of the mapping interface that will be interactive and parts that will not. Without careful consideration of the interface and the interactive elements of it, novice users might find such an application confusing and frustrating. Decisions on data models, classification and representation are critical to what is possible to ask from the application, and terminology such as layer names will end up in the interface. Thus, it is vitally important to ensure that the system is capable of presenting the user's mental model of the information. This does not mean that the system model should match the user's model. Only that it is capable of providing the users with their understanding of the world.

During the development process, the developers can refer to usability guidelines and design principles that have been learned in other projects or provided in the literature. For example, there are many accessibility guidelines that will define the level of accessibility of the application to a wide range of users. Another example is provided by the US Federal Government Usability.Gov website, where guidelines are provided covering issues from the design of the homepage to context organization. There are also emerging guidelines for certain types of geospatial technologies, and some are covered in the following chapters.

During geospatial technologies applications, there is a need to consider carefully which parts of the application should be open for customization by end-users and which should not. For example, as was covered in Chapter 3, there are many aspects of cartography that can have far reaching impacts on the way an application is used and how to integrate different data sources – for example projections. Yet, for most users, the issue of projections is not well understood and can lead to serious errors. Even for very experienced GIS professionals the issue of projections can be challenging. Therefore, the decision of projection should be considered carefully, in a way that minimizes the potential for user errors. Moreover, in most cases it will be wrong to expose the information about projections to the end-user. Colouring of thematic maps is another issue that can be handled through pre-set cartographic schemes that experienced cartographers set for end-users, and not to allow the end-user change them.

In addition to the use of usability engineering guidelines, a limited evaluation should be included during the development phase. As noted in this section, there will be some issues that relate to the core of the application (data structures, core functionality) that by necessity will have an impact on the interface. It is wise to test them at such an early stage, maybe by using paper prototyping. This is a technique in which the interface is evaluated through a paper mock-up. By discussing the mock-up with end-users, it is possible to make informed decisions about the directions that the development will take.

6.5   Evaluation and deployment

Once the application is in advanced development stage, it is common practice to test the application to see if it is answering the requirements that were set out for it. Some of these tests will be about the functionality of the system – using our geocaching device this will include the accuracy of the location information that it produces. In large GIS projects, these technical requirements sometimes play centre stage – for example, validating that the system has implemented the Open Geospatial Consortium standards fully. As noted, usability engineering should be integrated throughout the development process. However, it is critical to integrate it within the testing stage. The reason for this is that even if the application was developed without detailed consideration of usability and UCD principles, there are some things that can be fixed in the interface and that might improve the experience of the user while using the application.

Of all the areas of usability engineering, this is probably the most developed, in terms of the number of techniques that are available and the longevity of its use. The aspects that are covered by usability testing can be at the very detailed level to a more holistic and generic evaluation of the interaction.

The level of detail can be very high. For example, one of the first Usability Engineering methods that was suggested by Card, Moran and Newell in 1983 is Keystroke Level Modelling (KLM). This method is based on very detailed analysis of all the actions that the user is expected to carry out to perform a given task. The analysis includes counting every keystroke, mouse movement, and the time that it takes to switch between the keyboard and the mouse. For example, if the task is to find the direction between two locations using a Web-mapping site, the analysis will calculate the movement of the mouse to the 'directions' search box on the interface, the click of the mouse button that is required to activate it, the number of keyboard strokes that are required to type the two addresses etc. Using some standard measures for each action (such as about 0.20 seconds for keyboard press and release, or 1.10 seconds for mouse movement) it is possible to calculate the total time that it takes to achieve the given task. This can be compared to other existing systems, or used within the evaluation process itself, to see if some actions take an unduly long time.

At the other end of the scale there are the holistic methods, where the overall usability of the application is being evaluated. For example, heuristic evaluation is a method where a usability expert is using an existing checklist and their knowledge to review the interface of the application and identify usability errors that should be fixed. For example, such an evaluation can identify inconsistencies in the interface, where the same activity is termed differently in different parts of the interface.

As in the first stage of application development, usability engineering techniques cover the range from qualitative techniques, for example in asking a user to use the application and interviewing them after the event, to highly quantitative and structural, such as KLM. Some of the most common techniques are reviewed in Chapter 10.

Usability engineering techniques also differ in the complexity and cost that they involve. They can range from Nielsen's Guerrilla Usability (1995) where the testing is done for minimal cost in terms of equipment, number of users who are involved in the testing (usually about five) and the time that is dedicated to the testing. The heuristic evaluation that was mentioned before is a technique that Nielsen associates with guerrilla usability. An evaluation of a public mapping website that was carried out using this technique showed that it is indeed an effective method for geospatial technologies. At the other extreme are the expensive, laboratory-based experiments where a specialized usability laboratory is used. These can be rooms with unidirectional mirrors that allow the researchers to observe users while they use the system. Equipment such as eye trackers, which show what the user is currently looking at, can also be used. Simulators for real world environments, such as driving or flying can also be used. This type of evaluation is expensive, not just because of the cost of equipment, but also because it requires the participating users to travel to the laboratory and spend time on the experiment. In addition, the analysis of the results from such experiments takes longer.

Case Study 6.2: Usability evaluation of public mapping websites

An example of the effectiveness of Guerrilla Usability is the test carried out in 2005 on an evaluation of public mapping websites. The test includes providing users with a simple task of locating a common tourist attractions: Can a tourist find a famous landmark easily by using them?

The reasoning behind raising this question was that tourists are an obvious group of users of public mapping sites such as Multimap, MapQuest, Yahoo! Maps, Bing Maps or Google Maps. Market research information presented by Vincent Tao from Microsoft in a seminar around that time confirmed this assumption.

During the usability evaluation, we gave the participants the instruction 'Locate the following place on the map: British Museum: Great Russell Street, London, WC1B 3DG'. Not surprising, those participants who started with the postcode found the information quickly, but about a third typed 'British Museum, London'. While our participants were London residents, a realistic expectation from tourists is that they would know the postcode when searching for such a landmark.

In the summer of 2005 when we ran the test, the new generation of public mapping websites such as Google Maps and Bing Maps (Microsoft Virtual Earth) performed especially badly.

The most amusing result came from Google Maps, pointing to Crewe, a city in north west England, as the location of the British Museum (!).

The most simple usability test for a public mapping site that came out of this experiment was the 'British Museum Test': find the top ten tourist attractions in a city/country and check if the search engine can find them. Here is how it works for London:

The official Visit London site (http://www.visitlondon.com/attractions/culture/top-ten-attractions) suggests the following top attractions: Tate Modern, British Museum, National Gallery, Natural History Museum, the British Airways London Eye, Science Museum, the Victoria & Albert Museum (V&A Museum), the Tower of London, St Paul's Cathedral and the National Portrait Gallery.

Now, we can run the test by typing the name of the attraction in the search box of public mapping sites. Interestingly, as late as 2007, none of the common search engines managed to pass the test on all the top ten attractions, which are visited by millions every year. There is a good reason for this – geographical search is not a trivial matter and the semantics of place names can be quite tricky (for example, if you look at a map of Ireland and the UK, there are two National Galleries).

On the plus side, I can note that search engines are improving. At the end of 2005 and for most of 2006 the failure rate was much higher.

6.6   Usability engineering in research

In addition to the use of usability engineering techniques in product development processes, the range of methodologies and techniques that were developed within this area are widely used in research. Within GI Science research, these techniques are widely used by researchers who focus on HCI aspects of geographical information, as expected. In addition, they are widely used by researchers in geovisualization who are exploring how novel visualization methods of large scale data sets can contribute to knowledge construction and sense making by end-users. A third area where the techniques are widely adopted is the development of spatial decision support systems. These are computer systems that assist their users in solving spatial problems, which were covered in Chapter 4. In this type of situation, it is important to ensure that the interface provides the user with information to enable making a correct decision.

Finally, researchers who are developing applications in the GeoWeb are also utilizing usability engineering, since in Web applications there is generally more awareness of usability aspects. In all these cases, usability engineering is used beyond its usual role of testing that the application is usable for the audience for whom it is intended. In research settings, usability engineering methods provide tools to collect information about the users and the way that the geospatial technologies are used. The ability of usability engineering techniques to provide robust quantitative and qualitative evidence of interactive sessions is the reason for their use in this context. While this is a different goal from the way in which usability engineering techniques are usually deployed, they require very little adaption in terms of data collection protocols or the detailed methodology of transferring the techniques from research context to application evaluation context. An example of the use of usability engineering techniques in research that focused on the use of geovisualization for knowledge construction is provided below.

Case Study 6.3: Usability study of geovisualization

C. Tobón

Geovisualization is in important respects a cognitive activity, the purpose of which is to gain an understanding of a geospatial dataset by representing or encoding data in some graphical form that can reveal otherwise hidden information (MacEachren and Kraak, 2001; MacEachren, 2004). Geovisualization tools or systems are frequently used to explore and sift information from large or complex geographical datasets. Geovisualizations have at least two components: dynamic and interactive computer environments; and users whose domain knowledge is key to the data exploration and hypothesis formulation process. The computer system must therefore allow users to manipulate and represent data in multiple ways in order to enable the investigation of 'what if' scenarios or questions that prompt the discovery of useful relations or patterns. In this manner, the geovisualization environment is intended to support a process of knowledge construction which is 'guided by [the user's] knowledge of the subject under study' (Cleveland, 1993: 12).

Nevertheless, the nature of data exploration and knowledge discovery in geovisualization that aims to 'discover patterns within the data while simultaneously proposing a hypothesis by which the patterns might have come to be' (Gahegan, 2001: 275), implies that tasks and goals are often ill defined, making them difficult to formulate and even measure. Formal usability evaluation techniques usually require structured tasks or scenarios to assess user performance or success in accomplishing them. Experiments were therefore designed in an initial usability evaluation that was carried out using a commercially available system to ascertain the priorities for supporting different geovisualization tasks. Findings from this evaluation guided the formulation, coding and analysis of user tasks that aided in the understanding of whether a novel geovisualization environment supported them. In summary, a combination of analysis methods was explored as a means of understanding the extent to which potential users could utilize computer-based tools effectively and with ease to accomplish their work.

The first usability evaluation was carried out using a commercially available tool, which comprises a visualization system (DecisionSite Map Interaction Services by Spotfire®, Boston, MA) and a lightweight GIS data viewer (ArcExplorer from ESRI®, Redlands, CA) coupled as a plug-in to DecisionSite. Figure 6.1 illustrates the environment. To the left hand side of the image is the GIS data viewer with the map at the centre, a context window and layer control at the bottom, and a toolbar at the top to link the two systems. The main limitation of this environment is the lack

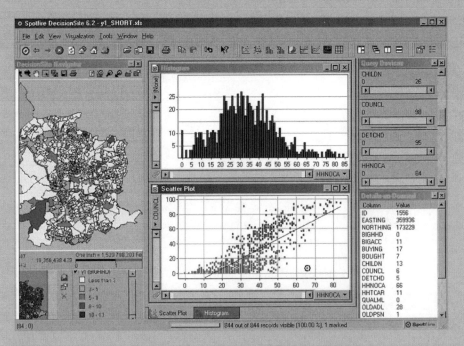

Figure 6.1 DecisionSite Map IS. (Copyright © 2003 ESRI. All rights reserved. Used by permission) (Image courtesy of Carolina Tobón)

of a dynamic link between the two components. Hence, when data are selected on the map in ArcExplorer, the operation is not reflected in the visualizations in DecisionSite or vice versa. Instead, once a visual operation such as a selection is made, for instance in DecisionSite, a button in one of the tool bars has to be pressed for the selection to be reflected in the map.

It was anticipated that users would have difficulties in exploring spatial data using this environment. Thus a first usability evaluation was designed to obtain as much information as possible about the ways in which the limitations of the environment would restrict user understanding and exploration of the data. For this purpose, nine participants were asked to perform four tasks. The first two tasks were open ended questions designed to allow the users to become familiar with the environment, while revealing how they attempted to solve the task. Users were allowed to ask questions, and were encouraged to explain problems encountered and to engage in discussions about the information they were obtaining. In the last two tasks, users were not allowed to engage in discussions and were required to obtain more precise answers about the dataset being explored. Despite the aforementioned limitation of the environment, the evaluation was nevertheless successful in gathering information about how users attempted these tasks, their difficulties or successes in doing so, and discovering other problems that required system support.

The participants who were invited to take part in this study were expected to have some experience with handling data in a GIS or in other systems where graphical displays are used for spatial data manipulation. Participants included GIS professionals and students of various ages and nationalities, as well as planning students with different levels of GIS experience. The lack of a dynamic link, however, made it difficult for users to understand the full functionality of the coupled system. At least two participants did not understand that the data represented in all views were from the same dataset. It took a further two users the whole session in order to grasp this. Two more failed to comprehend the information that the environment was providing and therefore did not see the utility of the linked software. Nevertheless, seven out of the nine users commented positively about the advantages of combining flexible and interactive tools for attributing value exploration with the mapping capabilities of a (lightweight) GIS, and were excited about the possibilities for hypothesis formulation that tools of this nature might support.

A second environment, in which DecisionSite was coupled to ArcMap in ArcGIS^{TM}, was therefore created in order to solve the problem of dynamic linkage described above. Figure 6.2 shows the environment developed for the second evaluation (DecisionSite on the right, ArcMap on the left), where the selection or highlighting of features in any view is reflected in all others. It was an objective of this study to obtain further information from the users about the functionality that environments for the visual exploration of geographic vector data should include, rather than defining a priori what such functionality should be. For this reason, DecisionSite was coupled to a fully fledged GIS, and the evaluation was designed to give participants as much flexibility as possible in tackling tasks and in using the functionality they deemed necessary to solve these tasks.

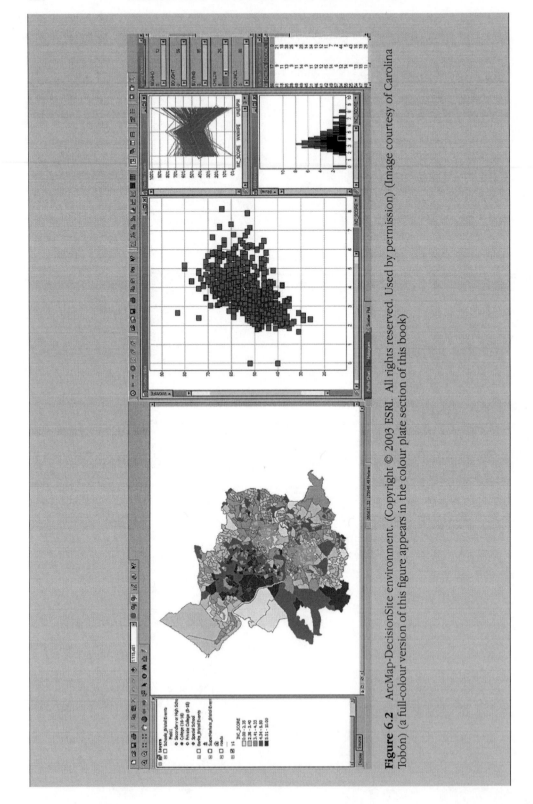

Figure 6.2   ArcMap-DecisionSite environment. (Copyright © 2003 ESRI. All rights reserved. Used by permission) (Image courtesy of Carolina Tobón) (a full-colour version of this figure appears in the colour plate section of this book)

Tasks for this evaluation were defined according to three factors: the extent of the geographical area that needed to be investigated, the number of attributes or variables to explore, and the type of visual operation users had to perform. The first two factors had two levels: either one or many areas or attributes to be explored. The visual operations required users to identify geographical area units with particular characteristics, locate where a particular phenomenon occurred, uncover associations between attribute values, and compare geographical areas where some relation between attribute values occurred. An experiment was designed so that combinations of the three factors (spatial extent, number of attributes investigated and visual operation) at their various levels defined 16 tasks of differing complexity. Twenty participants with varying levels of GIS expertise completed two sessions on two different days when they had to solve these tasks. Although the tasks they were presented with in each session were different in terms of the particular data attributes that they explored, the types of tasks were the same. The experimental design allowed two observations to be recorded for every task type. This made possible the investigation of learning the effects of the software environment, as well as comparison of user performance before and after they gained some confidence in using the software or acquired knowledge about the data.

Performance measures, such as the time it took respondents to complete a task, were therefore recorded. Factor analysis was used in the analysis of the measures, in order to explain differences in the performance variables (time and success rate) in terms of the main effects of each of the aforementioned factors in the outcome (or performance measure) of the experiment. This technique also allows the detection of interactions between the factors, or in other words, the effect of each factor at varying levels of each of the other factors that account for over-all user performance.

Both the main and the interaction effects were likely to be important. This is because the time required to complete a task (or the degree of success in performing it) was expected to depend on the types of visual operation attempted, the spatial extent of the problem area and the number of variables involved in solving the task, as well as on the combinations between these factors. The results of the statistical analysis confirmed this to be the case, and further qualitative information gathered during the experiments helped explain the particular needs of users for supporting each type of task. Thus the statistical evidence was complemented with interviews and questionnaires which provided information about user perceptions and experience of the environment. Interviews also made it possible to obtain detailed explanations of how users went about solving tasks or about the problems that they encountered. This adds weight to, and enriches, the quantitative evidence. Users also answered a questionnaire designed to obtain information about the perceived usefulness and ease of use of the environment.

Questionnaires provide a quick and inexpensive measurement instrument in this kind of study. They also generate useful quantitative data, particularly if measurements are recorded on a Likert scale, which can be used as an appropriate yardstick with which to measure each user's perceptions. Questionnaires administered in this second study made it possible, for example, to establish that although the environment was perceived as being very useful to obtain relevant information and speed up user response once it was mastered, the software was not seen as easy to use, at least initially.

To summarize, a number of usability techniques were tailored and combined to learn about some of the types of tasks that geovisualization applications need to support. It was shown how such methods can be used to evaluate not only a system or prototype in terms of some aspect of its usability, but also to demonstrate research concepts, such as how well a particular visualization works for certain data types.

(Adapted from Tobón and Haklay, 2003)

Summary

Usability engineering provides a set of tools that can be deployed and used throughout the product development life cycle. The integration of the specific methodologies and techniques allows for the translation of the UCD principles into practical measures and matrices. These, in turn, provide practical tools that can be integrated into the workflow that is used for the specific application or product.

Usability engineering techniques range from the inexpensive and simple to those that require expensive equipment and experiment design. Many of them can be used even by a single developer who is working under time and budget constraints.

However, most usability engineering techniques require careful adaption to geospatial technologies. For example, some heuristics for website design recommend reducing the amount of graphic files that are being sent to the user, to speed up the page load. Translating this to a Web-based map is counterproductive – the map should use a significant proportion of the screen. In addition, there are techniques that are especially useful for the analysis and study of geospatial technologies – such as the screenshot study.

Further reading

There are several websites that provide detailed information about usability engineering and methods. Jakob Nielsen, who is the author of one of the first books on usability engineering (1994a) has an extensive website with hundred of columns covering mostly Web usability issues (useit.com).

As was mentioned earlier, the US Government usability.gov provides a set of articles for public sector websites.

For more generic discussions on usability engineering and usability methods, see Usability Net (http://www.usabilitynet.org/). The site provides a comprehensive list of methods and techniques, with detailed instructions on how they can be used.

In addition to Nielsen's book, Cristine Faulkner's *Usability Engineering* (2000) is valuable, as it does not just discuss the cases where usability engineering works, but also presents challenges to existing methods and caveats that must be considered when usability engineering is used.

Revision questions

1. What are the main aims of usability engineering, and why do we use the term 'engineering' to describe this activity?
2. In the three case studies that are described in this chapter there are several usability engineering methods that were integrated in them. Can you identify them?
3. In a few points in the chapter we have sketched the development of a geocaching specific GPS receiver. Suggest how you would develop it using usability engineering principles.
4. The website useit.com provides a set of heuristics for website evaluation. Use these principles to evaluate a Web-mapping site of your choice and suggest how you would change these guidelines for mapping based systems.

Section III

Practicalities and technique

7 Application planning

Jochen Albrecht and Clare Davies

One of the mantras of software development is the need to always keep the application's user in mind; it is they for whom we make all our efforts. This is the core principle of User-Centred Design as we have discussed in Chapter 5. We need to understand what it is that the user is trying to accomplish, what their experience or skill level is, and how the task to be solved fits into a larger workflow and purpose. As such, UCD application planning is very much like systems analysis where the different elements of the system are carefully designed, except that here the user is at the centre of it all. One of the core differences between general information technologies application and traditional GIS applications is the steep learning curve that is required from the users due to the complexity of the software and the fact that this complexity is exposed in all its glory (or ugliness) on the interface. Even when it comes to other geospatial technologies, such as PND or Web-based mapping, beyond some basic operations it is common to see that the system is becoming complex to operate, requiring the user to learn several disjointed steps to accomplish a specific task – such as finding a hotel. This chapter's first two sections discuss some of the issues of GIS interface and the ways researchers address this issue, and we focus especially on task analysis. While Lanter and Essinger (1991: 1) noted that 'some of the most successful user interfaces are complete illusions outwardly bearing no resemblance to the data processing happening inside the machine' almost two decades ago, most GIS interfaces – from commercial to open source – require the user to have significant understanding about the way that the system is implemented to operate it. Users have mental models about the tasks they accomplish with a system – this is their internal representation of what they want to do, what the computer does and the way the system lets them accomplish those tasks. These models are defined by the user's prior experience, existing knowledge, and preconceptions about tasks (Rauh *et al.*, 2005). For both a task and the way to accomplish the task to make sense, they must correspond to existing knowledge the user already has. The third section will dwell on this transition between the developer's view and the user's view of a GIS application. Formal specifications of both views are seen as one possible way to create a consistent user interface, and can be used in the design process – especially of major systems where there is a large team of developers and there is a need for formal and accurate definition that the whole team can use. For many years, researchers have been gathering user requirements, but as the fourth section mentions, this is becoming increasingly difficult with the proliferation of contexts in which geospatial technologies are used, and especially with the emergence of user-generated content websites and compositions of Web services through mash up. Clues may be drawn from another domain, that of workflow management in the business community as well as in scientific process composition. The fifth and sixth sections discuss different approaches in these two domains. We also describe the

Interacting with Geospatial Technologies   Mordechai (Muki) Haklay
© 2010 John Wiley & Sons, Ltd

importance of the research body on GIScience ontology which has some very practical value in GIS application development based on the semantic Web. Recent efforts in this realm are presented in the seventh section, while the remainder of the chapter provides pointers to what technical and scientific developments to watch.

7.1 GIS interface complexity

As noted in Chapter 1, GIS combines the functionality of database management systems and computer aided design software, with cartographic design and statistical analysis added for good measure, so it should not be surprising that it is one of the most complex widely used applications around. Not only does it typically consist of over a thousand functions but it requires the user to combine different metaphors and concepts borrowed from separate disciplines in order to accomplish even a basic task (Driber and Liles, 1983; Mark and Gould, 1991; Traynor and Williams, 1995). In consequence, it usually takes a combination of multiple courses at university level and years of experience to become a proficient user. A lot of this has to do with the multi-pronged origins of GIS itself, which was developed by surveyors, landscape architects, and census statisticians, to name three prominent origins. These disciplines have little in common, but contributed approximately equally to the conceptualization of what we now know as GIS. The National Center for Geographic Information and Analysis (NCGIA) conducted a series of workshops in the early 1990s that brought together experts from a variety of disciplines to tackle the complexity of GIS user interfaces (Frank, 1993; Nyerges, 1993; Turk, 1993; Volta and Egenhofer, 1993; Medyckyj-Scott and Hearnshaw, 1993). In this early literature, the issue of understanding and identifying the tasks that the user performed with the system was recognized as central to the design of well functioning and usable GIS. This is based on the understanding that unless we can clearly express what are the tasks that the user is trying to perform with the system we cannot develop a system that supports this work. Importantly, the tasks are not about what the user will do with the system, but rather what the real world transformations are that should take place. In most cases, the computerized system will support some task that the user can perform without the use of computers, and it is this type of activity that we want to capture. For example, the most basic operation with a GIS is 'create a map'. This can be done without computers, but requires careful drafting and ability to trace base maps, transpose their coordinate systems, etc. Thus, the use of GIS makes such a task faster than carrying it out manually. However, if we are setting out to analyze the user's needs, we can develop a task description that focuses mostly on what the user is trying to achieve without any consideration to how the system will be implemented.

Task description and task analysis are, therefore, central to the process that we describe here of hiding the complexity of the system from the user.

7.2 Task analysis in GIS

There are many well-documented methods of describing work tasks, both within and outside the context of computer use (Kirwan and Ainsworth, 1992). Most of these methods focus on the qualitative or quantitative description of tasks, knowledge requirements and errors, without any specific methodology for extracting prescriptions about individual actions, for example differentiation between tasks that directly contribute to the outcome of the task (selecting the colour of a line on the map) and enabling actions, such as switching the computer on. An exception is the task-action grammars (TAG) proposed by Payne and Green (1986). TAG splits a task into individual actions, mostly described using terms that suggest a preparing/performing distinction (e.g. 'select object', 'modify'), but the grammars are aimed more at uncovering inconsistencies in the patterns of actions than at identifying individual actions as productive.

This emphasis on analytical description is not surprising, given that task analysis may be performed for many different purposes, ranging from risk calculation to the design of training courses. The prescriptive requirements naturally vary with the goals of the analysis. The work/enabling distinction proposed by Whitefield *et al.* (1993) should make it possible to identify portions of user-system interaction that involve many enabling actions or a particularly long time or high amount of effort spent on such actions, for each work action. As a result, specific aspects of the interaction may be identified which would particularly benefit from a redesign if we are examining an existing system or evaluating a prototype; the purpose of such an exercise may be either to reduce enabling actions, to reduce or alter them for some users whose expertise or tasks cause the actions to be unnecessary or tedious, or to change the nature of the enabling actions so that they more closely support the user's intentions within the task. Ultimately, this is just one more tool intended to create systems closer to users' real needs.

Existing analysis methods already allow us to quantify work using the number, duration and apparent physiological workload of actions, as well as the percentages of these spent in making or recovering from errors (e.g. Zapf *et al.,* 1992).

Task analysis can be linked to ideas which arose much earlier in the history of analyzing work. Any book on time management, of which there must be hundreds, will attempt to show the readers that they spend vast amounts of their time on non-productive activity. It is probably true that no form of human activity, from cookery to librarianship, is free from what Whitefield *et al.* (1993) labelled 'enabling' behaviour. Meetings, completing forms, searching for keys and placing paper in a printer tray could all be viewed this way. We should be cautious in assuming that the presence of enabling actions within a computing task is in itself a signal to redesign the system: it may be that we can never eliminate, nor want to, some enabling actions from computing tasks.

The existence of a variety of actions and procedures within a task may in fact benefit workers' or personal users' motivation under many circumstances, as was demonstrated in the revolt against Taylorism and the move towards job enlargement/enrichment earlier in this century (Kelly, 1982). Taylorism, a way of thinking which centred on the improvement of industrial productivity through the implementation of 'the one best way' to perform each task, became outmoded as work analysts began to focus on the social aspects and non-financial motivations of the workplace. Greif (1991) compared Card *et al.*'s (1983) GOMS and Keystroke-Level models, which were covered in Chapter 6, with Taylorism in their assumption that it was possible to identify the one 'best', most efficient, method of performing any given task. Greif (1991) argued that such assumptions ignore the importance of considering human motivational needs, especially the need for 'growth'. Current interest in user experience, which extends the analysis beyond efficiency per se, can be seen as a way to include these motivational needs.

According to Greif (1991), German work psychology, and particularly the work on 'action theory', favours systems which are adapted to an individual user's understanding of the tasks; thus the choice of available actions and the level of detail inherent in those actions should be made to fit the individual's conceptual model. Those conceptual models in turn, according to the German theorists, are based on users' motivational needs. The motivational need for 'growth' and fulfillment does not mean that all non-goal-achieving activity is desirable; nor does it prevent us from identifying such activity even if we decide not to remove it. The work/enabling distinction is related to the concept of 'idle time' in production lines, which consists of balance-delay time (time spent waiting for another person or system to complete a process), non-productive time (time spent handling or moving materials, or in checking errors and defects) and waiting time (time wasted due to machine breakdowns or interruptions in supply) (Kelly, 1982: 71). Such periods are clearly non-productive, and the worker in a poorly designed production line has no choice but to 'perform' the waiting actions in order to continue the task. Additionally, some method study techniques distinguish within an operation between actions which 'make ready', 'do' and 'put away' (Kanawaty, 1992: 95), of which the first and third are clearly analogous to 'enabling' actions.

However, care should be taken in analyzing office-based work with techniques developed for industrial production: often the aims of the techniques are different as well as the context of use (e.g. they may aim to identify a standard production rate or assess workers' personal performance). Additionally, in manufacturing contexts workers' activities are largely physical and thus subject to physical limitations. Thirdly, the goal of the activity is always clear: making (or physically altering) something. Finally, the worker has not chosen her or his goal: the task is fixed and usually does not involve mental creativity.

Nevertheless, even in the domain of creative work within office-based computing tasks, Green (1990) proposed a conceptual 'dimension' that may be linked to the work/enabling distinction. This is the 'cognitive dimension of viscosity'. Green defines viscosity as the degree to which a system makes it difficult for a user to change their mind – the difficulty of altering an entity after it has been created. Thus a program written in a disorganized way could require extensive rewriting if the programmer later decided to incorporate extra functionality; a road marked within an image editing package ('raster GIS') may not be easily changed if the user decides to alter a junction. Again, the viscosity dimension has limited applicability – not everything one does is creative, and not every source of inefficiency is due to difficulty in altering created artifacts – but nevertheless the extra effort expended in a highly 'viscous' system certainly entails 'enabling' activity and is arguably unnecessary.

Another conceptual framework for tasks was offered by Don Norman in his exploration of 'stages and levels' in HCI (Norman, 1984). Norman's four-stage concept of a task meant splitting it into intention, selection, execution and evaluation. Recognizing that the execution of a task to fulfill a given goal was usually likely to involve a complex sequence of actions, Norman (1984) argued (at a time when most computer users were not yet using graphical user interfaces) that systems requiring the user to 'point' to select an action (practically, a GUI-based system) were likely to be more efficient, but more restricted, than those requiring the user to name the action, for example by typing its command name. Norman (1984) also went further and argued that the difference between users who preferred GUI systems such as menus, and those who preferred 'name'-based systems such as command-line user interfaces, was to a large extent attributable to the emphasis placed on the selection of actions as opposed to their rapid execution. This is turn, he argued, reflected the type and extent of a particular user's knowledge: some users knew all the possible actions already, and wished to execute them as rapidly as possible; others wished to be visually reminded of what it was possible to do. In the work/enabling terminology, one could argue that the emphasis of the latter group was on being supported in enabling (selecting) the action, while the former group focused on minimizing enabling actions and just working (executing). Both the work and enabling actions within a given task will fall largely within Norman's (1984) third 'stage' of a task, i.e. the 'execution' stage, although some will be part of the earlier stage of 'selection' or the later stage of 'evaluation'.

Many of these task analysis frameworks were proposed for systems where tasks are fairly easily identified and defined, such as word processing. In contrast to these systems, GIS are sophisticated products with some of packages placing more than 10 000 commands or functions at the user's disposal. Nevertheless, the most common tasks performed in reality are more mundane ones such as preparing maps for printing or plotting on paper, querying and updating database records, data entry by digitizing the map data, checking for and correcting errors, searching through the system to find particular information, and summarizing data (such as sales totals or highway accident statistics) into printed reports. GIS differ from text editing software not just in the type of data handled, but also in their use for information storage, retrieval and analysis as well as editing. Any attempt to improve GIS usability must start with the recognition that users interact not only with a computer system but also with a map, and thus the analyst should be familiar with cartographic literature on map design and use as well as ergonomic prescriptions about the user interface (Wood, 1993). Maps, and hence the software to handle them, require decisions to be made and recorded about projection, scale, symbols, colours, relief representation, generalization and labelling. With this added complexity, for the task analyst as well as for the user, we should not be surprised that little published work has analyzed GIS tasks in detail.

Some authors (e.g. Nyerges, 1993; Gould, 1994) have called for the gradual development of detailed task taxonomies for GIS, but to be feasible such taxonomies would have to develop from a relatively high-level definition of 'task'. Nyerges (1993) adopted a view of GIS tasks based on the idea of tasks as transformations, and the 'transformations' in question concentrated on the transferral of information between mental, real-world, graphical and digital forms. It could be argued that such changes of form constitute a subset of the transformations a user may have to perform within the overall work domain.

Albrecht (1994) argued that the GIS user faces a fundamental problem in having to translate high-level job requirements into long sequences of low-level system-specific commands. Therefore, he proposed the development of a high-level language for translating between these two task levels, to facilitate GIS use. While this idea focused on the use of GIS to solve analysis, modeling or decision-support problems, the paper also included a genuine attempt to present a hierarchical breakdown of sample GIS-related tasks. This notion of a high-level 'language' is one possible solution to the problem of simplifying user system interaction, and the design of such a language could be facilitated by applying a task analysis coupled with an evaluative tool such as the work/enabling distinction. Other possible ways of describing or taxonomizing GIS tasks have been discussed in Davies (1995).

At a much finer level of detail, Haunold and Kuhn (1994) used the GOMS 'Keystroke-Level Model' of Card *et al.* (1983) to analyze manual digitizing of property data. Their study illustrates the usefulness of such a low-level analysis in identifying potential productivity improvement, through the reduction of the number of actions necessary to fulfill a low-level task goal. By way of an example, Haunold and Kuhn (1994) described the steps required in the system they tested to enable the user to zoom in to an area of the displayed map, and suggested that the ideal number of actions would be only two: a single button press by the operator and the system processing time required for redrawing the screen. Not all operations are as obviously open to redesign as that zooming action; in addition, when faced with a rich assortment of operations it is not clear where an analyst should focus attention in order to maximize gains in productivity. The potential value of the work/enabling distinction lies in its distinction between the actions that really do the job, and those which do not, a distinction made implicitly by Haunold and Kuhn (1994) in their zooming example. Making this distinction may allow us to see which tasks have a relatively high proportion of enabling actions, and decide whether these are the tasks that would benefit most from redesigning the user system interaction. In other words, distinguishing 'work' from 'enabling' may add a prescriptive dimension to the descriptive procedures of task analysis.

Case Study 7.1: Development of a land use-based transportation model

Albrecht *et al.* (2008) describe a rather complex but fairly traditional GIS workflow (Figure 7.1 depicts the program's logic; the database schema is too deep to be represented in a single figure). The goal of this spatial decision support system (SDSS) is to model existing flow in a multi-modal network (that is, a system in which people are using several modes of transport such as car and public transport) and to find optimal locations for a transit hub based on existing parcel-level land use. The procedures of the SDSS combine many hundred GIS operations for single model iteration.

The very model *structure* is dynamically changing as the model is run. For instance, the choice of a commuter's mode or route is influenced by how many other commuters crowd into one's first or second option. Many commuter train passengers, for example, are familiar with the increased number of passengers when flooding closes roads during rain or snow storms. The same happens on a smaller, but cumulatively as effective scale every day in metropolitan areas around the world. From an abstract perspective, the phenomenon is well studied and has become a popular example in complexity theory. From a user interface development perspective, it is almost impossible to create an easy to use and intuitive interface that hides this complexity from the user. Of course, sometimes it is quite useful, especially for the scientifically inclined user, to see this complexity.

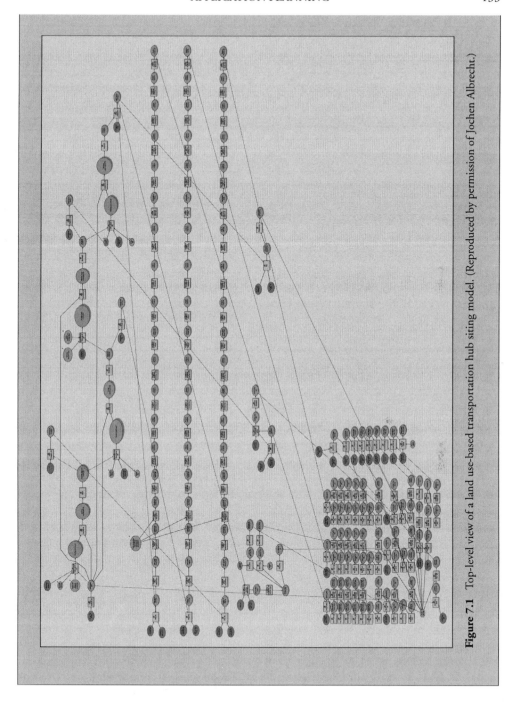

Figure 7.1 Top-level view of a land use-based transportation hub siting model. (Reproduced by permission of Jochen Albrecht.)

As Case Study 7.1 clearly demonstrates, there are situations where the inner working of the system should be hidden from the user. In addition to task analysis, which we have carried out, we should also note the pioneering work of Andrew Frank and Max Egenhofer at the University of Maine in the early 1990s. Volta (1993) separated the formalization of the problem domain – that is identifying the objects a user manipulates and their pertinent operations – from its visualization, through interaction elements such as windows and dialog boxes. Bruns (1994) tried to bridge the cognitive dissonance between table top and map space by creating so-called direct manipulation interfaces, where the user combines iconic representations of GIS objects aiming at creating a visual map algebra, which is a common process of integrating different layers in a GIS through algebraic transformation. We will revisit the notion of visual representations of analysis in the section on GIS workflows and process models. Jackson (1990) suggested a series of mappings between three domains: the source domain, the target domain, and the user interface domain, or visualization. In his thesis 'Visualization of metaphors for interaction with GIS' he developed formal specification methods to describe the domains involved and identify similarities and dissimilarities between them. Golledge's (1995) approach to the same dissonance problem was to identify spatial primitives as building blocks of higher-level GIS operations and subsequent tasks. Albert and Golledge's (1999) thorough investigation of the cognitive aspects of map overlay and Riedemann's (2005) parallel work on topological operators seem to have brought a closure to the topic, to some extent.

7.3   Formalized analysis of GIS user interfaces

With the advent of GIS as a potentially interesting research object in computer science, more formalized approaches to the study of GIS user interfaces began to appear. Kösters *et al.*, (1996a,b) took well-established methods from classic object-oriented systems analysis (OOA) and applied in parallel to (what developers believed are) the domain of GIS tasks and to the analysis of the user interface itself. Use of the same tool was supposed to guarantee that the 'mapping' referred to above could take place. It was a formidable effort that eventually failed because of the complexity of desktop GIS, especially since they have tried to tackle a complex traffic simulation system, and the subsequent size of the formal analysis model. Li and Coleman (2005) use a similar instrument (now based on the unified modeling language) to describe workflows, not on the level of individual GIS operations but on that of user tasks. Their goal is control quality and the formalized approach certainly goes a long way to assuring that.

Oliveira and Medeiros (1999, 2000) use considerable effort to formalize the static components of the user interface as well as a procedural view of how these are utilized. While their study objects are GIS and spatial decision support systems, this line of research of a group of information scientists who we will encounter in other parts of this chapter is decidedly not about user tasks or functional GIS components. The result is a dissection into software components (as in COM or CORBA modules) that are great from a software engineering perspective but are entirely void of any user experience considerations.

7.4   User experience considerations

Given the steep learning curve of GIS software, the user interface needs to be adaptable, i.e., work for the novice user as well as for the expert one. According to Mark (1992: 551), 'a main objective of GIS is to allow the user of the system to interact vicariously with actual or possible phenomena of the world. If this is so, then the system which mediates between the user and the world should be as unobtrusive as possible'. He goes on to distinguish between kinds of users (geographic knowledge), their computing skills, frequency of software use, cultural and personal background, and only then the purpose of the user's application. Davies and Medyckyj-Scott (1996) studied the first half of this list and concluded what seems obvious but had not been proven before: experience matters. Here, experience is the sum of educational level, familiarity with the software, and most significantly the match (or lack of) between the developer's and the user's conceptual model of the task at hand.

A user's experience is obviously strongly influenced by the culture she grows up in and the language they speak. Mike Gould (1989) built his early career on exploring the different spatial metaphors in English vs. Spanish and the effect that this has on GIS usage. David Mark (1992) extended this to look at a range of other languages such as French, German, or Croatian. Following the work of Lakoff (1980) and Norman (1990), spatial metaphors and image schemata became a fertile research topic (Gould and McGranaghan, 1990; Kuhn and Frank, 1991; Mark, 1992; Nyerges, 1992).

Of particular interest is Nyerges' contribution, as he tries to identify cognitive primitives that can be used as building blocks representing elementary spatial relationships such as container, part-whole, linear order, link, center-periphery, etc. We will revisit this theme in Section 7.6. Bunch and Lloyd (1996) give credence to the fact that managing the cognitive load experienced by learners is the key to representing geographic information.

Promising as this work looked back then, it has since only made advances in linguistics and cognitive science but had no lasting effect on the design of GIS applications. As to cultural differences mentioned by Mark above, the notion that culture matters was picked up by social critics of GIS (Pickles, 1995; Rundstrom, 1995), but in the long run, market forces have proven to be stronger: GIS applications still look very much the same, no matter whether they are used in a London finance office, a school on a Navaho reservation, or a Nepalese erosion prevention program.

7.4.1 User requirement gathering in the age of Web-Mapping 2.0

Research on the material presented in the previous sections has not been very active since the mid 1990s. However, because of the transition of GIS applications from traditional desktop environment to distributed Web services, there is a renewed interest in descriptions of the system and bridging the cognitive gap with the user. Traditional application development methodologies such as Joint Product Design, Joint Application Design, Rapid Application Development, and more recently agile methods, assume that we can foresee the type of geographical questions to be asked and the characteristics of the application's user. More important, textbooks (Downton, 1991; Whitten and Bentley, 2005) advise us that we work with the user in reviewing the requirement list. However, in the new range of application that emerged with the growth of Web 2.0 environment, these parameters are not available anymore (Kazman and Chen, 2009). Baranauskas et al. (2005) tried to address this problem by moving onto the Web and conducting a public participation process to solicit input from potential users of a Web-based GIS. Subsequently, and notwithstanding Brewer's (2002) call-to-arms, the only recent user requirements gathering for spatial applications are about mobile devices (Queiroz and Ferreira, 2009).

7.5   Task analysis as the basis for workflow management

With the user as a criterion out of the picture, we need to refocus on the set of functionalities available in GIS and related software. In the 1980s and 1990s, Goodchild (1988), Lanter (1994), Jung and Albrecht (1997), and probably many more compiled lists of GIS functionality. In the age of Google's spatial products, the incorporation of spatial data structures and functions in the statistics packages, and numerous open source tools with varying degrees of analytical functionality, the notion of what constitutes a GIS is a bit more fuzzy. The challenge is to match functionalities with generic user tasks, and most importantly to find means of communication that allow for that match to happen not just in theory but in reality, during a product design process. Albrecht (1998) did this for desktop GIS, and Klien *et al.* (2006) updated this line of work for Web-based applications in disaster management. The following is a review of task analysis for geospatial applications.

Still on the desktop side, Oliveira *et al.* (1997) developed a processing environment, which is both more generic than Albrecht's (1998) VGIS framework and because of that less directly applicable to a GIS implementation. As such, it serves more as an Application Programming Interface (API) for rapid prototyping of workflows in a multitude of application areas that require a lot of fine-tuning to result in a user-friendly application. The advantage of such an approach is the possibility of linking applications from very different domains, though there is virtually no use of standards that we expect nowadays from interoperable software. Originating from the same research unit, Weske *et al.* (1998) describe possibly the same system, now called WASA, which becomes the foundation for two diverging developments. On the one hand, they describe something akin to the scientific workflow environments that will be discussed in Section 7.6. On the other, WASA is a precursor to a distributed, Web-based toolset that is now at the heart of application developments worldwide covered in Section 7.7.

Workflow management has become a major focus of business-related information system research. It is a suite of applications and procedures that assist in managing internal workflows inside a corporation and deal with high level tasks, while allowing the monitoring and management of their progress. There are obvious gains in the rationalization of workflows, which have not yet been implemented on any significant scale with GIS packages. One aspect that links with past research is the issue of quality management. Li and Coleman (2005), for instance, conduct their task analysis with the goal of focusing on quality control aspects of workflow management.

Geographers have maintained for generations that (geospatial) context matters. Next to location and scale, context is at the core of geographic thought and theory (Spedding, 1997; Massey, 1999; Gertler, 2003) and so it is a little surprising that this theme is receiving more attention within computer science rather than in GIScience. Cai (2007) was the first to formally acknowledge the role of context in the specification of GIS tasks. Cai's article ties together several strands of research described in different sections of this chapter. For one, he picks up Volta's formalization of mappings from the developer's view to the user's view, albeit now with modern ontology tools. Cai also employs user input to fill gaps in the specification of context that are to be expected in a distributed Web-Mapping 2.0 environment. As such, he manages to keep his ontologies vague, with the system adding constraints on the fly as the user assembles the application. Liu *et al.* (2007) is a first report by a new research group at the National Center for Supercomputing Applications (NCSA) that picks up the context theme and a range of tools and software environments that help to build task ontologies (in particular for seismic disaster management) and end-user ready decision support systems.

Case Study 7.2: NASA's Earth Science Gateway

Bambacus *et al.* (2007) introduce the Earth Science Gateway (ESG), a geospatial Web portal designed to support prototyping applications and to reuse data by sharing Earth observations, Earth system modelling, and decision support tools. ESG's interoperability provides easy integration of systems and components through open interfaces for rapid prototyping. In their article, Bambacus *et al.* (2007) report on a rapid prototyping session during a GEOSS demo in May 2006. As do other countries and agencies, NASA and NOAA collaborate on modelling and predicting Earth science phenomena, such as global atmosphere circulation and wind speed. After the simulation results are produced, they are put into services that comply with the Open Geospatial Consortium (OGC) Web Coverage Service (WCS) and Web-Mapping Service (WMS) and the services are registered into different catalogs, such as the Earth Science Gateway. In their workshop session, the participants used the ESG to rapidly prototype an application to identify locations for building wind farms to produce electricity in Hainan province, China.

The workflow consists of three main components:

1) Search ESG using 'wind' to find services of interest from tens of wind-related services, such as G5FCST Wind Shear.
2) Add the G5FCST Wind Shear service that was delivered to the viewer to get the desired application.
3) The application shows some wind situations but no detailed wind speed information. Other services, such as the Mean Wind Speed service with rough and fine resolution are then added.

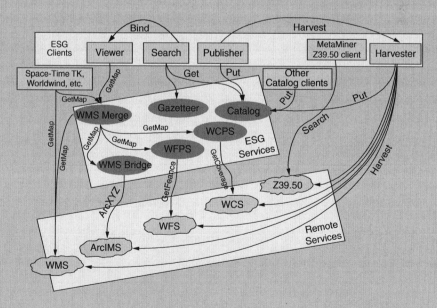

Figure 7.2 Web services chaining for the Earth Sciences Gateway (a full-colour version of this figure appears in the colour plate section of this book) (Reproduced by permission of Phil Yang)

> This application can then be saved, and whenever a user brings up the application, updated observations and simulations will be integrated.
>
> Through this process, an application can be prototyped quickly with the support of the ESG. In the process, users do not have to know who provided the observations or simulations, or who provided the external WMS. Professional users only focus on the application logic by searching available services and selecting needed services. Public users only need to bring up the application through an Internet hyperlink prepared by a professional. In this example, the legacy system of collected observation data and simulations of wind speed are leveraged in the find and bind process.
>
> The ESG has the capability of bringing up 3D and 4D visualizations and is targeted at serving the communities in sharing global earth observation data and simulations. Therefore, the ESG better serves the purpose of rapidly prototyping national applications and GEOSS applications than other more generic portals.
>
> Services found through the ESG discovery functions are limited by the availability and the quality of the service, which depends on several factors, such as 1) the accuracy of observed data; 2) the quality of simulation models; 3) the quality of post-processing of information to provide the service; and 4) the reliability of the service.

In the context of the Alexandria digital library project, a research group around Terry Smith (1995) developed in the mid 1990s a first geoscientific processing workbench. Alonso (1999) continued this work and established with OPERA a programming environment that has become the tool of choice for the next ten years. While Smith and Alonso focus on the concatenation of functions (not necessarily GIS functions but also shell scripts, Java applets, etc.), others like Bennett (1997), Marr et al. (1997), and Jung and Albrecht (1997) focused on the development of process libraries. These are modules that represent physical or social spatial processes, encoded in some standard formalism such as a Stella™ model. Similar to Oliveira and Medeiros (2000), Liu et al. (2005) use a component-based approach to the definition of geo workflows. Theirs is a more sophisticated implementation of the kind of tools that Marr and Albrecht used to define their process libraries. The problem with these implementations is that they employ highly customized, non-standard tools that decrease the likelihood of widespread use.

7.6   Geo-scientific workflows and process models

To address this problem, a coalition of researchers from the universities of Santa Barbara and San Diego developed a workflow system that now has the flexibility we were missing from Oliveira's earlier work. Based on a profusion of standards, the Kepler system (http://kepler-project.org) allows for the specification and documentation of a wide range of scientific workflows in distributed environments. Their work is in direct logical continuation of Smith and Alonso; Jäger et al. (2005) provide an application of Kepler for geoscientific workflows.

Case Study 7.3: Kepler

In scientific workflows, each step often occurs in different software (ArcGIS, R, Matlab, RePast, C#). The Kepler scientific workflow system contains a wide variety of analytical components (e.g., spatial data functions and support for external scripts), allows direct (real-time) access to heterogeneous data, and supports models in many science domains.

Figure 7.3a illustrates some of the system's components using a trivial example. The Director controls the sequence of actor execution. Each actor takes data on its input ports, processes that data, and sends results to its output ports. Actors transform input tokens into output data tokens which then get passed to the next actor under control of the director (Figure 7.3b). Each workflow component is self-documenting as it is defined; the specification becomes automatically part of an ever growing component repository (http://library.kepler-project.org/kepler/ and Figure 7.3c). For scientists, this becomes a boon because such models can not only be easily shared but also referenced in published papers (Figure 7.3d).

Kepler uses hierarchies to encapsulate complexity (compare this with Case Study 7.1). Composing models using hierarchy promotes the development of re-usable components that can be shared with other scientists.

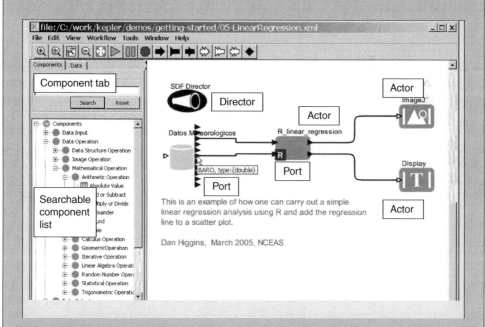

Figure 7.3a Important components of the Kepler user interface (Reproduced by permission of Dan Higgins)

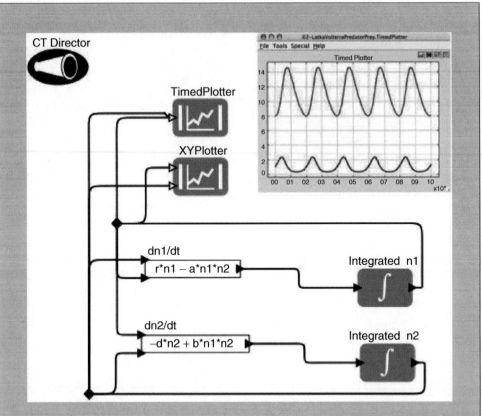

Figure 7.3b   Kepler execution is similar to Stella™ (From the Kepler Users Manual (BSD Copyright))

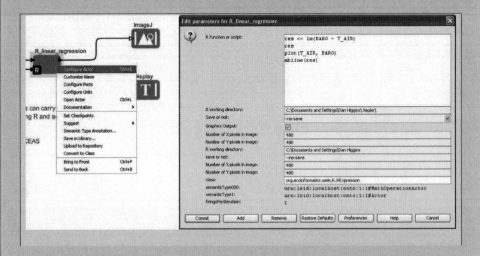

Figure 7.3c   Configuring a Kepler component may involve setting a number of parameters (From the Kepler Users Manual (BSD Copyright))

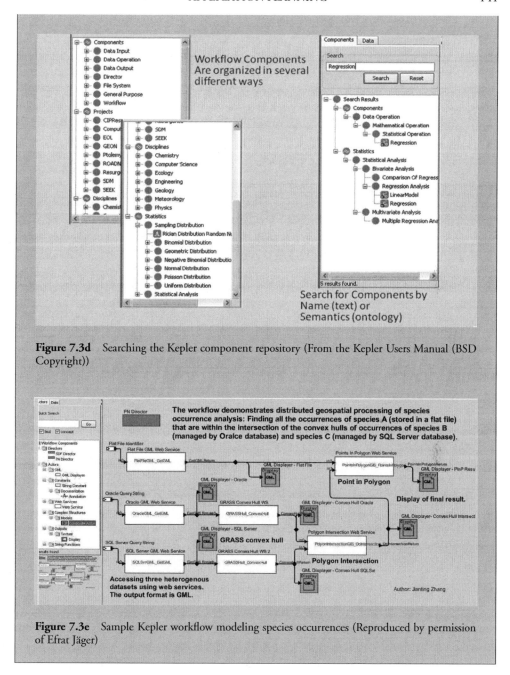

Figure 7.3d   Searching the Kepler component repository (From the Kepler Users Manual (BSD Copyright))

Figure 7.3e   Sample Kepler workflow modeling species occurrences (Reproduced by permission of Efrat Jäger)

7.7   Ontologies in support of application planning for the semantic Web

One of the characteristics of the Kepler system is its adaption to large scale distributed computing (Grid computing). Complementary to the work described in Section 7.4.1, most academic GIS applications move now into an open source-based, distributed processing realm. The key to the success of such endeavours is the use of computer-based ontologies that allow formalizing conceptual models of spatial data and processes, which in turn are a prerequisite to the mapping from one domain to another (Timpf, 2001). At a fairly basic level, using a simplistic use case from a homeland security working group, Wiegand and Garcia (2007) develop a task-based ontology to search and combine data from different sources.

Medeiros's group, too, has moved towards Web-based user interfaces. We already mentioned Baranauskas *et al.*'s (2005) attempt to solicit input from potential users of a Web-based GIS. Schmiguel *et al.* (2004) contributed the analysis of user interfaces for Web-GIS and Medeiros *et al.* (2005) describe a fully-fledged workflow-based spatial decision support system in a Web environment.

As mentioned before, ontologies are seen as key to the discovery and the combination of Web-based functionality. Lutz (2007) describes the use of the Web Ontology Language variant OWL-S to formalize Web services that can then be chained into workflows. His article illustrates how cumbersome it still is to develop user interface service descriptions, overcome the inflexibility of reasoning implementations, and map between domain ontologies. At the same time, he shows the way to develop robust Web-based GIS applications without having to resort to Kepler, which in the end is still a tool for the academic community and thus assumes a high level of technical and domain knowledge rather than for end-users.

Summary

GIS application planning has changed dramatically since the early 1990s. Where traditional work involved cognition and reduction of user interface complexity, the emphasis is now on ontological tools that help to bridge between the inherent complexity of the software and the task-oriented conceptual model of the user. However, for many applications the core demands for bridging the cognitive gap between users and internal representation are still key. The same tools are also used to modularize GIS functionality to repackage it as workflows in a Web-based environment. The use of advance workflow and ontological research might change the face of geospatial technologies over the coming years, though most implementations are still focused on research rather than practice.

Further reading

Cai (2007) provides breadth of discussion into the issue of application planning and Lutz (2007) methodology and level of detail. Davies (1998) provides a detailed discussion of task analysis with the differentiation between work and enabling task, including an analysis of GIS users' tasks. For a detailed discussion of task analysis in general, see Sharp *et al.* (2007). The very readable PhD thesis of Lemmens (2006) provides a detailed explanation and information. You may also want to keep an eye on the continued development of Kepler and its European counterpart Taverna. Finally, bookmark the website of Claudia Medeiros (www.ic.unicamp.br/~cmbm/); her group has developed by far the largest body of work on HCI and GIS in the area of the semantic Web and application development.

Revision questions

1. In a distributed Web environment, what are the tasks that are common to all applications?
2. Does the phrase 'spatial is special' still make sense when we look beyond desktop GIS?
3. What is the role of the user in GIS application planning? Given the limitations outlined in this chapter, how can they be accommodated?
4. What is the most important skill set for (a) a developer, (b) an academic interested in twenty-first-century application planning?
5. How do mobile devices change the context for GIS application development? For a first glimpse, browse the spatial application on the iPhone^{TM} apps store. What works, what doesn't? Can you identify major gaps in the provision of applications on this particular hardware platform?

Figure 3.2a-b Hierarchy with Harmony – same but different (Data source: OpenStreetMap.org)

Figure 3.3a-b Simplicity from Sacrifice, navigating route in Central London
(Data source: OpenStreetMap.org)

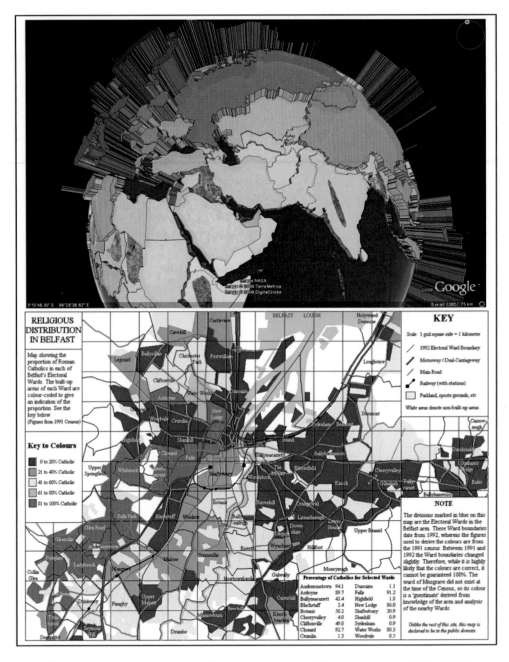

Figure 3.4 Examples of map junk. (Source top: Sandvik, 2009; source bottom: CAIN, 1996)

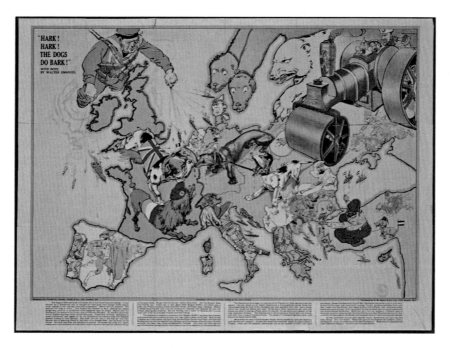

Figure 3.6 First World War propaganda map (Zimmermann, 1914)

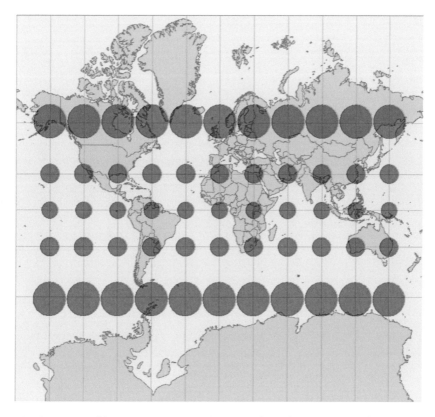

Figure 3.8 Distortions of the Mercator projection (Source: WikiMedia under license of Creative Commons: http://creativecommons.org/licences/by-sa/3.0/)

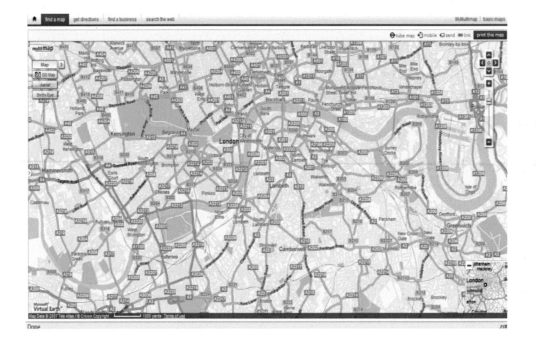

Figure 3.14a-b Multimap maps at two different scales (Data source: Multimap.com)

Figure 4.2 Face-to-face discussion using Luminous Planning Table (Image © Eran Ben-Joseph)

Figure 4.4 High resolution map from OSM of the area near University College London (Data source: OpenStreetMap.org)

Figure 5.1 Policemen with the new information system installed (Image courtesy of Aaron Marcus)

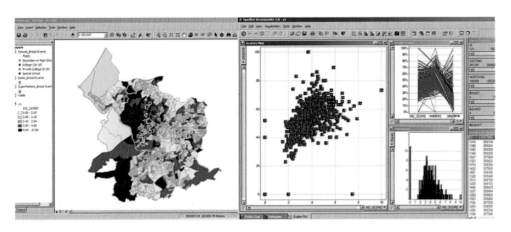

Figure 6.2 ArcMap-DecisionSite environment (Copyright © 2003 ESRI. All rights reserved. Used by permission)

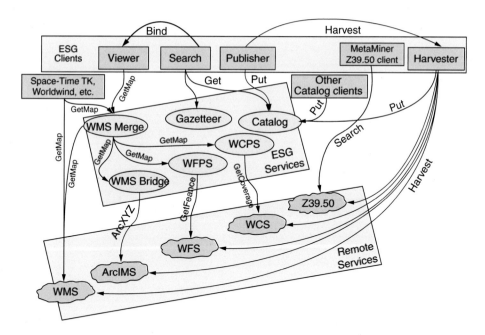

Figure 7.2   Web services chaining for the Earth Sciences Gateway

Graphical Element	Primary Visual Variables						Secondary Visual Variable	
	Shape	Size	Orientation	Hue	Value	Saturation	Texture	Orientation
Point							For point objects this variable is not appropriate - the objects will be too small to see the texture.	For some point objects this variable is not appropriate - the objects will be too small.
Line			For some line objects this variable is not always appropriate				For some line objects this variable is not always appropriate	For some line objects this variable is not always appropriate
Polygon / Area	For area objects the shape changes rarely – when a cartogram is created	For area objects the shape changes rarely – when a cartogram is created	For area objects the orientation changes rarely					
Uses and notes	Qualitative differences	Shows magnitudes or variations	For some point objects this variable is not always appropriate.	Qualitative and Quantitative differences	Qualitative and Quantitative differences	Qualitative and Quantitative differences	Qualitative and Quantitative differences For some point objects this variable is not always appropriate.	Qualitative and Quantitative differences For some point objects this variable is not always appropriate.

Figure 8.1   Different visual variables for symbols

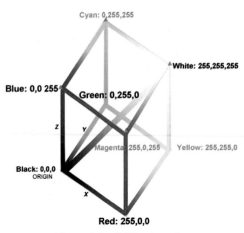

Figure 8.2   RGB color cube

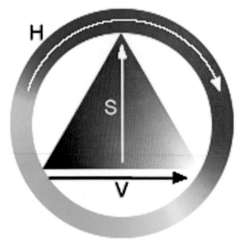

Figure 8.3   The HSV system (Source: WikiMedia Commons – licensed under the Creative Commons Attribution ShareAlike 3.0 License)

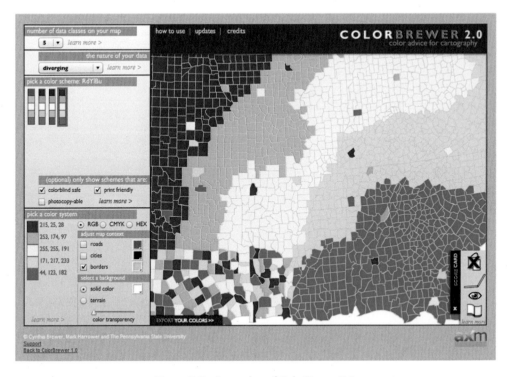

Figure 8.5   Screenshot of ColorBrewer 2.0

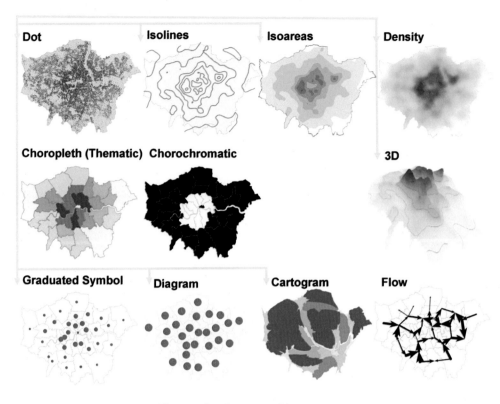

Figure 8.6 Nine types of thematic maps

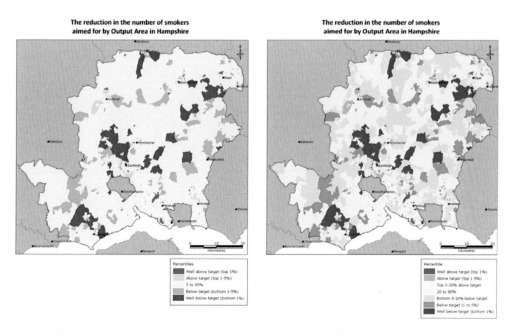

Figure 8.7 Two visualizations of the same data, with five (left) and seven (right) classes

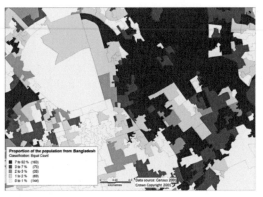
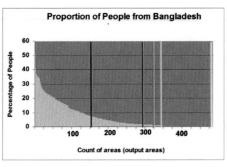
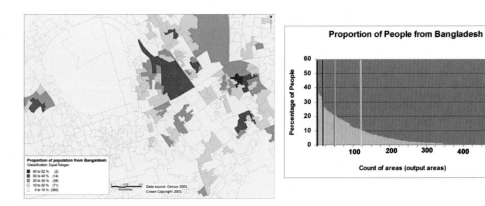

Figure 8.8a-b   Same data, same spatial units, same number of classes, different classification method

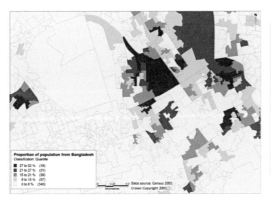

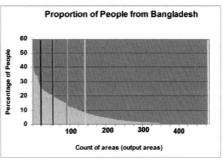

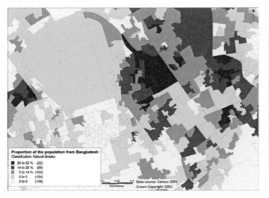

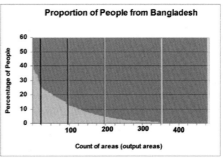

Figure 8.8c-d Same data, same spatial units, same number of classes, different classification method

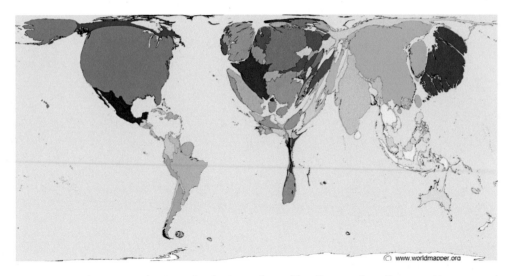

Figure 8.9 Cartogram showing distribution of wealth (Source: http://www.worldmapper.org/display.php?selected=170)

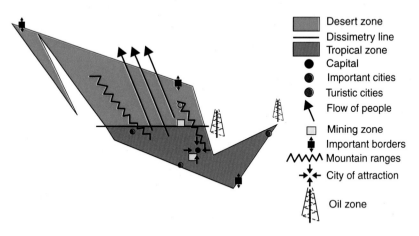

Figure 8.11 Mexico – a set of chorems showing dynamics among some main places

Figure 8.12   England Noise Mapping site

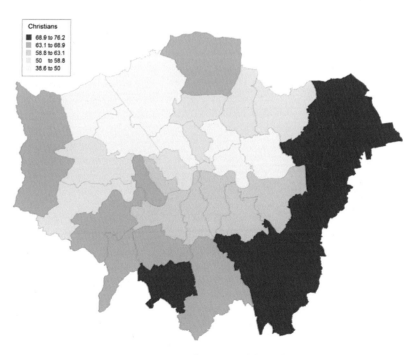

Figure 8.13a   Map elements and their placement

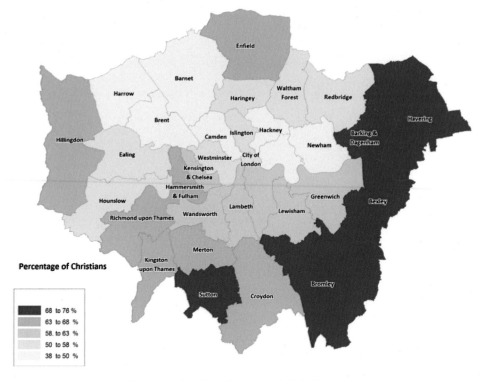

Figure 8.13b Map elements and their placement

Percentage of Christians residing in London Boroughs, 2001

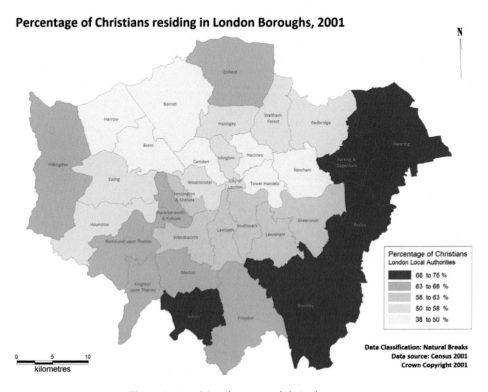

Figure 8.13c Map elements and their placement

Figure 11.3 A mini A–Z for London, showing the area covered by a digital map (Data source: Geographers' A–Z map company)

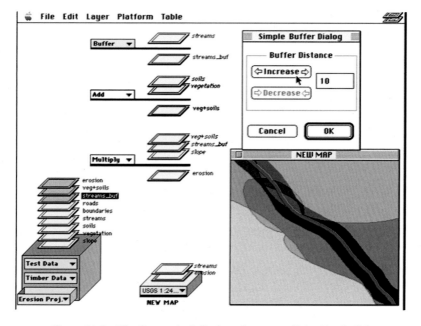

Figure 11.6 The Geographer's Desktop (courtesy of Max Egenhofer)

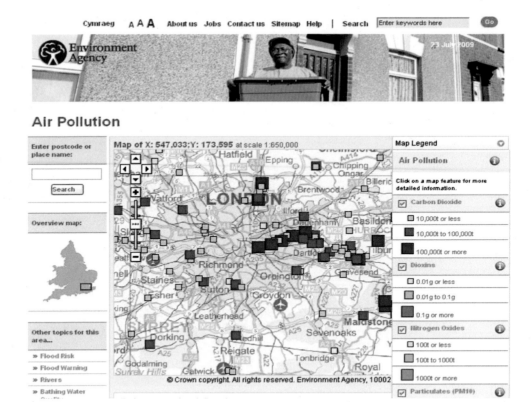

Figure 12.1 Results of air pollution in the 'What's In Your Back Yard' website provided by the Environment Agency (EA)

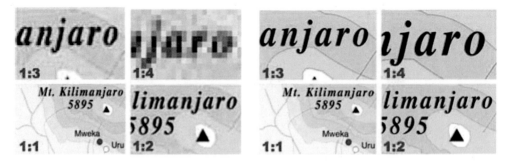

Figure 12.5a-b (a) Raster Zoom-in Function (Shand, 2002); (b) Vector Zoom-in Function (Shand, 2002)

8 Practical cartography

Catherine (Kate) Emma Jones

In Chapter 3, we discussed some of the high level principles to help guide the production of a well designed map that is both simple to use and easy to interpret. With the emergence of digital mapping applications there has been a tendency to neglect the effort required in designing the map, assuming that because the user can set the visual display properties, the details of the map are unimportant. Instead efforts are focused on the advancement of technical capabilities. However, by following the simple cartographic guidance and principles outlined in Chapter 3 and by considering the design guidelines in this chapter it should be possible to provide comprehensible maps in a way that matches the user needs of the application.

To create a well designed map that is both useful and usable it is essential to adopt a systematic approach. Plan carefully and evaluate what works and what does not: map prototyping enables you to hone design constructions. It is important to remember that making good maps takes time and it is a task that, at the outset of a project, to which time should be allocated. With attention to detail and ensuring the fundamental map elements are included alongside appropriate *symbology* (the type, colour and graphic appearance of symbols), *typology* (the different fonts) and colours you can improve the likelihood of developing a map that successfully engages the user.

In this chapter we cover the use of colour in maps, data classification and the role of the various graphical variables in the symbology. However, there is a very important point that should be considered before everything else – is a map the most appropriate form for portraying the information? In many cases, geospatial technologies can use location information and digital geographical information to provide the user with a useful application but the delivery of the information to the user does not always need to be in the form of a map. In-car navigation systems can be effective by providing audio instructions or by displaying arrows. For many situations, these alternative representations are often more effective than a map. This is a very significant aspect of these technologies that should be highlighted when cartography is discussed. Now, if you have decided that maps are the appropriate mode of visualization for your application – read this chapter.

8.1   The role of symbology in map making

Symbology – the shapes, patterns, symbols and colours that are used to depict objects on a map – is one of the most powerful tools used in designing a usable map. The symbology of a map provides very important information through the use of visual variables to mark features, places and other attributes on the map (Monmonier, 1996). The most important consideration is the choice of symbols and their perceptual impact, so that users must clearly be able to discriminate between different map symbols (Robinson *et al.*, 1995). The different types of symbols should be unambiguous to the user, which means there must be sufficient contrast and differentiation between the different symbols. It is not simply enough to make them larger, because if they are too large the map will look cluttered and the different symbols will overlap each other.

The logic behind map symbols is based on viewing them as graphical elements and understanding their different dimensions: size, shape, texture, orientation and colour. The different dimensions of map symbols can be grouped together into three key components: the basic graphic elements, the primary visual variables and the secondary visual variables (see Figure 8.1). The basic graphical elements that are appropriate for symbols are point, line and area (polygon) each of which conveys a slightly different message. The point symbolizes a location in the geographical space whereas the line represents a continuum indicating directional location and a sense of movement through that geographical space. The area, on the other hand, outlines a more defined geographical space – an extent or cohesive space (Robinson *et al.*, 1995; Monmonier, 1996).

Contrast can be added to each of these elements by changing either the primary or secondary visual variables. The primary visual variables are associated with the shape, size, orientation, hue, saturation and value (the variables of colour which are explained in Section 8.2). The secondary visual variables are also used to provide further contrast to the graphical element but in some circumstances they can actually cause objects to become too visually confusing. The shapes of a graphical element can either be geometric and abstract or utilize icons which often use image representations that fit with real-world perceptions (see the weather map case study 3.5). For example the locations of restaurants are illustrated with an icon containing a knife and fork or a pub represented with a beer or wine glass. These icons provide an intuitively encoded view of the world that is familiar to the users, and can be interpreted more easily with less impact on the user's cognitive load matching the user's intuitive interpretation of its meaning.

Some sectors have standardized icon sets, and in this case they should be followed. Such common sets of icons are used by the oil or transport industry, as well as weather and geological maps. The connotative meaning attached to these icons is often obvious to the user but their practical use and legibility very much depends on the scale of the map. A small scale map showing a very large geographical extent does not easily support an icon symbology set. This is because the icons cannot be drawn small enough to retain any sense of clarity and legibility. One further point to note is that the geometric symbols rarely contain an intrinsic cultural meaning and so absolutely, without any exceptions, they require a legend to be included on the interface (van den Worm, 2001). The legend must be provided in an efficient and user friendly manner to ensure short-term memory is not negatively impacted and time is not spent looking for the legends instead of exploring the patterns, therefore when developing geospatial technologies remember not to hide legends or make them very difficult for the user to find.

8.2   The role of colour in map making

Colour is one of most exciting aspects of design for map making. Monmonier once described it as, 'a cartographic quagmire!' (1996) but if used carefully it can invoke implicit

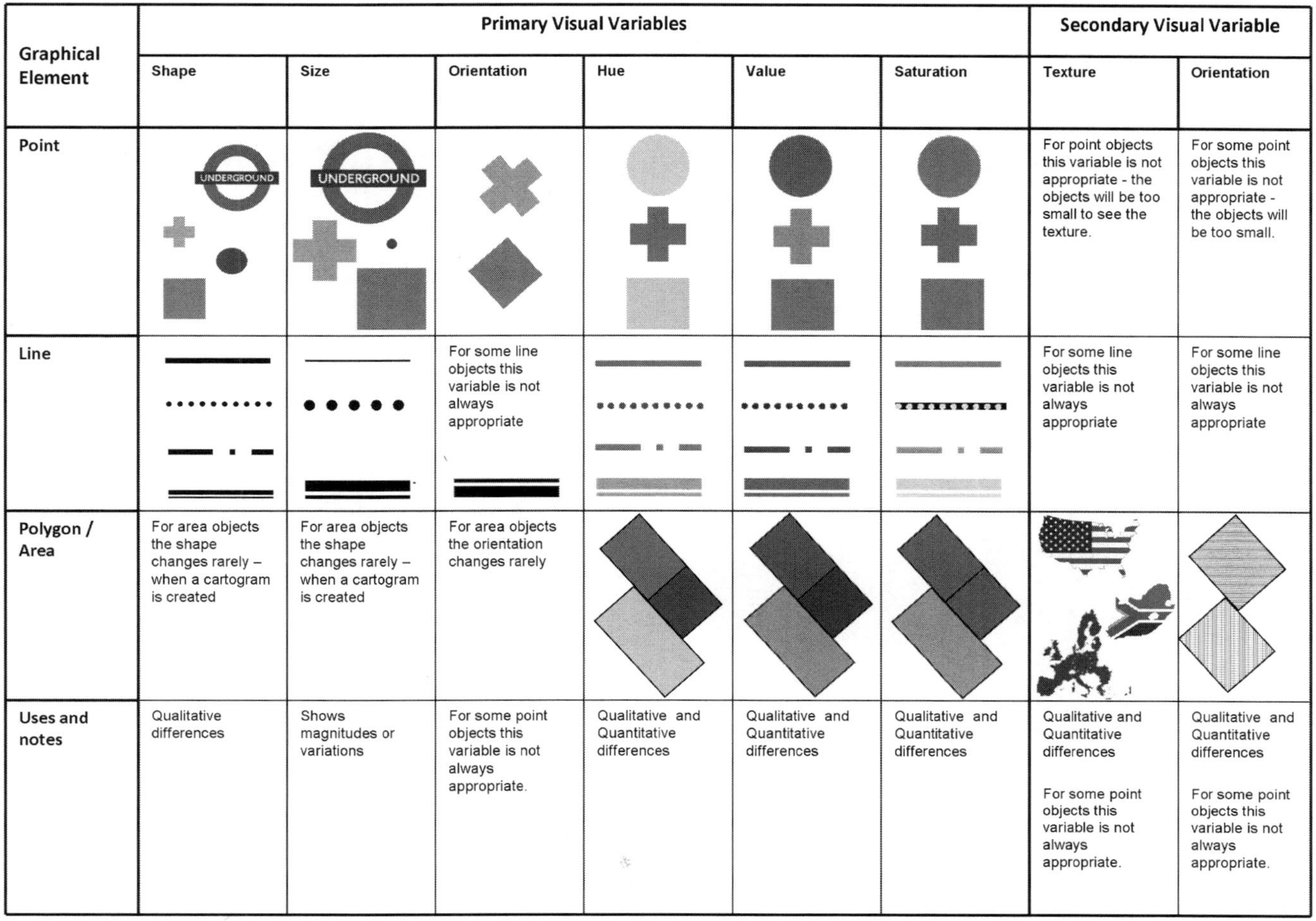

	Primary Visual Variables						Secondary Visual Variable	
Graphical Element	Shape	Size	Orientation	Hue	Value	Saturation	Texture	Orientation
Point							For point objects this variable is not appropriate - the objects will be too small to see the texture.	For some point objects this variable is not appropriate - the objects will be too small.
Line			For some line objects this variable is not always appropriate				For some line objects this variable is not always appropriate	For some line objects this variable is not always appropriate
Polygon / Area	For area objects the shape changes rarely – when a cartogram is created	For area objects the shape changes rarely – when a cartogram is created	For area objects the orientation changes rarely					
Uses and notes	Qualitative differences	Shows magnitudes or variations	For some point objects this variable is not always appropriate.	Qualitative and Quantitative differences	Qualitative and Quantitative differences	Qualitative and Quantitative differences	Qualitative and Quantitative differences — For some point objects this variable is not always appropriate.	Qualitative and Quantitative differences — For some point objects this variable is not always appropriate

Figure 8.1   Different visual variables for symbols (a full-colour version of this figure appears in colour plate section in this book)

connotative meaning by harnessing user's own perceptual and spatial ability. Using colour effectively can and will enhance the ability of users to identify features, objects and patterns (Brewer, 2004; MacEachren, 1992; Monmonier, 1996; Robinson *et al.*, 1995). The application of colour is not simply an exercise in decoration. Colour can present and enhance maps whilst also supplying the aesthetic qualities which secures the user's interest and attention.

Careful selection of colours can create harmonious hierarchies accentuating the visual significance of selected features and objects whilst enabling the user to see visual hierarchies in the data. There are many different aspects for the use of colour when designing maps, so it is best understood in terms of understanding how we see colours.

8.2.1   Colour perception – how we see colour

Colour is not a constant variable. There are many components influencing how colours are perceived including the type of light source available and the medias/surfaces upon which the map is created. Colour is a product of our mental processing. It is perceived by the eye when required conditions are met: there must be a **light source**, an **object** and an **observer**. Usually, we assume that the light source appears to emit white light (electromagnetic radiation with wavelength) but actually it is a series of colours that comprise the visible part of the electromagnetic spectrum. If the full range of the visible spectrum is emitted by a light source such as the sun, our eyes will perceive this as white light. If the wavelengths of the light are separated by objects and reflected, the light continuum that enters our eyes is then transformed by our mind into sensations of hues comprising red, green, blue, yellow etc. (Robinson *et al.*, 1995: 342).

Therefore if maps are viewed in different light sources the colours will be perceived differently. If you are viewing the same map in sunlight or in a room with fluorescent lighting, they will look different due to the method in which light is absorbed or reflected (Krygier and Wood, 2005). This is true also for electronic screens which emit light to create maps. They will look different under differing lighting conditions. In the same way as a light source alters how colours are seen, objects also modify light. Therefore the choice in the type of paper, printer, ink, projector or monitor all affects how the colours of the map are perceived, see Table 8.1.

Finally, the observer also has an important bearing on colour perception due to the individual variability that exists with our ability to see and recognize symbols, shapes and colours. A significant proportion of the population have some level of colour deficiency (MacEachren, 2004: 54) and maps need to be designed accordingly.

Colour deficiency seems to occur in men more than women, 8% of all men are assumed to have difficulties in perceiving colour but these issues only extend to around 0.4% of women (MacEachren, 2004: 63). One such colour deficiency is known as colour blindness. People who are colour blind cannot differentiate between certain colours, such as red and green, so a map compiled with objects in both these colours will hinder knowledge construction. Spatial units comprised of these colours will look the same and will therefore be interpreted as such. Colour deficiency also occurs in older people, and if you are creating mapping applications for older populations you should take account of the fact that: (a) blues are difficult for them to differentiate and (b) the colours used need to be more saturated (Krygier and Wood, 2005).

Table 8.1 Influences on colour perception (adapted from text by Krygier and Wood, 2005: 258)

	Type	Consequence	Action
Light source	Low intensity light	Reduces intensity of colours	Use more intense saturated colours
	High intensity light	Increases intensity of colours	Use less intense saturated colours
Map Media	Matte paper	Reduces intensity of colours	Make colours more intense
	Glossy paper	Increases intensity of colours	Make colours less intense
	Projected on wall	Depending on type of projector or and the material on which the image is reflected will increase or reduce intensity	Test out map colours and change accordingly. **This is one of the most common mistakes that are made by map designers, as the map 'looks good' on the computer's monitor**
	Computer monitor	Makes colours more intense and vibrant	Test out map colours prior to use and change accordingly to the media/surface map it is being created for. Consider what will be the common characteristics of the monitors of the users.
Observer	Colour blind	Cannot perceive difference between red and green	Do not use red and green in the same map
	Older people	Difficulty in seeing difference in blue colours	Do not use sequential blue colour schemes for maps to be used by older peoples
	Older people	Difficulty seeing colour	Make colours more saturated

8.2.2   Colour systems and definitions

How is colour created on screen and print?

Creating colours is complicated. The differences in displays and printing equipment will lead to obvious disparities in colours of the same map made available through alternative mediums because each colour model is based on different assumptions of the light sources. Print and screen will mix colour according to their light source – coloured ink in a printed map is reflected whereas colour on a screen is produced by monitors emitting light. The standardized colour specification model for commercial print is the CMYK (Cyan, Magenta, Yellow, Key – Black) whereas RGB (Red, Green, Blue) is the common model for screens and display monitors. In both systems the spectrum of colours is created by mixing together small dots of base primary colours, be it Cyan, Magenta, Yellow and Black or Red, Green and Blue. Sometimes translation from one colour system to another does not return exactly the same colours. Note that some RGB colours are vivid but these colours are not always directly transposed into the CMYK colour spectrum because of impurities in the different light sources (Brewer, 2004). Thus, when designing a map you need to beware of these differences and design separate sets of maps for printing versus screen use. For the development of geospatial technologies it is expected that most of the time the maps will be displayed through the computer monitor so the RGB colour system is the one you will encounter the most.

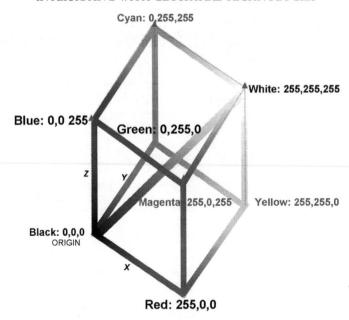

Figure 8.2 RGB colour cube (a full-colour version of this figure appears in the colour plate section of this book)

The RGB colour system is an additive system which adds mixtures of three primary colours together to create all other colours in the spectrum of visible light. A colour consists of various quantities of each of the three primary colours represented by a number corresponding to their position in the colour cube. (In many cases there is a range between 0 and 255). Figure 8.2 shows how additions of the primary colours will result in an array of colours visualized as vectors in a 3D array containing X, Y and Z coordinates, Red, Green and Blue respectively). If the maximum values for red, green and blue (255,255,255) are added together in equal measures then the result is white light.

In the RGB colour specification model the corner position of the origin is always 0 in red, green and blue which equates to the total absence of any colour and therefore returns black (no light). The three primary colours are located as vectors adjacent to the black. If pairs of primary colours are then added together it is possible to achieve the secondary colours of yellow (255R, 255G, 000B), magenta (255R, 000G, 255B) and cyan (000R, 255G, 255B) (Bunks, 2000). When amounts of the additive primary colours increase the colours get more intense. Shades of grey are made up of equal amounts of red, green and blue and are represented by the horizontal vector between black to white (the neutral axis) (Brewer, 2004; Robinson *et al.*, 1995; Krygier and Wood, 2005).

The CMYK colour cube is somewhat similar, but it is based on a subtractive model where the mixing relationships are reversed (Brewer, 2004). In this model the primary colours are cyan, magenta and yellow and when you add them together it forms the colour black. What this means in reality is that printers use ink and when all the inks are mixed together they will make the colour black. In this case the baseline colour is white (as it is expected that white paper will be used) which becomes coloured when ink is added to it. It works differently to the RGB model because the ink is reflecting light and not emitting it and so the paper takes away wavelengths of reflected light to produce the required colour (Peterson, 2009). When the CMYK colour vectors are increased, unlike the RGB scheme that increases intensity, it produces a darker colour. To create white the primary colours are subtracted from each other. If you are creating maps that will be used simultaneously on the screen and on print it is always sensible to test them out by printing as you design – note that the colours will be treated inconsistently for different printers.

8.2.3   Our perceptual colour system – HSV

The RGB or CYMK colour schemes do not align perfectly with the colour spectrum that our eye perceives. This is because we perceive colours differently and human perception is usually understood by the shade, brightness and paleness or vividness of a colour. Hence perceptual colour systems try to model the mental model that we create. The Munsell colour system, defined in 1905, is one of the perceptual colour systems that is widely used and is internationally recognized. It is also a popular system with map and cartographic designers. It comprises three attribute values known as hue, value and chroma and it enables colours to be perceptually scaled so as no two colours are visually perceived to be the same. A detailed description of this system is in Robinson *et al.* (1995: chap. 19).

Another perceptually scaled colour model is HSV which stands for hue, saturation and value and equates to the artist's use of tint, shade and tone. This is widely used in GIS software. In this colour model *hue* refers to what we commonly call the shade of the colour, whereas *saturation* is the purity of the hue which equates to our mental description of whether a colour is vivid or pale and the *value* refers to how bright the colour is and what we describe as the lightness of darkness of a colour. Note that saturation is sometimes called *chroma*.

In a 12 step colour wheel, hues are the basic primary colours defined by the spectrum: red, orange, yellow, green and blue. The HSV colour model is a conical shape with the upper outer edges of the cone representing these saturated hues. The sequence of the hues is according to the longest wavelength (red) to shortest wavelengths (blue) (Brewer, 2004). Like in the RGB system, a set of three numbers are used to define the levels of hue, saturation and value. Therefore a saturated hue is pure and has no white or grey in it. To set the hue we use a range of number between 0 and 360. To make the hue paler grey is added by changing the saturation. If grey is then added to the pure hue the saturation of the colour changes so whilst the hue is still the same it will become paler and less vivid. The amount

Figure 8.3   The HSV system (Source: WikiMedia Commons – licensed under the Creative Commons Attribution ShareAlike 3.0 License). (A full-colour version of this figure appears in the colour plate section of this book)

of grey in the colour increases and is defined by a number between 1 and 100. Therefore saturation represents the percentage purity of the colour. A hue that is completely distorted will be neutralized, in effect becoming grey and will have a zero in the percentage of purity. A colour is also described by how bright it is which refers to the amount of light derived from the colour. Value represents the amount of light that is derived from the colour. If the value of the colour is high then less light is coming from it and it will be perceived as lighter and brighter. Likewise colours with low colour values are perceived to be darker since less light is emitted from the colour. The value of a colour is set according to the percentage of lightness where 100% equates to the maximum proportion of white in the colour and is the lightest it can be, look at the hue, saturation and value variables in the graphic of Figure 8.1.

We have to be careful when we choose colours because of the different ways in which colour is made, compared to how we perceive it. The careful selection of each of these HSV colours will improve the usability of a map, and in most mapping systems the HSV settings use automatic mathematical conversions to relate them directly to RGB settings (Robinson *et al.*, 1995).

There are some general tips about the influence of colour scheme in map use, collated from the key texts:

- Lightness can used to rank data – (see sequential colour schemes later in this chapter) and can be used with a constant hue – increasing the percentage value of a colour will make it lighter.
- Saturation reinforces the lightness changes to help differentiate map symbols.
- Differences in hues can indicate types of features on the maps and this can be useful for classes of qualitative data, where the differences between different types of information are not along a scale.
- For categories with small area objects (polygons) saturation can add emphasis by making them more vivid. Likewise large areas with highly saturated colours will dominate the map (rightly or wrongly).
- Saturation on its own is not easy to distinguish and is difficult to adjust.
- Implicit differences are accentuated by using pure hues such as red, green and blue.
- Certain colours are easier to see – so aim for maximum visual differentiation. Use pure hues with colours of shorter wavelengths for the background but change the amount of light to help it recede even more.
- For more guidelines look at the work of Cynthia Brewer.

8.2.4   Colour design and decisions for map making

The HSV colour model is very useful for map making as it acts a catalyst for knowledge construction and helps us make sense of the information portrayed on the map. Most GIS software will provide options to edit both the HSV and the RGB colour settings, so conversion between the two is made easier. With this in mind a good choice of colour will support the depiction of information and help the user to understand the meaning of the map and poor colour choices and overall design may even lead the user to think the information is untrustworthy. This is where the art of map making comes to the fore. Many technology-driven GIS maps will be coloured according to system defaults that fail to convey the implicit connotative meaning of the map. Therefore, the most important rule for digital cartography is: Never accept the system defaults without careful consideration and do not underestimate the time it takes to produce a well designed map.

Kraak and Brown (2001) have a useful discussion on colour choices for Web maps. They point out that it is always important to remember that the users of Web-mapping interfaces never have control over how the colour of maps will appear on their monitors. One solution to this problem is to use the 'Web safe' colour palette which has 216 colours, examples of this colour palette and the RGB equivalent are available on many websites.

8.2.5   Figure-ground and visual contrast

An important feature for consideration when designing with colour is the relationship between the foreground and the background, the relationship known as the figure-ground. This relationship builds on the principle of hierarchy with harmony which was discussed in Chapter 3. When an image or map is viewed, the human eye and the mind will react spontaneously to perceive a visual array often identifying an inherent order and structure to it. This order comprises the figure and the ground (Robinson *et al.*, 1995). The figure-ground relationship in a map exploits this ability of the human eye to visually separate elements where the foreground figure is the point in the image where the eye will naturally settle. The advance and retreat of different map objects and layers arises from the physical adjustment our eyes make when they perceive longer or shorter wavelengths (Krygier and Wood, 2005). For this reason there are certain colours that naturally stand out. The red colour range is a suitable example of this. These colours will always appear closer to the user whereas blue colour palettes tend to recede into the distance.

The choice of colours for the foreground or background can change the emphasis within the map so users perceive objects differently. The selection of colours for the foreground will aid differentiation between it and the background – this promotes the figure-ground relationship. The background should be recessive and so the colour selection should send the objects to the back of the visual hierarchy, so it is the last object to be seen. The objects recede into the distance rather than advance to the front. Good map design will accentuate the natural figure-ground organization and so discriminate the objects you want the user to identify most clearly by setting them as the figure, whereas less important objects in the hierarchy are set as the ground. To test your figure-ground relationship within your maps it is useful to stand back from the screen, look at the map and note what you see first and ask the person sitting next to you. Is it the layer you want your users to see first?

Colours must be selected to enhance the differentiation between the ground and the figure. Some techniques will help place the emphasis on the figure making it advance to the front: the use of detail promotes an area to the figure; place the figure in the optical centre of the map (see Figure 8.14) this draws the eye to it; smaller areas naturally emerge as the figure where larger areas recede into the background; and objects that are familiar to the reader invariably promote themselves to the figure (see Figure 8.4). Techniques that help enhance the priority of the figure include highlighting borders between objects and using

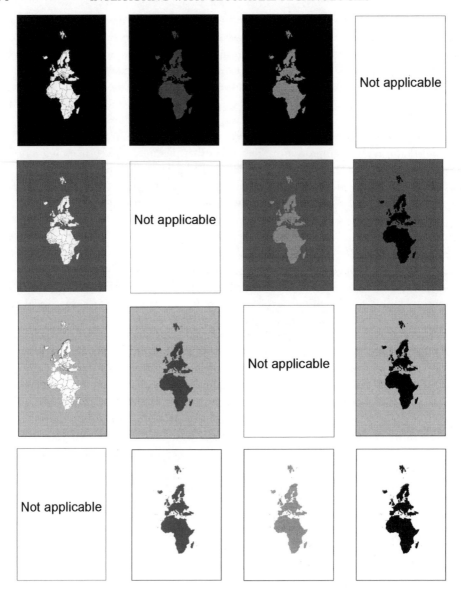

Figure 8.4   Figure ground maps using different contrasts

haloes to distinguish and place emphasis on objects – often used in labelling if you want them to stand out.

There are some simple rules to remember the influence different colours will have on the figure-ground relationship. If setting the hue (colour shade) of objects, colours with longer-wavelengths will proceed to the foreground of the map whereas shorter wavelengths spontaneously fall into its background. These hot–cold maps, which use blue and red, accentuate the objects coloured in red. Think of temperature maps where the areas with the highest temperatures stand out the most. For the purity of a colour, one that is totally saturated and at its most bright and vivid will come to the front and be perceived by the eye as a foreground object whilst a pale and unsaturated colour with lots of grey added to it will give way to other more saturated colours, look at the point and area symbols in Figure 8.1. The same hierarchy can be emphasized by using lightness, if the value is very light then the colour will advance. Figure-ground relationships can be further improved by careful use of colour contrasts.

Contrast is the basis of seeing (Robinson *et al.*, 1995) and it describes the method in which two or more colours interact and consequently influence each other. Maps will be more visually powerful if contrast is used effectively, because all colours are influenced by the colour(s) surrounding it. This is known as *simultaneous contrast*. In figure-ground maps contrasting colours can help the map reader to distinguish objects more easily and aid the natural order of a map. Consider the colour palette of Google Maps and how main roads are more obvious to the eye than the residential roads. Go back and look at Figure 3.2a and b, Figure 3.2a was designed to have subdued colours with little contrast so that it could be used as a background map, all objects in it are recessive. Whereas in Figure 3.2b the hierarchy of the data is illustrated to the user by manipulating the contrasts of lines, shapes and colours to achieve an implicit visual structure (MacEachren, 2004).

Figure 8.4 illustrates how changing the contrast alongside the background/foreground palette will alter the visual hierarchy of the map emphasizing the figure-ground relationship. Which map do you think has the most obvious visual hierarchy? Some of the figures have more contrast and more effectively emphasize the visual variable of the countries. The use of the same colours (in this instance grey) groups all the countries together – so colour provides an excellent tool for grouping objects together based on similar attributes. This type of grouping is often referred to as Bertin's selective and associative principles whereby similar objects with the same attribute data values become one unit. A technique that is very common is to think about how points of interests are grouped together on a detailed reference map.

In Figure 8.4 when the continent is coloured grey and surrounded by black here the contrast fails to accentuate the hierarchy and it is not as easy for the eye to naturally identify the figure from the ground. When using a colour, warm colours (reds, oranges and yellows) are better for accentuating figure objects in the figure-ground hierarchy as previously mentioned. This is because of the very strong visual impact that colours in this palette have. They also have a tendency to attract the eye and invoke excitement, but you should avoid giving neighbouring objects a similar colour otherwise they will not stand out and may be perceptually grouped as the same by the reader. Cooler colours such as blues, green and violets (those with the shortest wavelengths) cause objects to recede into the background and so are more useful for ground objects. One tip is to remember that complementary colours on the colour wheel, which are opposite to each other, are useful for mapping the figure and ground perspective and produce nicely balanced colours, but neighbouring colours on the colour wheel are not always successful (see Figure 8.3) and so will interfere with the inherent order of map objects. A figure-ground map made up of an orange foreground and red background will not be as effective, these neighbouring colours offer poor contrast between themselves. Nevertheless a palette comprising of a blue background with a green foreground will be visually effective. Think about the natural colour palette of green for land and blue for sea. This colour scheme always brings the land to the front of the map as this is what users wants to investigate first; the sea coloured in blue always fades into the background. Maps that invert this relationship and put land in blue and sea in green will always confuse the organization and interfere with the connotative meaning of the map.

8.2.6   Colour schemes

As we have already identified, colour is an important and supportive feature of map design. If used appropriately colour can be applied to a range of different data types including: binary (see chorochromatic map in Figure 8.6), qualitative (see diagram map in Figure 8.6) and quantitative (see thematic map in Figure 8.6), but the colour choices will depend upon the data types of the attributes being mapped. Colour is most useful because it can be applied to all the different map objects to help enhance their intrinsic meaning and inferred interpretation. By making careful colour choices, spatial patterns in the data can be illuminated, and in a case of an application aimed at letting users explore the data, new patterns can be found.

There are many types of colour schemes and palettes. Two of the most important colour schemes for quantitative data are the sequential colour scheme and the diverging colour scheme. They are used to provide colour to thematic (choropleth) maps.

In a sequential colour scheme the lightness (value) of a colour is used to exaggerate the intrinsic sequence of data distribution (see thematic maps in Figure 8.6). For this reason it is a useful scheme for data which can be ordered or ranked. It is a useful scheme if you are mapping population densities, percentages of attribute etc. The lightness of a colour is used to imply the inherent order. Most often a single colour (one hue) is used and its value (its percentage of light) is altered incrementally to create a palette that uses lightness to differentiate high or low data values. The lighter the value of the colour the lower the data values it represents. Using this type of colour scheme, magnitudes or quantities can be deduced by the users as the colour's hue progresses from light to dark.

In sequential colour schemes research has shown that if the colour value of each class has less than 10% differentiation then it is difficult to tell them apart (Brewer, 2004; Krygier and Wood, 2005) due to insufficient colour separation. For this reason it is recommended that a sequential colour scheme with only one colour should not have more than seven classes, or better still five classes. When using just one colour, experiments have shown that yellow is generally not an effective colour choice? If you are mapping more than five classes, remember that the optimum number of classes is seven plus or minus two. With seven or more classes it may be necessary to use a sequence with two colour progressions and vary their lightness, otherwise you may not have effective differences between the colours. Remember for hue progressions in sequential maps the colours do not need to contrast significantly (unlike in figure-ground maps). For a two hue sequential colour scheme select colours next to each other on the colour wheel, for example yellow to red or green to blue.

There are a few limitations of sequential colour schemes. They should not be used with a spectral colour palette going from blue to yellow to red. There are two reasons why this is not appropriate: first, a spectral colour palette does not produce a natural order in the mind of a user and the inherent ranking of the data will be obstructed and, second, in a spectral colour system yellow is often the colour of the mid range values with reds representing high values and blues low values, but this location of the yellow may confuse the reader by emphasizing the difference around the mid range. In a sequential colour palette the mid range is not used to display critical values – it is for representing data within a continuous sequence that is the priority (Brewer, 2004; Robinson *et al.*, 1995).

In contrast, a diverging colour scheme does work with a spectral colour system. This is because the colour palette for a diverging colour scheme sets out to place significance on the

Table 8.2   Colour rules for map making

Never accept the default settings without checking their suitability
Grey tones are distinguished as a sequence from dark to light and the eye can only really differentiate five greys on one map
Use well known colour conventions such as land in green and sea in blue
Use colour to facilitate perceptual grouping of like features with the same attributes
Colour should enhance the hierarchy of objects
Colours should be discrete and not garish
Always design specifically for maps to be produced in grey scale – never just convert a coloured map to grey scale.

values at the ends of the data range so the extremes are highlighted. An example of a diverging colour palette is used in the maps of Case Study 8.2. Such a palette is useful because it enables the comparison of data ranges according to critical values – giving readers more information. A diverging scheme, sometimes known as a bi-polar colour scheme, uses a double ended colour progression diverging away from a shared lighter colour used to represent a critical value or mid-point in the dataset (Brewer, 2004). If you want to convert a map into grey scale – for example, if you need to use them with devices which have monochrome screens – diverging colour schemes should be developed specifically, they do not photocopy or convert effectively since both sides of the progression will appear the same.

A diverging colour scheme can be effective, for the hues at either ends of the range use contrasting colours to highlight the extremes, yet they also need to complement each other, a careful balance is required on both sides of the critical value. Changes in value of the colour can be used to create the progressions but the same gradation change of value should be used for each hue progression. This ensures the progression has a symmetrical quality in its divergence pattern, which is imperative. There are some standard conventions that will make the colour choice easier in certain circumstances according to the specific nature of the variables being mapped. For example temperature maps are commonly depicted using a diverging blue to red scale with 0 degrees representing the critical point.

Qualitative colour schemes are applied to data that do not have any implied order and as such are generally useful for categorical data. For this reason it is important that the colours must not imply any sense of order to the data to avoid misinterpretation. Qualitative schemes are best represented using different colours which have very small variations in saturation and values but no wild variations (Brewer, 2004). Careful use of hues by varying the lightness and darkness can be used to emphasize specific categories in the map that you want to stand out to the user – for areas with categorical categories you want readers to immediately find select hues with longer wavelengths, warmer reds or oranges and make them just a little darker than other categorical hues.

Robinson *et al.* (1995: 386) suggests that in qualitative maps, between 8 and 15 different hues can be easily distinguished by the reader as long as they are sufficiently different from one another. Beware, if you are changing the texture by giving the hue a pattern, the patterns might irritate the user by being very distracting.

Case Study 8.1: Using ColorBrewer to select useful colour palettes for mapping

Cinthia Brewer from Penn State University is an expert in Cartography. She has designed several interactive tools that can help you to design better looking maps. Together with Mark Harrower from the University of Wisconsin Madison, she has developed an application that is one of the most useful aids to learn about colour selections for maps. The tool 'ColorBrewer' is very effective: http://www.colorbrewer2.org. It was originally developed in 2002 and has been improved and relaunched in 2009. It is a flash-based application that runs in an Internet browser. The tool helps users select colour palettes appropriate for the data you are visualizing and also recommends if it is useful for the media you will use to deliver the map. The Application will provide you with RGB, CMYK or Hexadecimal values for the colours you select.

While ColorBrewer does not provide many colour palettes, and it is very likely that your GIS package includes many more colours and options, using the application can teach you the principles of how to apply colours. One of the major advantages of ColorBrewer is that it includes a 'score card' that will allow you to evaluate the suitability of the colour palette to a range of uses – from paper to computer screen.

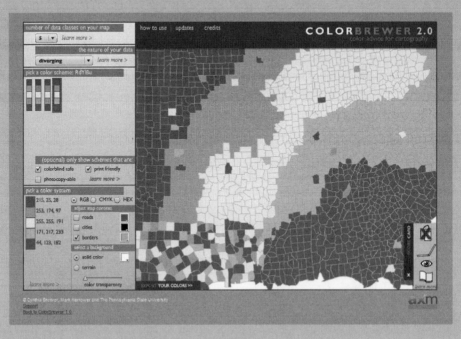

Figure 8.5   Screenshot of ColorBrewer 2.0 (a full-colour version of this figure appears in the colour plate section of this book)

8.2.7 Colour connotations, cultural interpretations

So far, we have discussed physical and technical aspects of colour use. Many of them are universal. However, in different cultures colour can have different meanings, therefore certain variables might carry with them a habitual inferred meaning emanating from the colours selected to represent them. We have previously mentioned the use of colour to represent land/vegetation and sea/water using green and blue respectively. For these types of representation white equates to snow and light brown can be effectively used to characterize desert areas. To familiarize yourself with these type of cultural associations of colour typical in topographic and geological mapping have a look through the portfolio of maps available on the National Geographic website (http://maps.nationalgeographic.com/). These connotative meanings of colour provide very effective conventions for these types of map products.

Colours carry meaning and colour choice can evoke immediate emotional responses based on cultural specific norms. In Chapter 3 we mentiond the possibility of making the map more appealing to the user by harnessing an emotional response – remember the propaganda maps. Colour provides a means to achieve this, particularly if the cultural meaning of each colour is considered. For example 'traffic-light' maps shaded using colour gradients from red – orange – green are immediately interpreted by users in the West. Of course, such use may not be good for people with colour blindness as this is a palette not suitable for people with such conditions. Red being the most advancing colour will encourage the users' eye to settle on any spatial units coloured in it. Furthermore in the West the colour red is equated with all things dangerous, for example, traffic symbols use red circles containing speed limits denoting that if you are driving faster than the speed indicated then you would be driving dangerously. Conversely, using the same cultural notions green generally means everything is OK – green on a traffic light means go, and in the UK it also appeals to the rural idyll of a green and pleasant land. Therefore data categories coloured in green indicate the values are nothing to worry about. Finally, orange represents values in the middle of the range.

For other parts of the world the cultural interpretation of using the colours in a traffic light map may differ. In Russia, and for maps of the Cold War, red is clearly associated with Communism, but in China red is a symbolic colour for good luck and celebration. In Ireland orange is associated with the Protestant community, so using a traffic light map might be religiously insensitive.

8.3 Data classification – types of maps and thematic mapping

8.3.1 Types of representations

As noted in Chapter 3, there are two general types of maps. These are reference maps and thematic maps. It is useful to distinguish between them since the different types of maps will require different design approaches. Reference maps describe a location and are aimed at answering questions about location, distance or directions. They provide spatial facts and of include such maps as general topographic maps or street maps.

Thematic maps show at least one variable for data presentation, communication or exploration and are now common in the pages of academic journals, newspapers and reports. The theme that is presented can for example be derived from social, economic, political, epidemiological, cultural, environmental or physical data variables.

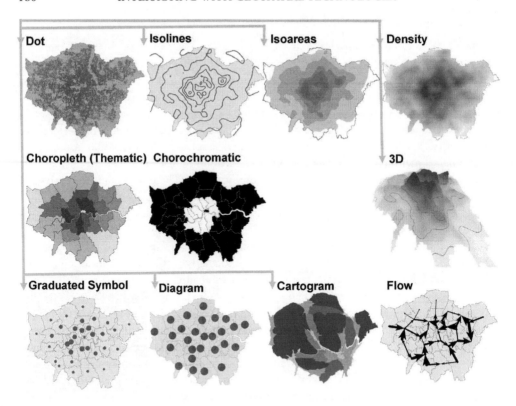

Figure 8.6 Nine types of thematic maps (a full-colour version of this figure appears in the colour plate section of this book)

Thematic maps have at least three purposes: they provide information about a specific place or location; they enable the exploration of spatial patterns and they facilitate the comparison of patterns for different places. Unlike reference maps, thematic maps do not provide a source for navigation, and within this context the reference information provided is for situational understanding. There are many different types of thematic maps and by presenting attribute data relating to spatial units in an intelligible way, thematic maps can be very useful (see Figure 8.6).

Creating maps is all about choices. These choices relate to the selection of the mapping method and the most appropriate graphical variables used to visualize your data, all of which depend upon the purpose for which the map is being created (Kraak and Ormeling, 2003). Data visualizations for thematic mapping can be comprised of many different forms in association with the three different types of objects: points, lines or areas (polygons). The different object styles are used to depict different types of information and the choice of thematic map will depend upon the form of the variables being mapped, for example qualitative versus quantitative, ordinal versus categorical data.

Figure 8.6 identifies nine types of thematic maps that can be used to graphically transform data into useful representations using various aggregation methods.

A dot map uses a point to represent a data value for different geographical areas. In this type of map a dot can be used to represent one object or several, for example the total number of deaths from cholera. Dot density maps can be problematic. The dots on the map rarely represent a specific location and there is a risk of misinterpretation. Issues arise in dot density mapping when the dots represent variable values of more than one for a particular area, also the dot can be placed arbitrarily within the area and thus obscures the real locations. The result will be distortion of the data distribution and might lead to a random pattern representing a concentration of dots that is simply not representative of the phenomena being mapped (Krygier and Wood, 2005).

Graduated symbol maps use a range of symbol sizes from small to large where the sizes of the symbols vary proportionally to the underlying data values. The symbols are usually centred at a specific object or within a geographical area. Often circles are used to show variation in population for different cities and the location of the circle on the map depicts the location of the city. As the symbol changes in size it represents changes in magnitudes and so it is suitable for mapping total counts. Note that graduated symbol maps can be confusing: if the gradations of the symbol size are too large they lead to overlapping symbols which are difficult to interpret or if the gradations are too small there is insufficient differentiation between the ranges of symbols.

Diagrammatic maps use charts to represent quantitative attribute data or counts of incidents in a qualitative group. The charts are commonly in the form of pie charts or bar graphs used to convey multi-variate information. Bar graphs are useful for displaying relative amounts and pie charts visually depict proportions. Pie charts can show the sub-groups of a variable, for example proportions of the population per age group. Like the graduated symbol map the charts are visually associated to an area by using its central point (known as centroid). These maps can be visually problematic: if the symbols are too close together they will overlap, so the scale used to produce the maps is important. Inclusion of too many sub-groups reduces the readability of these maps as they become cluttered and if very small proportions or values are included then they will be illegible.

Maps using *isolines* are lines on a map that connect data of the same value together, which are continuous lines that are never broken. A well known type of isoline is the contour connecting together points with the same elevation above sea level which is common in topographic maps.

Flowline maps represent the movement of objects from one location to another with the thickness of the flow line. The line is often varied to indicate the magnitude of the flow. In the past these maps were quite difficult to produce but they have been made easier with specialized computer algorithms such as Tobler's flow mapper. They are useful for mapping population flows. Note that too many flow lines will make the map very busy and may reduce its readability.

Density or surface maps create a continuous statistical surface from point data to identify concentrations of the points or the attribute data values of the points. They are useful for providing a representation of how attribute values change across a geographical area as they are a method for generalizing and smoothing data. A useful technique if you need to make point data anonymous. The technique produces output in the form of a grid (raster) made up of a series of cells, with each cell representing the estimated spatial density of events or attribute values (Fotheringham *et al.*, 2002; Martin, 1996). It is a useful representation technique because, as noted the result is a smooth estimate of probability density from an observed set of inputs (Bailey and Gatrell, 1995). They are commonly coloured using a hot-cold colour scheme where blue represents low density values and red high density values, hence they are sometimes called hot-spot maps.

Density surfaces are influenced by the distribution of the points, for locations where there are no points the values are *interpolated*, but this means the number of sample points and how they are concentrated will determine the pattern of the surface. They are also prone to suggesting the existence of equal density for all areas which have been classified in the same way, but this may just be the result of the data smoothing.

Cartograms are advanced visualizations which use the areas of a map as a symbol, these symbols are then used to create a type of proportional symbol map where the symbol shrinks or grows (the geographical area) according to the data attributes associated to the area symbol (Cuff and Mattson, 1982), for more details see Case Study 8.3.

Chorochromatic maps provide a useful method of representing non-numerical or categorical data where different geographical areas are shaded in different colours, it is also useful for binary data. The name chorochromatic is derived from two Greek words 'choros' meaning area and 'chroma' meaning colour. The colour palettes used in this type of thematic map do not represent any intrinsic hierarchy in the data but are used to identify the presence or absence of a given variable for a particular area. They are often used to represent geological, land-use or vegetation mapping. There are problems associated with this type of map when social data is being mapped. If these maps are not properly understood the reader might interpret both the colour of the area and its size as being significant when it is not – so the size of the area may incorrectly be interpreted as proportional to the number of people representing the categorical variable mapped (Kraak and Ormeling, 2003).

The final type is the familiar *choropleth* map, but the term is likely to be unfamiliar. The origin of the term is a Greek derivative from 'choros' meaning place and 'pleth' meaning value. These types of maps use different shading for geographical areas according to the values of the underlying data; they are data maps for discrete spatial units and are often created using data aggregations in the form of ratios or percentages. The boundaries of the areas are often derived from administrative datasets such as census wards or countries and these types of maps are the spatial equivalent of a histogram or bar chart.

Another alternative label for describing a choropleth map is the generic term *thematic map* and in this context it relates to the attribute variable being represented on the map. The latter term is fairly common and therefore more suitable for descriptive purposes. All sorts of thematic maps can be created with modern geospatial technologies which highlight patterns within data distributions although creating any thematic map requires a number of choices based on how the data is classified. For most types of thematic mapping the spatial units to be classified are discrete and distinct and can only be assigned to one class/interval at a time (de Smith *et al.*, 2006).

8.3.2   Data classification for thematic maps (choropleth maps)

A cartographic form or map is not just a means of communication but, if used appropriately, can be an effective tool for acquiring spatial understanding and knowledge. However, this raises the question of the statistical basis on which to classify ('colour-up') a thematic map and the number of interval ranges to adopt. The types of cartographic representations should be chosen carefully according to the values in the dataset and rules conforming to data classification for mapping (Alçada-Almeida *et al.*, 2009; Jenks and Coulson, 1963; Jenks and Caspall, 1971; Monmonier, 1996; Evans, 1977). The colouring of phenomena using numerical classification can be regarded as *standardizing by numbers*.

The look, feel and interpretation of any thematic map are influenced by three elements: the size and shape of the spatial units, the number of interval ranges (classes) and the method of classification (Robinson *et al.*, 1995). These elements are part of the *univariate* classification process whereby each element influences how the map is displayed. Spatial units themselves influence thematic maps since their size and shape have significant visual impact, spatial units that are not uniform in shape and size are likely to mislead the users' interpretation of the map. A thematic map created using spatial units that vary in shape and size leads the user into thinking that the larger areas are more significant because they have a bigger visual impact than the smaller areas. If you encounter such a problem, it is possible to use graduated symbols at the centre of each area to overcome this problem.

The classification process organizes data so it makes more sense to the user. It simplifies data through a process of generalization – from the wide range of values that exist in the data it reduces the amount of information that a user consumes (Harvey, 2008). In order to make the pattern obvious it is necessary to carefully select the number of classes (sometimes called intervals or groups) and the classification method. The guiding rule within the cartographic literature for thematic maps advocates seven classes, plus or minus two, as we previously discussed. The simple premise behind the principle is related to our ability to visually perceive objects. Too many categories make it more difficult for the user to process and retain the information in memory so it is more arduous to comprehend the map. The rule is derived from an early paper in cognitive psychology which suggests that 'the magic number 7' is significant (Miller, 1956) in relation to memory, attention, imagery, language comprehension and speech. Miller noted that people can recall around seven chunks (pieces of data) of information from the stimuli presented to them which are retained in the short-term memory (Sperling, 1960). More recent research suggests that for certain situations the magic number should be as low as four plus or minus two (Cowan, 2001). Despite the lack of consensus, the aim is to find the number of categories most cognitively efficient for the map user and remember that as the categorization becomes more detailed the effort required by the user to process them will increase (MacEachren, 2004).

Case Study 8.2: Selecting ranges for health data

A recent study carried by Wardlaw and Haklay at UCL (Wardlaw *et al.* 2008) investigated the types of cartographic representations that are useful to health service professionals in strategic decision making. A cartographic survey of health professionals focused on reviewing different health maps, originally drawn to emphasize a number of different cartographic aspects from data classification, number of ranges, colour schemes, representation of point data, raster or vector data for the background mapping to the mapping of multiple datasets.

One of the survey questions asked the respondents to comment on a univariate map showing the number of people who had given up smoking. The map was drawn to accentuate pattern differentiation according to performance to show areas where the quitting rate was well above the national target and where it was well below using a standard deviation classification to make use of a diverging colour palette. To solicit

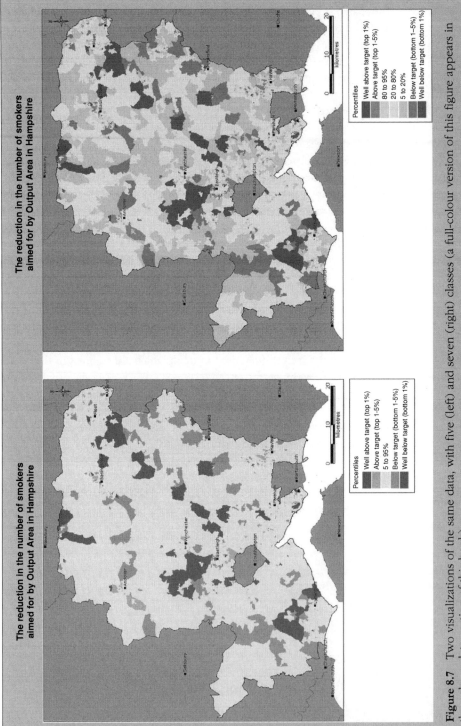

Figure 8.7 Two visualizations of the same data, with five (left) and seven (right) classes (a full-colour version of this figure appears in the colour plate section of this book)

a response relating to the preference of number of intervals and the corresponding amount of detail, the same maps were drawn twice but one had five interval ranges (figure left) and one comprised seven (figure right).

Feedback from the survey varied, with some commenting that 'Less ranges made it easier to identify the problem areas – Map b had too many ranges, with the areas of concern harder to pick out' (Research & Statistics Officer) and others saying 'Fewer bands made it easier to interpret but to inform decisions would need more detail.' These comments show preference for a number of intervals, but are inconclusive, and the decision should be influenced by the purpose for which the map is created. Use the following points as a simple guide to help you make your decisions (adapted from Krygier and Wood, 2005; Monmonier, 1996; Brewer, 2004; Robinson *et al.*, 1995).

- If your data is binary – representing a variable with values such as yes/no, true/false, or positive/negative – use two interval groups.
- To differentiate distinct patterns use between four and nine intervals/groups.
- It is hard to distinguish more than five shades of grey.
- A map with no interval groups and one which uses as many groups as there are unique values is difficult to interpret and identify patterns and values, and is best avoided.

Once the spatial units have been selected and the number of interval groups chosen, the next step is to identify what classification method should be used to group the data. The classification aggregates data, by grouping data within a particular range of values into a single interval group. Each group is represented by the colour of the line (Monmonier, 1996). The aim is to find the most appropriate classification to match the distribution of the underlying quantitative data. As with the size and shape of the spatial units, the choice of classification influences the 'look' of the map. With the same data and spatial units, different classification methods will result in radically different representations (Monmonier, 1996).

The most common classification methods in geospatial technologies are: quantiles (percentiles), equal interval, equal count, natural breaks (sometimes called Jenks) and standard deviations. Usually, there is also a custom classification that enables the map maker to create custom classification. To select the correct classification method you need to match the classification to the data distribution – this is very important. Exploring the distribution of the data either algorithmically or in a form of an ordered histogram is useful for this. Classifying data without having first explored the data distribution will fail to reveal geographical patterns in the data (Krygier and Wood, 2005) and may result in the inappropriate use of a classification.

Figure 8.8 gives examples of four types of data classification together with the break points which create the classification – shown in the histograms. The *equal count* (Figure 8.8a) range assigns interval values so each group has an approximately equal number of spatial units allotted to it. This distribution produces a map which appears to have a balance in the number of units but on many occasions has no bearing on the data distribution and is quite unhelpful in most cases.

In contrast, the map classified using the *equal intervals* classification (Figure 8.8b) assigns interval values so each interval group contains the same range of values and the data is divided

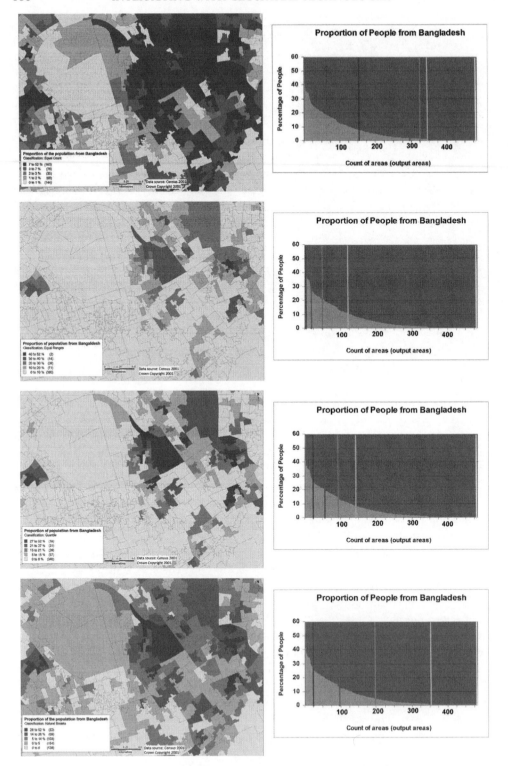

Figure 8.8a-d	Same data, same spatial units, same number of classes, different classification method (a full-colour version of this figure appears in the colour plate section of this book)

into equal parts. It is easy to understand and this makes the legend easy for users to interpret (Slocum, 1999) but it often fails to represent the underlying distribution of data, especially when a significant proportion of the data are concentrated in a few of the intervals. It is especially useful for mapping data that are uniformly distributed (Monmonier, 1996).

A *percentiles* or *quantiles* map (Figure 8.8c) is useful for exploration. It ranks the data and then divides them so that all intervals contain approximately the same number of spatial units. This classification produces a visually balanced and distinct map because it never has empty classes or classes with too few or too many values. It does force some values into the same interval when it may not be appropriate and the consideration of ranges does not always make sense.

The final classification in Figure 8.8d uses the *natural breaks* method. The natural breaks method is a statistical technique to identify clusters of values that are inherent in the data distribution. Each cluster is then put into a different interval or group for the purpose of classification. Natural breaks seem like the most relevant classification for representing the structure of the underlying data distribution because often the interval groups maximize differences to ensure similar values fall into the same cluster (Harvey, 2008). Maps classified by natural breaks minimize differences between values within the interval groups whilst simultaneously maximizing differences between the group intervals (Krygier and Wood, 2005). This classification is least likely to miss potentially significant data groupings and accounts for its distribution, but the accompanying legend is sometimes difficult for the user to interpret (Slocum, 1999; Monmonier, 1996).

The other type of map classification (not illustrated) uses the standard deviations statistical analysis. This classification will highlight how attributes vary from the average. The break values are set so that each interval group represents one standard deviation from the mean, therefore it provides a representation of how dispersed a dataset is, since the smaller the deviation the less dispersed the data will be. This type of classification, like the natural breaks method, reflects the data distribution and is easy for users to conceptually understand (Slocum, 1999). The rules of data classification are summarized in Table 8.3.

Table 8.3   Rules of data classification

Considerations	
Think and look at the data before you classify	Cover the full data range
Experiment with classification schemes to get the pattern you want to match the purpose of the map	No overlap between classes
Method of classification affects the 'look' of the map	Always include a fully detailed legend
Classification should 'match' data distribution	Make the legend meaningful to the user (e.g. rounded numbers, use short textual descriptions)
Number of classes is usually between five and nine	Ensure the legend and classification fully match
Beware of system defaults	
Maximize homogeneity within interval groups	
Maximize differences between interval groups	

In summary, data classifications facilitate the spatial organization of statistical data and an effective one will aid users to make useful interpretations about the geographical patterns. It may be necessary to try and highlight critical values in the distribution although if there are no critical values then the guiding principle should be to maximize homogeneity within classes and the differences between them. This can be done by always exploring the data distribution before it is mapped. To make the classification more meaningful it is also a good idea to consider if the interval groups should have any specific meaning in relation to the purpose for which the map is being created. By following these principles of data classification you will avoid one of the biggest pitfalls associated with the oversimplification of distributions which arise from the arbitrary selection of interval groups. Do not accept system defaults without due care and attention.

Case Study 8.3: Advanced cartographic representations – Cartograms and Chorems

Robert Laurini, Monica Sebillo and Giuliana Vitiello

In conventional planar cartography, since the Earth is more or less spherical, one of the objectives is to use projections which minimize the distortion from the reality. However, in some cases, the message being delivered is most important; and sometimes this message can be more understandable when a certain level of distortion is accepted. So, the goal of cartograms and chorems is to schematize some areas in order to maximize the understandability of the message.

Indeed, for effective decision-making, it is more important not to represent exactly everything with high level of details, but rather to show the more salient features or characteristics. By doing so, sometimes hidden relations can be more easily discovered.

Cartograms were invented by Tobler (Tobler, 1961, 1963), they are maps in which the particular distortion chosen is made explicit. Area cartograms are drawn so that areas representing places on paper are in proportion to a specific aspect of those places. In cartograms, the most important aspect is to adjust the area in relation to a given variable, so the graphic preserves the spatial relations between the different objects, but not necessarily the shape.

An excellent collection of cartograms is available from http://www.worldmapper .org/ from which Figure 8.9 is provided. This figure shows the repartition of wealth in 2002: we can easily compare the respective distorted shapes of North America and Africa.

Unlike conventional cartography in which territories are represented proportionally to their area, in cartograms areas are made proportionally from other data. As it is simple to transform a territory according to those numbers, the more important problem is to preserve spatial relations with the neighbouring territories (known as topology). For this purpose, special algorithms must be designed.

Another type of cartogram was developed by Danny Dorling (Dorling, 1991) based on circles which are set to be proportional to a data attribute. Figure 8.10 gives an example extracted from the 2004 presidential election in the US. For designing a

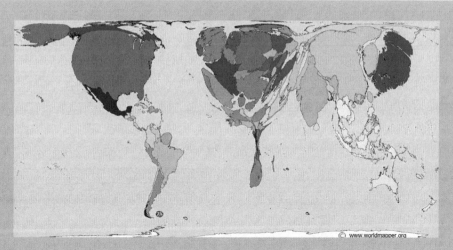

Figure 8.9 Cartogram showing distribution of wealth (Source: http://www.worldmapper.org/display.php?selected=170) (a full-colour version of this figure appears in the colour plate section of this book)

cartogram, one of the more popular methods is based on Gastner and Newman's algorithm (Gastner and Newman, 2004).

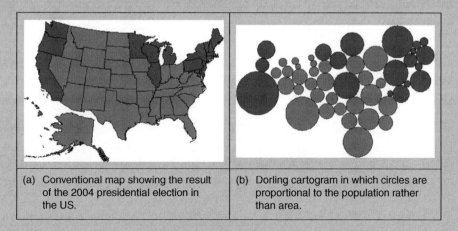

| (a) Conventional map showing the result of the 2004 presidential election in the US. | (b) Dorling cartogram in which circles are proportional to the population rather than area. |

Figure 8.10 Example of Dorling cartogram. Source: http://www.goldensoftware.com/newsletter/Issue53Mrb.shtml

The term chorem comes from the Greek word $\chi\omega\rho\alpha$, which means space or territory. According to the definition of the French geographer Roger Brunet (Brunet, 1986), *chorems* are a schematic representation of a territory, where objects and phenomena of interest are emphasized through proper visual notations, whereas details related to other aspects are discarded. Different to cartograms, where values of a single variable are shown at a time, chorems allow designers to assemble into a single (chorematic) map more than one thematic layer, thus representing the relative importance of a set of objects and phenomena related each other.

Figure 8.11 shows a set of chorems, related to Mexico, where some aspects are highlighted, such as:

- the geometric shape,
- the most important cities,
- the different climatic areas,
- the flows representing the direction of people migration.

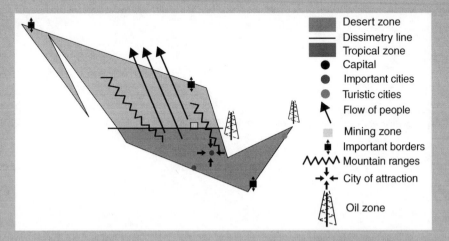

Figure 8.11 *Mexico – a set of chorems showing dynamics among some main places* (a full-colour version of this figure appears in the colour plate section of this book)

Since their first appearance, chorems have had a large diffusion into the geographical community (Lardon, 2003) where they have been widely investigated and developed (Cheylan *et al.*, 1990) due to their capability of conveying (possibly) complex information to (non-expert) users through a visually simplified reality. In fact, many application domains have benefited from adopting chorem-based cartography to sketch real-world situations, including applications related to political, economical, environmental and demographic aspects.

In recent literature, chorems appear to be a very interesting approach to investigating a number of issues, including way-finding (Klippel, 2005), providing a visual summary of spatial database contents (Del Fatto *et al.*, 2007) and providing a global vision of a spatial database (Laurini *et al.*, 2006). In fact, despite being a simplification of a territory with simple geometric shapes, chorems try to rigorously represent the structure and the evolution of a territory, by preserving both spatial and temporal relationships among elements. This capability of chorems can be exploited to provide decision makers with a means to acquire syntactic information (*what*, *where* and *when*), as well as semantic aspects (*why* and *what if*), useful to support human activity to model, interpret and analyze the reality of interest.

Given the well-established relevance that chorems may have for the geographical community, computer scientists working in the field of geographic information systems have recently focused their research on the development of chorem-based solutions for automatically extracting geographic knowledge from spatial databases and representing it onto maps.

When designing a chorematic map, the simplification rate should be high because the involved geometry should only resemble the associated concept. Then, a proper balance between the simplification level and the whole generalization task is necessary in order to preserve topological and directional relationships, which are fundamental to convey the right information towards users.

8.4   Mapping conventions – map elements and layout

8.4.1   Map elements – titles, legends, scale bars and north arrow

The final section in this chapter is dedicated to the most important elements that should be included in a map to ensure that it is useful and usable to the audience. Some of them are useful for any map, be it a digital map or paper map, and some are relevant to printed maps. One of the issues with geospatial technologies is that all too often the designers do not take into account the differences between the map that is displayed on the screen and a map that the user prints to use off line. This usually leads to inferior outputs. Some simple map conventions can improve the legibility of the map. Here we discuss titles and additional information, legends, scale bar and north arrow.

A map should always have a title. Deciding on a useful title is subjective but it must always make sense to your user, who is your target audience. The title should always say something about its topic. It is often useful to provide information about the geographical area of the map, although this can be problematic for a dynamic Web application since the user may be able to pan and zoom which means the location is always changing, and the title would need to re-align itself constantly. In these circumstances a dynamic overview/inset map provides the user with a larger areal extent locating the situational context of the centre of the map; this is particularly useful for small scale maps. The other important component of the map title is the temporal information about the map, as users can then make a value judgement about the timeliness of the data and how suitable it may be. Krygier and Wood suggest that the title should be two to three times bigger than the fonts used on the map itself. This is so that the title stands alone and does not appear simply as another map label (Krygier and Wood, 2005).

It will not be possible to contain all the details about the map topic in the title and nor should you try. For this reason the map should also contain some useful information to complement the map title. This can be in the form of text boxes, or in an interactive system an additional window that can be called upon request. The purpose of these is to provide more detailed descriptive information about the data sources, missing values, the author, the author's contact information, a copyright statement, the map projection and its coordinate system and other useful details. For Web-mapping this information can be documented on a separate page as long as the link to it is easy to locate within the map interface.

The next element, which was mentioned earlier in the chapter, is the legend, which is one of the most essential map elements. Providing the user with a clear description of the map layers will ensure proper interpretation of the map. It is absolutely essential in the knowledge construction process and without it the user will fail to understand the messages portrayed in the map. In many digital applications, the details of each layer are derived from file names but usually this will not explain to the user what the layer means. It is therefore critical to ensure that the naming of layers is in line with their knowledge. Legends provide a lookup key to the symbols used in a map and details for interpreting any data classifications and their associated colour schemes that are used.

There are some hints and tips for legend design. One possibility is to group symbol items together based on similarity, for example, or functional group or graphical element. This associative grouping will aid spatial cognition. It is useful to keep the legend's design as simple as possible. It is important to keep the legend in proportion to the map area. Do not create an oversized legend or else this will be the most dominant design feature of the map. There are many Web-mapping applications that neglect this fact and as a proportion of the page space available the map is very small which makes the map much more difficult to use and view (Haklay and Zafiri, 2008) – for example the Environment Agency case study in Chapter 3. Numbers in legends should be rounded to the nearest whole number, and if the numbers in a data classification correspond to something very technical and your audience are non-specialist then use textual description to add meaning. Furthermore, the convention is for the data values to be arranged vertically with the lowest values placed at the bottom.

A scale bar is especially useful if the maps are to be used for navigation or distance comparison. If a user is unfamiliar with an area it will help to get a sense of the areal extent of the map. Scale was discussed in detail in Chapter 3 where it was noted that for Web applications the only useful representation of scale is to use the scale bar. If you are developing applications for the UK note that some of the older users will be more comfortable with the imperial units of measurement whilst the younger users will be most familiar with the metric system. Likewise different countries use the different systems so check the national standard and if necessary present the scale in both unit systems.

For paper maps the north arrow used to be an essential component as it helped the user orientate it in geographical space. The north arrow is a directional indicator. With the advent of Web-mapping interfaces the north arrow is not quite so important as long as the map window always points to the north. If the application facilitates user controlled animated fly, like Microsoft Virtual Earth and Google Earth, then the north arrow once again becomes an important element of the map. Keep the representation of the north arrow simple so as not to advance it to the front of the map layers and prevent it acting as a distraction. The final map element that may be useful is a border which is known as the neatline. This element draws a box around all the objects in the map window to maintain a sense of grouping.

Case Study 8.4: Noise pollution mapping in the UK and its map elements

The noise mapping interface was developed for the Department for Environment, Food and Rural Affairs to provide information about road traffic modelling in major urban areas in the UK. The map uses a location marker showing the centre point of the location searched by the user. However, notice that there is a significant absence

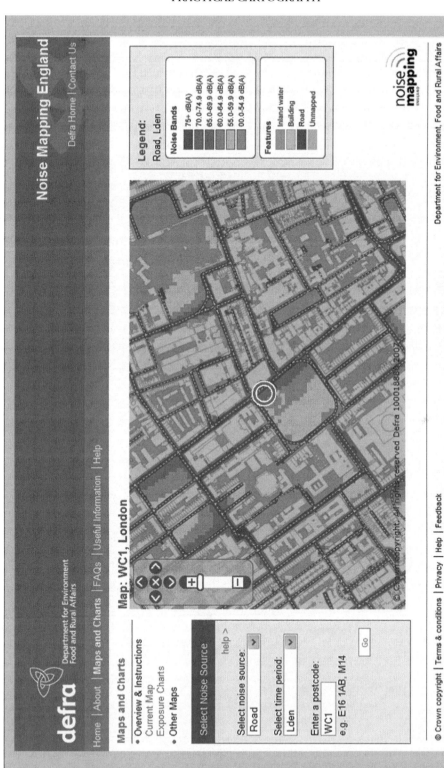

Figure 8.12 England Noise Mapping site (a full-colour version of this figure appears in the colour plate section of this book)

of any street names to help the user contextualize the map. Also the marker symbol is in colours which advance it to the forefront so this is what the eye settles on when viewing the map. Its size is also very large and it obscures the information in the data layers below (which is what the user is really interested in). The result is that this map will only be useful for people already with detailed local knowledge of an area with a comprehensive internal mind map.

There are some noticeable map elements that are missing – can you identify them? Firstly the map has no scale bar. There is no inset map so users cannot orientate themselves within the greater city area. This map shows a detailed view of central London. Are you familiar with London? If not does the location defined by the postcode in the title mean anything to you? Does the title help you to interpret the map? There is no north arrow, but the map does orientate itself to the north so this is not such a fundamental absence. Now look at the legend – the colour scheme uses a traffic light colour palette so the colours have cultural meaning (even though it is not a great palette for colour blind people). Look at the legend, the numbers are very precise using 1 decimal place but this does not add much to its intelligibility. To non-experts it would be useful to have a textual description alongside the numbers in simple words explaining what the numbers mean. Also note the orange design of the surrounding interface distracts the user.

8.4.2 Layout

There are five key components of usability which were first proposed by Nielsen (Nielsen, 1999). And they apply as much to Web-mapping interfaces as they do to standard Web design. They are efficiency, learnability, rates of errors, memorability and satisfaction (see Chapter 6 for more details). Consider the maps in Figure 8.13a-c and the five usability components which the inclusion of map elements onto the map can assist in working to achieve usability best practice. The addition of map elements to maps b and c helps to improve spatial cognition so assisting the efficiency of the knowledge construction but where they are placed is also relevant. The title, legend and scale, their size and placement and the use of a neatline all contribute to the overall user experience and ensure the user is satisfied with both the experience of using the site and the quality and usefulness of the output. Which of the maps in Figure 8.13a-c is more satisfying to look at? Which one is the most easily understood?

The placement of objects both on the map and in the interface is very important to the overall usability and usefulness of the map design. The aim of a good map layout is to seamlessly allow the user to scan the map and find out the information according to the purpose for which the map was created. Well designed map elements will reduce the cognitive load of the user and increase the ease with which the user can interpret the map. If the layout is successful the user will not notice it, the eye will naturally focus on the map and its content (Krygier and Wood, 2005). It will provide the mechanism for spatial cognition by being aesthetically pleasing. Conversely, a map with a weak layout, one that is cluttered, requires the users to consciously take time and search for information, or a layout which advances the map elements at the expense of the map layers will only frustrate the user.

Experiments carried out by Jacob Nielsen since 2006 show users scan Web pages very quickly and there is a common pattern in how they do it. Eye tracking experiments indicated

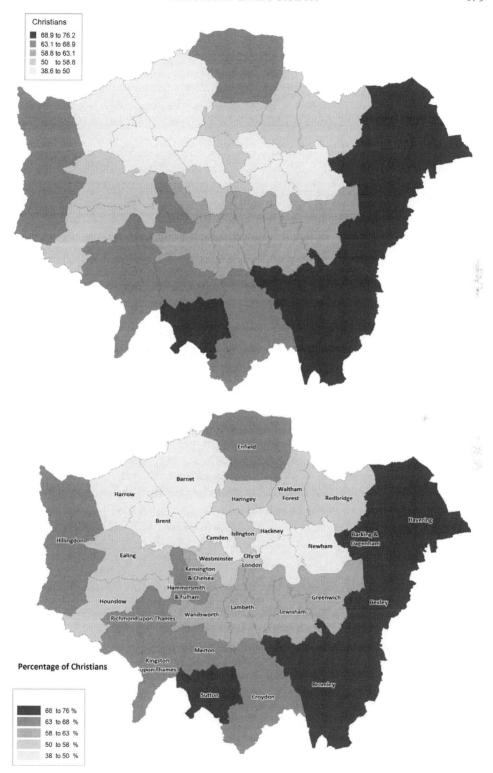

Figure 8.13a-b Map elements and their placement (a full-colour version of this figure appears in the colour plate section of this book)

Percentage of Christians residing in London Boroughs, 2001

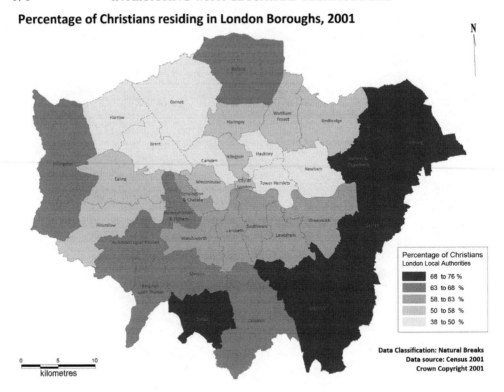

Figure 8.13c (*Continued*)

that users often read Web pages in an F-shaped pattern: two horizontal stripes followed by a vertical stripe as they move down the page. They begin using a horizontal scanning of the top part of the page and then move down and scan horizontally sequentially and end with a final vertical scan of the left hand side of the page. Essentially they scan with a top to bottom motion. The elements need to be positioned in accordance with how the eye will scan the page, see Figure 8.14, as this will also help to create the natural visual balance.

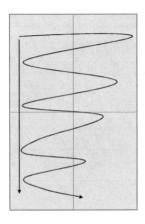

Figure 8.14 Direction of eye movement when scanning a Web page

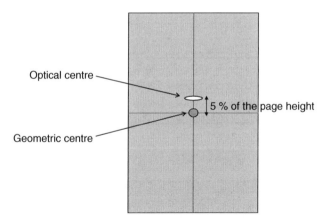

Figure 8.15   Optical centre of map/page layout

Robinson and his colleagues note that a well balanced design ensures: nothing is too light or too dark (this includes objects in the map and the interface), nothing is too long or too short, too large or too small, in the wrong place or too close to the edge of the mapped area (Robinson *et al.*, 1995: 334). The aim of achieving visual balance is to establish a harmonious arrangement of objects around the optical centre of the map (see Figure 8.15). This is best done iteratively, moving map elements around the map until it looks balanced. A composition that is unbalanced will be perceived by the user as random and accidental.

Summary

Cartography is a wide field of knowledge, and the knowledge that you can find within the literature is wider and deeper than what is covered in both this chapter and in Chapter 3. By highlighting the main issues of symbology, colour, data classification and map elements, we aimed to provide you with working knowledge about the issues that most frequently appear in map composition. In some geospatial technologies you will find yourself designing a map that will not be altered by the user, so you have more control over its look and feel, and have the ability to apply the guidelines that are provided by the cartographic literature. In other cases – for example when designing an interactive system where users will switch layers on and off and compose maps for themselves – the control is handed to the user, to some extent. However, even in these cases it is important to ensure that the interface encodes cartographic rules to assist the user in creating effective maps.

Because cartographic knowledge is wide, and also depends on practice, it is recommended that if you are dealing with a system or a situation where many types and variations of maps will be used then find a cartographic advisor. An experienced cartographer will help in setting out the rules, set up colour ranges and other attributes of the maps, thus improving the quality of the final product. Finally, never underestimate the time it will take to produce a well designed map.

Further reading

While the books that are mentioned in Chapter 3 will provide you with detailed knowledge of cartography, there is now a growing body of books and papers on practical aspects of cartography. John Krygier and Dennis Wood (2005) *Making Maps* is highly accessible and includes many more illustrations which emphasize both good and bad cartographic practice. Cynthia Brewer (2008) *Designed Maps* focuses mostly on the composition of maps for printing, but also includes many useful hints and tips on the design of maps in general. Edward Tufte (2001) *The Visual Display of Quantitative Information* is also great as a design baseline – it is also very aesthetically pleasing. For online design of maps, Stamen Studio (www.stamen.com) are active in designing effective maps, as well as Axis Maps (http://www.axismaps.com/), specialists in cartographic design. The International Cartographic Association (http://cartography.tuwien.ac.at/ica/), especially the Commission on Use and User Issues, is a source for current research and practice in cartography across the world.

Revision questions

1. What are the main ways of representing colours in a uniform way?
2. What are the core elements that should be included in a well presented map?
3. What are the impacts of colour, symbology and map elements on the different maps that are presented in this chapter? Look at the colour plates that are provided in this book and analyze the maps that are presented there.
4. In a similar fashion to the analysis of the UK noise map, in Case Study 8.4, choose a website with which you are familiar, and analyze the information that is presented in it.

9 Principles of interaction

Jessica Wardlaw

The aim of this chapter is to focus on the elements of the interface that are critical to the design of geospatial technologies for computer desktops or over the Internet, and we will focus mostly on Graphical User Interface (GUI). The most important message that this chapter advances is that consideration of the user experience is imperative for successful geospatial technology design. The chapter opens with an outline of the key elements of the theory of interaction for GIS, including some of the theory about what the GUI means to users in the real world and linking elements of the GUI to actual tasks being performed by the user (see also the discussion in Chapter 7), drawing particularly from Shneiderman's Object-Action Interface model. We then describe the main interaction styles and the main elements of an interface (windows, icons, menus, and pointers) illustrated with examples from desktop and Web-based GIS. The chapter discusses some alternatives to the traditional direct manipulation mode of interaction, for example through speech or gestures, which offer solutions particularly suited for designers of mobile geospatial technologies for collaborative projects for example. The integration of the theory and the GUI elements will finally be used to suggest some guidelines for designers of GIS interfaces, drawing inspiration from Jeff Johnson's (2000) checklist in his *GUI Bloopers: Don'ts and Do's for Software Developers and Web Designers*. This chapter should encourage a deeper consideration of the user in the design process for geospatial technologies interfaces.

More and more non-geographers, or 'accidental geographers' as Unwin (2005) has described them, are using geospatial technologies, especially thanks to Web 2.0 applications. However, little research has been done into how geospatial technology interfaces can be designed for non-experts. Ask a novice to carry out an analytical task using a popular GIS package (such as ArcGIS, MapInfo or AutoCAD Map) and it would be very difficult for them to know where to start without lengthy training. GUI design is critical if we are going to cut down on user errors in GIS, which seem to be the most common types of errors in these systems' operation. For example, in some of the most popular packages, a user can spend the whole day editing a data file but lose their work because in these packages the process of saving the workspace (that is the organization of layers, data sources and cartographic settings) can be easily confused with the saving of the data file itself. It is down to the software designers to make it obvious to the user what is required for each task. Software developers can profoundly influence the users' experience through the user interface.

So what do we mean by the User Interface (UI)? The UI is the conceptual link between human intention and what a computer can offer (Gould, 1994), above and beyond human capability. Designers construct user interfaces as illusions behind which all the internal software architecture is hidden, so that the user is able to understand and use it. Users need

Interacting with Geospatial Technologies   Mordechai (Muki) Haklay
© 2010 John Wiley & Sons, Ltd

illusions because the capabilities of computers are so abstract, and the user needs to map the software's internal operation, which is coded, to the task that they are trying to perform. Jeff Johnson compares user interfaces to human skin. Human skin isn't just a superficial surface layer whose purpose is to hide the internal working of the human body. It has its own structure and depth. Different kinds of skin serve different functions; for example the skin on the palm of our hand is distinctly different from the skin on our scalp or our lips because they all serve a different purpose. Skin has a range of purposes, including protection, transfer of moisture and tactile sensation. Each type of skin has a different structure so that it can carry out its particular function. In the same way, to make systems truly usable software illusions are built on top of the underlying functionality. These illusions make abstract things appear concrete and give the users the impression they are controlling real objects. The best example of this is the computer 'desktop', with a Desktop, Files, Folders, a Recycle Bin and Inbox. While the manipulation of files which the desktop environment supports can be executed using a Text Terminal through Command-Line script, non-computer specialists will find such a mode of operation tricky to learn. Great effort has been put into designing a Graphical User Interface that gives access to the same functionality but through the direct manipulation mode of interaction, which allows the user to carry out activities that are similar to actions in the real world. These include pointing to an object, tapping it, holding and releasing, dragging and dropping and other similar operations. While these operations are carried out by manipulating mechanical artefacts (for example, a computer mouse) and clicking on buttons, after very little training most people will intuitively grasp the metaphors that are being used and will be able to operate the application successfully.

Everyday life is often made more complicated by designers who don't consider the people for whom they're designing (Norman, 2002). Think of all the times you've pulled a door handle that needed to be pushed. Either a flat plate or a handle that makes it obvious that the door should be pushed would be more suitable. Consider the number of different tap designs, some of which look 'clever' but are hard to operate. Sharp *et al.* (2007) highlights remote controls (with their numerous buttons) and answering machines as also being examples of bad usability. Sometimes usability is intentionally inefficient. The QWERTY keyboard, which typists have become so accustomed to, was allegedly designed for inefficiency: the most frequent combinations of letters were on different lines to slow down the typist and prevent the typewriter's typebars jamming the machine (Harrower and Sheesley, 2005). The amount of pressure put on the key determined whether the letter was typed. By placing common letters on the left hand side the typist is encouraged to use the little finger on their left hand for the letter 'e'. This is a machine-centric solution, forcing users to work in a particular way due to the limitations of the technology rather than the needs of the user.

Case Study 9.1: A close up on flood maps

The UK Environment Agency *'What's in your backyard?'* (WIYBY) website, aims to provide information for the public about the flood risk of different parts of England and Wales. We have discussed the map on this website briefly in Chapter 3, and here we give it a more detailed examination. The way in which the website operates is by allowing users to enter their postcode into a text box to find out information about that location. However, when you type in the postcode and press enter, the map window does not zoom fully into that location as the user would expect – especially when considering that each postcode in the UK can be as small as defining an area of 25 homes. For example, type in a London postcode and you are taken to the 1:650000 scale which shows the whole of London. This is not much use to the user who is interested in knowing the flood risk for their property. It goes against the users' conceptual model of how the map should function.

You can see from the screenshot in Figure 9.1 how small the map window is in relation to the rest of the page. The size is fixed to 600 × 400 pixels regardless of the screen size. A large part of this is taken up by the legend of the map. This can be

Figure 9.1 The Environment Agency 'What's in your back yard?' website (Source: Environment Agency)

hidden away by clicking on an arrow button on the map since it becomes an obstruction to users trying to identify whether their property is in the flood risk zone. However, the user needs to discover that the arrow is the appropriate object on the interface that will lead to the collapse of the legend.

There are zoom and pan controls to the side of the map window which can be used to manipulate the map. However, there is no mechanism to zoom in on a point on the map. If the user wants to focus in on a particular part of the map they are forced to use the controls on the left side of the map. On Google Maps and similar Web-mapping applications a convention has emerged that by double clicking on the map window, the view of the map will zoom in on the map. Instead, on WIYBY double clicking opens a page full of text with information about the flood risk of the location that was double-clicked.

WIYBY is certainly not the worst example of a Web-mapping application. On the contrary, it is now in its fourth or fifth iteration and with each version it is improving. However, as we have discussed in Chapter 3 and here, we can see some significant usability shortcomings. Some of them are 'internal' and part of the way the application operates and presents information. For example, it does not zoom in far enough to match the users' expectation about the area that they should explore. Other problems are 'external', emerging from conventions that other, more popular websites, are using.

These two aspects of usability are significant for any GUI design. Applications should, as a general rule, follow the convention of other applications to ensure that they are useful.

What applications such as WIYBY show us is that the interface, the application and the user experience are all integrated in the mind of the user. Thus, we need to understand how to construct the interface from its elements to achieve a useful and effective interface. As was noted in the early 1990s, it is advisable to design the UI before writing the code behind any GIS software (Frank and Mark, 1991), and this rule applies to other geospatial technologies.

9.1   Key elements of the theory of interaction for geospatial technologies

In general, there are two main principles to be adhered to when designing any GUI: make things visible and provide a good conceptual model. A conceptual model is simply the model of a product that the designers want users to understand. An example from the real world of a simple conceptual model is a pair of scissors: it is clear that by moving the handles the blades move. This is compared to say a digital watch with four different buttons, with nothing to let the user know how to set the time. If you have a good conceptual model, errors made will be minimal and the user will be able to return to the appliance and remember how to use it without looking at an instruction manual.

Geographical Information Systems are designed around a suite of commands, grouped into classes of similar data manipulations (Shneiderman and Plaisant, 2005). The user then has to map the task they desire, e.g. which are the houses at risk of flooding along a river, to the commands available, e.g. creating a buffer zone of influence around a river.

Task mapping for computers is inherently challenging since computer functionality is so abstract from the real world. For example, when driving a car it is logical for the driver to turn the steering wheel in the direction they wish to turn. This is a very logical conceptual model for the driver. Steering a boat, on the other hand, is more difficult; if you want to turn right the rudder must be steered left. It is similarly unnatural to move the indicator in the car up for turning right and down for turning left, compared to turning the steering wheel in the direction of turn (Lanter and Essinger, 1991). On computers there are many examples of unnatural mappings. For example, to shut down the Microsoft Windows XP operating system users must click on the 'Start' button.

It is often impossible to provide complete natural mapping, and the different requirements from the computer interface require some level of compromise from the designer and the user. For example, the very successful desktop metaphor which is used in most personal computers includes within it an item which is illogical in the real world. This is the position of the 'trash can' or the 'Recycle bin' on top of the desktop. In the real world, the bin is kept well under the desk, and not on top of it. On the computer, it is necessary to position it on top, as the visibility of the object is central to its use.

Figure 9.2 provides an example for natural mapping between control and the controlled objects. These diagrams denote the configuration of hobs and their controls on a stove. The top two diagrams show stoves where the mapping of which dial switches on which hob is unnatural; the arrangements of the dials and the hobs are in direct conflict. Typically these arrangements require a diagram to tell the user which dial controls which hob, since the user is likely to forget the mapping next time they use the stove. The others are examples of

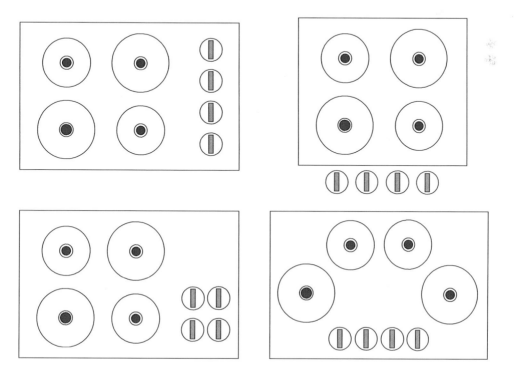

Figure 9.2   The mapping of stove dials (adapted from Norman, 2002 and Lanter and Essinger, 1991)

alternative arrangements where the configurations are matched, which would make it easier for the user to remember which dial controls which hob next time they use the stove.

In GIS this can be compared to moving the mouse forward if you want the cursor to go up on the screen. Web-mapping applications such as Google Maps have used this mapping for their panning control: moving the mouse forward scrolls the map down/south and moving the mouse backwards moves the map up/north. In this particular case users are able to learn quickly that the pan works by clicking on a location and dragging it to where they want it to go on the screen. As a result, users typically do not notice the slightly unnatural mapping required. However, it is up to the designer to ensure that wrong conclusions are not drawn from any unnatural mappings like this.

Shneiderman and Plaisant (2005) outline the **Object-Action Interface model**, which describes and explains the connection between the users' tasks and GUI. They advocate developing a hierarchy of objects and actions associated with each task, which helps to systematize the design of GUIs. In the same way that the natural world is classified into a hierarchy of kingdom, class, order, genus and species, this model decomposes the digital world into a hierarchy of actions and objects.

At the top of a computer's object hierarchy is the high-level concept that computers store information, which users generally understand. Moving down the hierarchy, stored information is organized into directories and files. In turn, each directory is composed of files or entries, each with a name, length, date of creation, owner, and so on. Likewise, each file is an object that has a lower level structure consisting of lines, fields, characters and fonts. Notice that in geospatial technologies the use of two formats for storing information can lead to confusion. At the top level, users should understand the concept of layers, and see the mapping between layers and files. However, some of the popular packages store one layer in several files, which is a cause for user errors and misunderstanding. In addition, the ability to store information as a raster or in a vector makes things more difficult. In vector file storage, the user can explore below the layer level to the objects that make the layer – and different systems have different rules about mixing points, lines and areas in one layer. In contrast, in the raster representation it is not possible to explore below the layer.

Interface actions also are decomposable into lower-level actions. The high-level plans, such as backing up a data file, may require selection, copy and save actions. The mid-level action of saving a file comprises selecting a folder and moving the file, providing a password, overwriting previous versions, naming the file, and so on.

There are also many low-level details including permissible file types or sizes, storage space, or responses to hardware or software errors. These settings, however, are infrequently used so are hidden from the main view unless the user specifically asks for the details. Finally, the users carry out each low-level action by selecting a button in a dialogue box or clicking on a pull-down menu. This can be aided by **progressive disclosure** so that the user does not see all the options at the same time. It is advisable to hide detail and complexity until the user needs it. For example, deactivate controls and don't display menu bar menus until they are relevant.

Defining the task domain for GIS software is challenging because the tasks are so wide-ranging (see Chapter 7) and the level of technical expertise required by the user to carry out the tasks is varied. Nevertheless, task centric design of GIS interfaces has many benefits for the users. Wood (1993) makes a distinction between users who are making maps and those who are using them for such tasks as:

- Map reading (cognitive model)
    - Level 1: The most basic distinction between individual symbols

- Level 2: Identification and comparison of the spatial patterns of symbol groups
- Level 3: The application of deep-structure information to decision-making and content-knowledge-building for problem-solving
- Selecting (e.g. from a menu)
- Positioning
- Orienting (e.g. a symbol)
- Generating a path (position/orientation) over time
- Quantifying (entering a value)
- Adding text

There are a range of cognitive issues that are being addressed in the field of Human-Computer Interaction for GIS that explain why establishing the task domain for GIS software is so challenging (Lloyd and Bunch, 2003 and see Chapter 2). A large number of cartographic studies have focused on the schematization of geographical information for efficient communication, and cognitive processes involved in the interaction between the map and the map reader. Such studies have included the use of colour on maps, visual search processes used in map-reading, learning processes associated with maps and graphics, and learning with maps and text. These studies have typically gathered, analyzed and compared subject responses (e.g. reaction times, per cent correct or confidence) to stimuli under various controlled experimental conditions (see Chapters 3 and 8).

9.1.1   Styles of interaction

Interfaces have evolved since the era of punched cards required to operate early computers. With the appearance of the interactive terminal in the 1960s, **command-line interfaces**, which required users to learn syntax in order to carry out tasks and actions, were common (Shneiderman and Plaisant, 2005). Command-line interfaces are efficient, precise and speedy in their response (Sharp *et al.*, 2007). They are still effective in some types of operations, and they can be very usable if they are well integrated with an ability to edit the text. For example, query languages can be very powerful and efficient in manipulating geographical datasets. They are also advantageous for the user because it gives them a strong sense of being in control, despite high error rates due to the precise nature of the syntax. This makes the syntax hard to remember and it requires a significant amount of training, restricting use of any programs requiring command-line to expert frequent users. Command-line is still found in GIS software. For example, ArcInfo is still available from ESRI and you can carry out commands via command-line from ArcMap. In MapInfo the user needs to know the Structured Query Language (SQL) to carry out advanced queries.

Command-line interfaces have been largely superseded by **Graphical User Interfaces (GUIs)**. This interaction style requires users to point at, or **directly manipulate**, visual representations of objects to carry out tasks and actions rapidly, with the results immediately visible, or an indication that the computer is processing and that the user needs to wait. The common yardstick is that the application should respond within two seconds. If this is not possible the user should be informed about the delay and the need to wait. Very sophisticated and complex command languages that emerged in the 1980s, and are still present in GIS packages, such as ArcInfo for the more advanced analyses, have given way to comparatively simple direct manipulations of visual representations of objects and actions (Sharp *et al.*, 2007). For example, actions might be represented by rubbish bins for deletion, or folder icons to represent destinations for file copying. There are syntactic aspects to direct manipulation, such as dragging the file to the rubbish bin or selecting a button in a dialogue box rather than clicking on a pull-down menu item, but this knowledge is simple compared to the level of detail needed for early **command languages**. In addition, the visibility of the items on the GUI reminds the users about the functionality.

Other interaction styles include **menu selection** and **forms**. Many people associate them with GUI, but they can be implemented on a text-based interactive terminal. Menu selection requires the user to read a list of tasks of actions, select the one most appropriate to what they want to accomplish and observe the effect. This enables the user to complete their task in very little time if the correct terminology is used and items have been grouped to make them easily accessible for novice and intermediate users. Forms, on the other hand, are useful for data entry. With this interaction style, the user must understand the field labels, know the permissible values and data entry method and how to respond to error messages. This interaction style is most appropriate for knowledgeable intermittent users or frequent users.

There has been increasing research in recent years on the usefulness of alternative interactions for GIS, such as **speech** and **gestures**. Research has focused on mobile GIS whose users are out in the field updating data in real time.

The GUI also has to fulfill the needs of the user (Nivala, 2007). For mobile GIS, for example, it is recommended that the number of buttons on the UI is restricted so that the users can use the application while wearing gloves, or holding the device with one hand. Take a look at the ArcPad console in Figure 9.3. Only users with nimble fingers are able to press

Figure 9.3 A mobile GIS application. Notice the space that is dedicated to icons and menus and the area of the map (Reproduced by permission of Trimble and ESRI)

the buttons on this interface and only users with excellent vision will be able to read the map from the small map window. In reality, the user must use a stylus and thus need both hands for operation. When the display is only 3.5 inches big, 'screen real estate' is extremely valuable. Users in the field using a Web-based application are also reliant on any edits they perform being stored in the database, regardless of any instability in the systems' Internet connectivity. It is also important that the device carried in the field is not simply a note tool, so that tasks can be finished in the field rather than back at the office.

Oviatt (1996) presented results that suggested many performance difficulties with speech-only map interaction – including increased performance errors, spontaneous irregularities, and lengthier task completion time – were evaded when people could interact multi-modally with the map (e.g. with speech and a pen-based input). These performance advantages emulated a strong user preference to interact multi-modally. The error-proneness and unacceptability of speech-only input to maps was attributed in large part to people's difficulty generating spoken descriptions of spatial location. MacEachren et al. (2005) argue that multi-modal systems are more natural and conducive to collaborative work. Furthermore, they argue that the multi-step, multi-participant nature of decision-making using GIS suits a dialogue-enabled, multi-user interface. However, more research and development is required in this area.

9.2   Basic elements of GUI

The ubiquity of the GUI usually means that developers feel familiar with the concept and its basic elements. However, it is valuable to understand the role of the different elements of the GUI and how they influence the user's interaction. The elements that we are covering here are windows, icons, menus, and pointers.

9.2.1   Window

A window is part of the screen that displays information separately from the rest of the screen. For example, when the 'My Documents' icon is double clicked in the Windows Operating System, a new window will appear on the screen to show the files that are contained in the folder. When you open a word processor, the area in which you are writing is also a window. They were designed to overcome the physical constraints imposed on the user by the screen; more than one window can be open at a time allowing the user to have more than one application open at a time to allow multitasking. A user can easily manipulate a window: it can be opened and closed by clicking on an icon or application, and it can be moved to any area by clicking in a certain part of the window – usually the title bar along the top – and keeping the pointing device's button pressed, then moving or 'dragging' the pointing device. A window can be placed in front or behind other windows, its size can be adjusted, and scrollbars can be used to navigate its different sections. Scrollbars provide an ability to view information that is larger than the area that is captured by the window. The system memory is the only limitation to the amount of windows that can be open at once. It is often necessary to have many windows open at once, which can be overwhelming for the user, but functions have been created that assist the user in finding a particular window. For example in MapInfo, by clicking on the Window menu a list of all the windows that are open is provided. Unfortunately because MapInfo doesn't automatically give the user the option of naming the windows it is often very tricky for the user to locate the window they want to work in – this is an important and rather tricky aspect of multi-window applications, as the designer needs to consider ways in which the users can identify the window in which they want to work. For example, in Apple Mac computers, by hitting the F9 key on the keyboard, all the windows or applications that are open will be made visible and spread around the screen (Sharp *et al.*, 2007). This makes it very easy for the user to find the window they need.

There are also many types of specialized windows:

- A container window – is a window that provides the limit for the application, and other sub-windows are opened with it. Some GIS are using this ability to allow users to have multiple views of the information, while being aware that all of the views are integrated within a single activity.
- A Browser Window e.g. Web browser or file browser such as 'My Computer' allows the user to move forward and backwards through a sequence of documents or Web pages.
- Text terminals are text-user interfaces within the overall graphical interface. Examples include MS-DOS and UNIX consoles. In GIS, as noted, ESRI ArcInfo is operated through a text terminal window.
- A Child Window opens automatically or as a result of a user activity in a parent window. Pop-up windows on the Web are examples of child windows.
- A Message Window, or dialogue box, is a special type of child window. These are small and basic windows that display information to the user and/or get information from the user. They usually have a button, for example OK or Cancel, that must be pushed before the program can be resumed. Notice that most users won't perceive these as windows but as dialogue boxes (which they are commonly called). They provide the illusion of conversation with the computer, and are also very useful in informing users when a long process is running, and the user is required to wait until its completion.

9.2.2   Icons

Icons are small pictures on the screen that represent objects such as a file, program, Web page, or command. When double-clicked, icons immediately execute commands, open documents, and run programs. In many operating systems all documents using the same extension will have the same icon. This makes them useful when searching for an object in a browser list, provided they have a meaningful picture on them for the user to be able to remember what it refers to the next time they want to use the icon. For example Microsoft Office has standard icons for each of the component programs, and the file icons match the icon of the programs themselves, creating a good conceptual model for the user.

Icons are a significant part of interface and are used many times in toolbars. There are significant problems with icons, and they should not be designed without due consideration. First and foremost, icons do not always convey the correct meaning to the user. The common solution for that is a 'tool tip', or a bit of information that explicitly explains the function of the icon. Secondly, it is very important to group icons by functionality so that all the icons that relate to similar actions are linked to each other visibly. Icons should also be linked to the users model of the system and their daily work life outside the system. For example, the original design for the desktop interface was carried out in an office environment in the Western world, where file binders are common, so icons for files and folders depict this artifact. In many geospatial technologies applications, the icons can be rather abstract and this can make the learning processes more difficult.

9.2.3   Menus

Menus allow the user to execute commands by selecting from a list of options, just like looking at a restaurant menu. The user's choice is selected with a mouse or other pointing device; a keyboard may also be used. Menus show what commands are available within the software. This limits the amount of documentation the user needs to read to understand the software.

A menu bar is displayed horizontally across the top of the screen and/or along the tops of some or all windows. A pull-down menu is commonly associated with this menu type. When a user clicks on a menu option the pull-down menu will appear. Each menu has a visible title within the menu bar. Its contents are only revealed when the user selects it with a pointer. The user is then able to select the items within the pull-down menu. When the user clicks elsewhere the content of the menu will disappear (Figure 9.4 shows a menu in a common GIS).

There are many different styles of menus to choose from, including flat lists, drop-down, pop-up, contextual, and expanding e.g. scrolling and cascading. With cascading menus, consider that the user can often make errors, such as overshooting or selecting the wrong options. It can be frustrating for the user also to lose the menu as they try to manoeuvre the mouse onto the second or third tier of the menu, particularly if users find controlling a mouse difficult in the first place. Thus, ideally menus should not have more than two levels. There are a few conventions developers should stick to when creating menus. You will notice that in the majority of software '. . .' is used to denote a menu item that opens up a new window and '>' to denote a menu item that cascades into a separate menu. Notice how the sub menu in Figure 9.5 appears in the opposite direction to the main menu. This make the menu difficult for the user to manipulate.

A context menu is invisible until the user performs a specific mouse action, like pressing the right mouse button. Menu extras are individual items within or at the side of a menu. A context menu can be tailored to the object that the user is manipulating, as the selection of the object will always precede the menu.

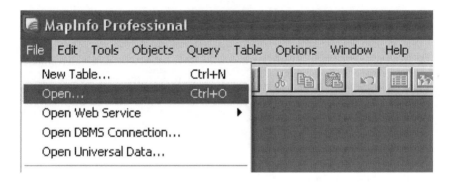

Figure 9.4   MapInfo file menu

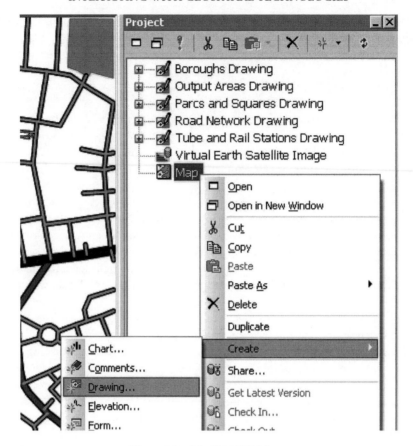

Figure 9.5   Manifold GIS Menu

9.2.4   Pointer

A pointer is a graphical image on a screen that indicates the location of a pointing device, and can be used to select and move objects or commands on the screen. It commonly appears as an arrow, but it can vary within different programs or operating systems. For example, word-processing applications use an I-beam pointer that is shaped like a capital I. Web browsers also often indicate that the pointer is over a hyperlink by turning the pointer into the shape of a hand with outstretched index finger, which informs the user what they have to do if they want to go to that page.

The changes in the icon of the pointer can be beneficial to the user; they indicate that the system is in a certain 'mode', so the next click will perform a certain function or activity. Often in GIS the graphical representation is associated with the particular function being carried out. This is useful feedback for the user. For example, in MapInfo a hand icon is used when the pan function is in use; however, when the user has pressed the 'Zoom In' button the graphical representation changes to a magnifying glass containing a + sign to indicate to the user that if they click the map will zoom in. The 'identify' operation uses the symbol for tourist information to help the user remember what that button does – display the information about an object.

When considering the relationship between the application and the user, it is important to take into account the pointing device that will be used. For geospatial technologies the main kinds of pointing devices are:

- A **mouse** moves the graphical pointer by being slid across a smooth surface. There are two main types of mouse:
  - The conventional roller-ball mouse uses a ball which, as it moves, rotates two small shafts that are set at right angles to each other. The rotation is measured by sensors within the mouse which transmit the distance and direction information to the computer. The computer then follows the movements of the mouse to move the graphical pointer on the screen.
  - The optical mouse is very similar but uses visible or infrared light instead of a roller-ball to detect the changes in position.
- A **trackball** is analogous to an upside-down mouse. It consists of a ball inside a socket containing sensors to detect rotation of the ball about two axes. As the user rolls the ball with a thumb, fingers, or palm, the mouse cursor on the screen will also move. Tracker balls are commonly used on CAD workstations, where there may be limited desk space on which to use a mouse.
- A **touchpad** or **trackpad** is often found on laptop computers. Users can move the graphical pointer on the screen by sliding their finger over the touchpad. It uses a two-layer grid of electrodes to measure finger movement and electrode strips which handle vertical and horizontal movement.
- A **pointing stick** is comparable to a touchpad, and is found on laptops embedded between the 'G', 'H', and 'B' keys. It operates by sensing the force applied by the user. The corresponding mouse buttons are commonly placed just below the spacebar. It is also found on mice and some desktop keyboards. The responsiveness of this device is slower than the mouse, so if this is the main pointer for an application, a suitable design should be used to allow the user to minimize their effort. A pointing stick is also common on mobile devices.
- A **graphics tablet** or **digitizing tablet** enables the user to enter drawings and sketches into a computer. A digitizing tablet consists of an electronic tablet and a cursor or pen. A cursor is similar to a mouse, except that it has a window with cross hairs for accurate placement, and it can have up to 16 buttons. This is typically used in GIS to trace and digitize paper maps. A pen (also called a stylus) looks like a simple ballpoint pen but uses an electronic head instead of ink. Electronics inside the trackball enable it to detect movement of the cursor or pen and translate the movements into digital signals that it sends to the computer. This is different from a mouse because each point on the tablet represents a point on the screen.
- A **touchscreen** is a device embedded into the screen of the TV Monitor, or System LCD monitor screens of laptop computers. It could consist of an invisible sensor grid of touch-sensible wires drowned in a crystal glass positioned in front of a real monitor screen, or it could consist of an infrared controller inserted into the frame surrounding the monitor screen itself. To prevent the screen from becoming dirty with fingerprints, modern touchscreens can be used in conjunction with stylus pointing devices, while those powered by infrared do not require physical touch, but just recognize the movement of hand and fingers in some minimum range distance from the real screen. Touchscreens

are generally considered the next step in human driven interfaces, particularly since the success of touchscreens in the mobile phone market.

- A **lightpen** is similar to a touch screen, but uses a light sensitive pen instead of the finger, which allows for more accurate screen input. When it touches the screen it sends a signal back to the computer containing the coordinates of the pixels at that point. It can be used to draw on the computer screen or make menu selections, and does not require a special touch screen because it can work with any CRT-based monitor. Some recent models of mobile phone use this technology.

9.3   Some guidelines for designing a GIS interface

So far in this chapter we have looked at the theory behind interface design and covered the foundations of Windows, Icons, Mouse and Pointer (WIMP), but now we need the tools for building your GIS system. The following guidelines will help you to design effective GUI application in a GIS and geospatial technology context.

9.3.1   Be consistent

One of the most important elements of a well-behaved interface is consistency. Make sure that the application has consistent sequences for actions in similar situations; identical terminology should be used in prompts, menus and help screens; and consistent colour, layout, capitalization, fonts and so on should be employed throughout. Exceptions, such as requiring confirmation of irreversible actions or no echoing of passwords, should be comprehensible and limited in number. Across GIS packages and Web-mapping websites, there is inconsistency between the symbols used on the pan/zoom/identify buttons. More importantly, there are inconsistencies between different operations – for example, zoom in might require drawing a window on the screen, while zooming out might require a single click. In other applications, the window-based zoom is separated. In the latter, the application is more consistent, as zoom in and zoom out follow the same operations. Inconsistency often results from anarchic development (as Johnson, 2000 calls it) – adding on requirements in a rush at the end of development. Detailed user requirements help to avoid such haphazard development.

Stick to conventions where you can but don't feel restricted by them. Notice that in GIS software, whether desktop or Web-based, there are conventions that can be used as a starting point. For example, it is now common to use check boxes to switch layers on and off. There is a need to separate toolbars for map manipulation and navigation, and the use of menus for higher level actions e.g. buffering.

9.3.2   Feedback

As Don Norman noted, it is critical to provide an obvious and immediate feedback for actions (Norman, 2002). This can be seen in applications such as Google Earth which highlights features that you can select as the cursor glides over them. Users often find trial and error learning frustrating and they do not like surprises, they prefer to be in control. More importantly, feedback reveals to the user what the software has done; some of the functionality in GIS packages has a permanent effect on the dataset in use and cannot be undone. While in word processing applications the undo function can be implemented fairly easily to compensate for users' errors, this is not the case with GIS, and therefore it is highly important to allow the user to know what has happened.

The time intervals users expect between actions and their completion is available in the usability literature and can be used as a guideline for the design and benchmarking. For example, users need to receive feedback within 0.1 seconds for successful hand-eye coordination (e.g. pointer movement, window movement and drawing operations) and feedback that a button has been clicked. Also 'busy' indicators should display within this time frame. Progress indicators should be displayed and most user-requested operations should be finished after 1 second e.g. opening a dialogue box. Otherwise this can have the same effect as a pause in a conversation and unsettle the user. However, in a multi-step task it is acceptable to allow the component steps to take 10 seconds, given that the user will see signs of progress and is informed about the length of the operation. In GIS software there is also a need to communicate very long operations and to provide an estimation of their length, because the processes take a very long time to complete due to their complexity and the size of the files.

Users need feedback when they click on a button, to let them know that something is happening. By the same token developers should try to avoid fake progress bars that do not relate to the state of completion of the action being carried out. It gives the user a false expectation of when they can expect the action to be completed.

9.3.3   Closure dialogue

When a long process such as a file conversion or a table join has completed, it is important to give the users feedback not only on the status of the process but also inform them clearly of its completion and its success or failure, so that they know whether they can continue. Some processes in GIS take a long time to complete, so feedback to the users to let them know the program is still running is essential.

9.3.4   Prevent errors

Make buttons big enough so that the user doesn't have to struggle to click on them with the pointer. Do not contribute to the 90% of computer errors that come from the user. Create a good conceptual model that users can easily learn. Hints include greying out menu items that are not appropriate and not allowing alphabetic characters in numeric entry fields. If the user makes an error, give clear, concise and constructive instructions for recovery.

9.3.5   Permit easy reversal of actions

It is important to make it easy for users to reverse actions they have performed and are not happy with. Conversely, make it clear to the user when changes they make will be permanent. For example in ESRI ArcMap when a user makes changes to the attribute table structure outside of the 'Edit' mode, the user is given a warning that any changes made are permanent. The inclusion of an undo button which makes it easy to reverse actions is also highly valuable. This is important as it gives users the confidence to try out tasks without worrying whether what they do will in any way impact upon a previous version of the file they are working on. When a function or task cannot be undone, the user should be warned, at least when the function is used for the first time.

9.3.6   Make the user feel in control

Effective interfaces use subtle methods to make the user feel in control of the computer, rather than the other way round. These methods are difficult to define, but through appropriate interface design it is possible to boost the confidence level of the user in the ability to use the interface. Here are some suggestions as to how this can be done with the design of GIS interfaces.

Remember that the screen belongs to the user (Johnson, 2002). The proportion of the map area in relation to the total interface can have significant implications on the effectiveness of the application. Jones *et al*. (2009) argue that zooming and panning are only necessary because the size of the map is constrained by the size of the screen. Reducing the need for zooming and panning saves the user time and frustration. Of course it is often not desirable or possible to fit the whole map on the screen. However, the other extreme is also unusable, as demonstrated in Figure 9.3. So much screen space is taken up by toolbars and controls that you can barely see the map. Contrast this approach with Google Maps and other recent Web-mapping applications where the map window is sized according to the size of the user's screen, so in general the map is the principal element of the screen.

More important than the physical characteristics and dimensions of the computer screen is its interactive role, linking the user's cognitive facilities to the map or information system. For example, zooming is likened to 'taking a closer look', but on a computer this is extended to display an enlarged part of the database (Wood, 1993). The screen display puts physical restrictions on the viewing freedom available to readers of paper maps. This provides the user with opportunities for exploration and discovery (Jones *et al*., 2009). Users of Google Earth are able to navigate the globe through a set of zoom and pan controls that are only visible when the pointer is located over them. Harrower and Sheesley's paper (2005) discussed the main types of zoom and pan controls and talked about which types are suitable for different kinds of system.

Speak to the users in their language, not in technical jargon. GIS software is overloaded with jargon which confuses users. Developers often include technical jargon in error messages so that when they are developing the software they can instantly identify what is causing the error. This, however, will not help the users to correct what they have done wrong. Talking to the users during the development process can help to establish their language and how different functions should be labelled.

9.3.7   Reduce short-term memory load

An indicator of how much the user is expected to remember in order to use a piece of software is the amount of help/support material; the longer the help section, the more information the user is expected to know, the less usable the software is. This is a very useful tip! Think now about how many times you need to consult the Help functionality in the various packages that you are using. Most GIS packages are extremely difficult for novices to learn because the interfaces make it so difficult to remember how to use them. If a command-line syntax is used, it is vital that the form, abbreviations, codes, and other information are provided in the help system and are consistent.

Make common tasks accessible and easy. Don't complicate the users' task or hide function-ality that users will want to use very often. The KISS principle is particularly important in the design of geospatial technologies interfaces: **Keep It Simple, Stupid!** Albert Einstein was one of the first proponents of this principle: 'It can scarcely be denied that the supreme goal of all theory is to make the irreducible basic elements as simple and as few as possible without having to surrender the adequate representation of a single datum of experience' (Einstein, 1934). The more complex and sophisticated the software functionality, the more likely its interface is to be random or arbitrary. This can be avoided by using some basic rules, such as ensuring that the layout should reflect the order in which people typically read documents and fill out forms. Controls that designers expect users to set or scrutinize frequently should be laid out prominently and placed where users will encounter them early at the top of the screen. Don't overwhelm the user with details and hide controls that you expect users to set rarely. Controls should be placed near any other controls or data displays they affect. However, developers should find out what users want, even if it contradicts the previous three recommendations.

A good example of a geovisualization tool that uses these principles, and how they can be achieved, is documented in Jones *et al.* (2009). The authors provide evidence through video analysis that a 'less-is-more' approach to interaction design can improve the efficiency, effectiveness and learnability of GIS. Giving the user full control over colour palettes, classifying ranges, moving around the map and trying to interpret complex statistics often distracts the users from their primary goal of gaining knowledge and insight from the data they are looking at.

9.3.8   Strive for universal usability

Cater for users at all ends of the experience and ability spectrum. Adding features for novices, such as explanations, and features for experts such as shortcuts and faster pacing, can enrich the interface design. This can include allowing users to rely on defaults where it is suitable. For example the auto setting on digital cameras is a great way for novices to use the camera without tweaking with all the settings.

Case Study 9.2: Google Earth interface

Google Earth is an example of a geobrowser or a virtual globe. It is a 3D viewer of geographical information equipped with the capabilities to rotate the globe, zoom in down to the level of a street and out to the level of the globe, fly over terrain, annotate points of interest, and hyperlink additional geospatial data (Andrienko *et al.*, 2007). Google Earth technology fuses imagery, terrain, and GIS data and delivers them to the user by means of a client-server architecture, where the application serves as a client that accesses the data viewing and navigation services from the Google Earth server over the Internet. Google Earth, however, was not designed as an analytical environment; it does not provide quantitative data analysis functionality but can be coupled with data visualization methods, as the example in Chapter 3 shows.

Google Earth is an infotainment application – it was designed to entertain and educate its users. In its basic version, the tasks that the user is expected to do with the application include searching for a location on earth, zooming to it and viewing information that is associated with it, such as video clips, photos or Wikipedia

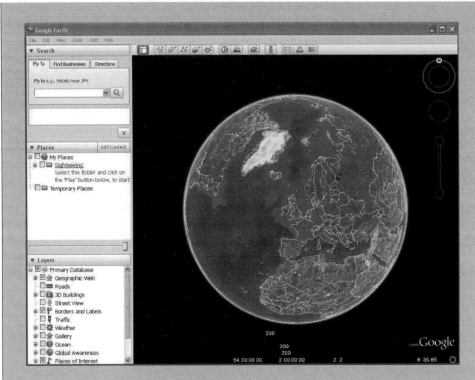

Figure 9.6 Google Earth interface

pages. At a lower level of tasks, the main actions that the user is expected to carry out are:

- searching for a location, possibly finding the directions between places;
- navigating in Google Earth for sightseeing, by panning and zooming;
- tilting and viewing hilly terrain;
- showing or hiding points of interest, as in some situations there is too much information on the screen;
- finally, limited ability of adding content, mostly by marking places.

When you examine the Google Earth interface, it is possible to recognize the decisions that were made by the designers and how they have utilized the basic elements of the interface to create a compelling and highly successful application.

The graphics are very impressive and, since they are in 3D, conform more to how they see the world than standard GIS packages, which are in 2D. The use of detailed satellite imagery is also increasing the accessibility, as research shows that for lay users, aerial photographs are more accessible than topographic maps. The zoom and pan controls match the response of the system i.e. you can zoom in at a fast speed and the image updates at the same speed, as recommended by Harrower and Sheesley (2005). This is done by progressive rendering, which means that as the user zooms, the client communicates with the Google Earth server and brings data back. The data is returned rapidly, and progressively becomes more detailed and clear. Special attention is given to the area at the centre of the screen, as this is the area that the user is focusing on.

The area that is dedicated to the globe (or map) is maximized, and occupies 68% of the screen space on a 1024 by 768 monitor – nowadays the minimal resolution of a computer monitor.

The search box is prominent at the top left area of the screen with clear tabs for 'businesses' and 'directions', which are now common across Web-mapping applications from Google, Yahoo! and Microsoft. This means that while learning the interface of Google Earth the user can employ knowledge gained from other systems.

The different aspects of information are confusing for new users, as there are 'places' and 'layers', though the differentiation between the two can be learned through exploration. The layers option uses some jargon, such as 'primary database' and 'layers'. The user can switch layers on and off through the check boxes that are presented next to each theme. By default, some of the layers are switched on and others off, giving the user a visual feedback that some things are marked as selected and some are not. This encourages exploration with the interface.

Some of the tasks such as marking a place and controlling the interface are provided through a single toolbar at the top of the window. The toolbar groups functions together – for example, adding point, area, line and video are all in one logical group.

The zoom and pan controls are outlined when the user doesn't hover over them, and only completely visible when the user has the cursor pointing to them, which increases the size of the map. However, this is one of the weakest parts of the interface, which confuses even experienced users as the same controls are used to zoom in and out as well as tilting the scene.

Summary

Without detailed conceptual design, taking the users as its primary focus, system developers may be implementing faster, better, more powerful solutions to the wrong problems (Winograd and Flores, 1986; Norman, 2002). Developers must step back and determine the motives and values of the users, and what they want to achieve. Successful user interfaces consider how the user thinks and works (Lanter and Essinger, 1991). Users do not think in terms of using algorithms, functions, stored procedures, databases or subroutines. Instead the user selects commands using buttons, menus, manipulating controls or typing.

In this chapter we have provided a review of the main elements of the interface and how they can impact geospatial technologies. We covered the main principles of interactions and how to create interfaces that convey the meaning of what they execute. We then briefly outlined the main elements of the interface that need to be taken into account in a model GUI interface, and finally, we looked at the guidelines for interface development, which the reader will find useful when designing their own GIS interface.

Revision questions

1. What are the main elements of the Graphical User Interface?
2. What should be the balance between the area of the screen dedicated to the map and other controls and elements of the interface? More importantly, why is this the case?
3. Explore the interface of a Web-mapping website and identify how the elements of the interface influence the interaction with the user. Can you identify the decisions that the designers made in terms of interaction with the user?
4. Assume that you are required to redesign the data collection device that is presented in Figure 9.3. What changes will you make and how will the new interface look?

10 Evaluation and deployment

Stephanie Larissa Marsh and Mordechai (Muki) Haklay

An important part of the User-Centred Design process is evaluation. As noted in previous chapters, there are very few universal rules that work with every interface and application. Even definitions such as 'an application should respond within two seconds to provide users with a feeling of interactivity' cannot be applied universally. For example, it is impossible in many situations to acquire the location from GPS satellites within this time limit. In other cases, calculating a complex route may require a processing time of longer than two seconds. These cases do not make the application unusable, or not interactive. Thus, each application and product will develop in their own specific way with a mix of aspects that can be handled through the use of guidelines and principles such as those that we have visited in previous chapters, combined with functionality and operations that were developed specifically for the product and to which there are no simple guidelines. Therefore, there is a need to check if the application or the product is answering the requirements that were set for it. Many times, especially in large scale GIS projects, the focus of the evaluation is on the technical aspects of the application. Questions such as the level of accuracy of data capture are tested thoroughly, while consideration of the user in the design process for geospatial technologies interfaces are ignored or sidelined. This will mean that the users of such systems are not as productive as they should be.

Thus, evaluation is a necessary part of the product design process. It provides an empirical evidence for how end-users are using the application, and, most importantly, to problems that are caused by difficult to use interfaces. In this chapter, we turn to the issue of evaluation, covering the frameworks, methods, data gathering and analysis techniques. We will mention briefly the methods that have been used with geospatial technologies, noting their advantages and disadvantages. Before turning to detailed descriptions, we need to look at the wider context of evaluation within the design process.

10.1 Evaluation options – from usability laboratory to guerrilla usability

In general, the evaluation of a product should be integrated in a reasonable way into the design process. One of the reasons for resisting the integration of User Centred Design principles into product development is a false notion that usability testing and UCD are time consuming and expensive operations. This notion of expensive studies might be based on early studies of HCI which followed many research protocols in psychology and behavioural studies. This would mean that the study was carried out in a special usability laboratory, where the participants will sit in front of the computer without anyone else in the room. A two-way mirror (which allows people to see through it) would be set in the room, to allow the researchers to observe the participants. In addition, video cameras, tape recorders and other devices would be set to record the interaction between the participants and the computerized system. This type of usability laboratory is a very powerful tool, and indeed can provide 'laboratory conditions' for testing and exploring some fundamental aspects of Human-Computer Interaction. At the same time, they are expensive to construct and maintain, and require specialized personnel to maintain and run the equipment that is used, and also require participants to travel to the lab, and thus increase the costs of an experiment as the participants might spend their time travelling to and from the laboratory. The use of such complex settings also means that if you want to include the use of a usability laboratory in your product development process, you need to consider carefully the costs of doing so and what you should do with the results.

However, this is an incorrect perception of usability evaluation and its costs within the development process. Even with the use of specialized usability laboratories, the cost of not working closely with end-users and testing the applications are much higher. Some applications are not accepted by their intended users, and therefore the whole cost of the development is going to be wasted. In other cases, major errors in design can lead to ongoing reduction in the productivity of users throughout its lifetime. A simple example of this with geospatial technologies can occur in enterprise applications that are built to be used by many users. If the application is used by 100 users, and they are expected to log into it twice a day, then a delay of 30 seconds in the startup time, which can be caused by lack of design of proper spatial indexing, will cost the organization the equivalent of two working months of a person every year. If the application is not central to the work of these users, then this time lost represents an additional cost that the organization will pay in every year that the product is used. While this example seems trivial, notice that it is very common for users of GIS in a corporate environment to wait 15 to 60 seconds every time they zoom or pan the map, and these add up to significant loss in productivity.

The development of usability engineering (Nielsen, 1994a) and cost justifying usability (Mayhew, 1999) provide further support to economic and operational justification of integrating usability within the development process. Nielsen provides an analysis of the return on investment from the integration of usability principles in the development process. His suggestion for guerrilla usability (Nielsen, 1994b), in which the methods are simplified, tests are carried out in the work place of the participants and the number of test participants is kept to a minimum, provides a blueprint for realistic integration of usability studies in development projects.

For example, Nielsen has shown that most of the issues of interface design can be identified by testing with five users, so the testing can be completed in a day. Using methods such as 'think aloud protocol' with the evaluator taking notes instead of recording, transcribing and analyzing the text can also reduce the costs and time of usability evaluation. In some contexts, especially as part of research projects where there is a need for statistically significant results, there is a need to use more participants and the use of more detailed analysis, but if the focus of the research is not on HCI, spatial cognition or other subjects that require robust experimental settings, guerrilla usability can be very effective. The most important aspect is to consider carefully which frameworks, methods and data handling techniques are the most suitable for the specific project, and are achievable within the given time and cost budgets.

10.2   Evaluation techniques

There are many existing HCI techniques for performing evaluations; each will be particularly suited to particular evaluation aims and objectives. It is likely that more than one technique may be suitable for a specific project or research question, though importantly these techniques are not mutually exclusive and combinations can be used. Each technique has its own advantages and disadvantages. This section gives a short description of the most widely used techniques available, following a comprehensive review of the literature in Human-Computer Interaction, usability, and usability in GIS and geovisualization, where geographical representations are used to reveal an underlying pattern in large and complex datasets.

A variety of different experiments are performed in the realm of HCI, for example rigorous controlled experiments (*in vitro*). Independent variables include tools, tasks, data and participant type. Dependent variables include accuracy (precision, error rate, number of correct answers), and efficiency (time taken). There is also usability testing which identifies and solves user interface problems. Other experiments include the use of metrics, heuristics, models, longitudinal and field studies (Saraiya *et al.*, 2004) all of which will be detailed below. There is also ethnographic-based research conducted 'in the field' (*in vivo*), for example in the workplace where a geospatial technology is used, observing 'real' users and 'real' use. Such experiments give invaluable insight into how tools are actually used to support a user's work, however, variables are often uncontrollable and data collection may become ad hoc.

Each technique described here and those applied in this research have advantages and limitations. Some of these will be inherent in the methodology; others will be specific advantages or problems encountered when applying the methodology to geospatial technologies. Table 10.1 shows the various methods that can be used. Frameworks are structures in which different methods can be applied. Various techniques can be used within different methods, and data analysis techniques describe some of the methods that can be applied to particular types of data recorded with certain frameworks, methods or techniques.

Each of the frameworks and methods described here can be implemented at different stages of the development process. Frameworks/methods will be more appropriate at certain stages of this process are highlighted in Figures 10.1 and 10.2. Figure 10.1 shows where the frameworks that are covered here fit within the development process, and Figure 10.2 is

Table 10.1   The most commonly used HCI methodologies

Frameworks	Methods	Data collection techniques	Data analysis techniques
Formative	Usability testing	Questionnaires	Content analysis
Summative	Field studies	Interviews/demos	ANOVA
Quick and dirty	Predictive evaluations	Focus groups	Severity rating
Longitudinal	Heuristic evaluations	Verbal protocol analysis	Problem frequency
Convergence	Cognitive walkthroughs	Onscreen capture	Performance
Case study	Co-discovery	Diary/note keeping	Subjective analysis
Remote study	Task analysis	Scenarios	Discourse analysis
Participatory design		Affinity diagramming	
		Card sorting	
		User defined tasks	
		Product defined tasks	
		Paper based prototyping	
		Eye tracking	

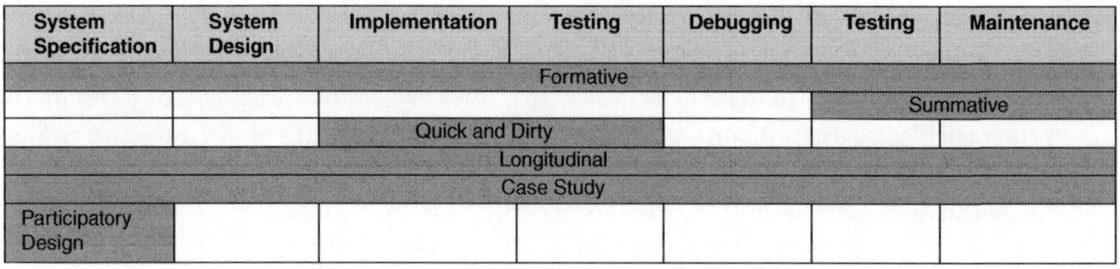

System Specification	System Design	Implementation	Testing	Debugging	Testing	Maintenance
Formative						
					Summative	
		Quick and Dirty				
Longitudinal						
Case Study						
Participatory Design						

Figure 10.1   The role of frameworks in the application development process

doing so for the methods. Each method can be implemented in a number of the frameworks, however, Figure 10.2 shows that particular methods will be more appropriate in certain frameworks, depending on the stage of software development. The data collection and analysis techniques implemented will depend on the aims and circumstances of the specific development process and evaluation.

10.2.1   Frameworks

Experimental frameworks are structures in which methods and techniques are placed to yield the appropriate data. Eight main frameworks are described here; it should be noted that these frameworks are not mutually exclusive, for example a remote study may be a formative study at the same time. The value of considering the different frameworks is to consider which one will provide the most suitable data for the problem that you are trying to solve, or provides the best answer to your research question. We are going to cover the frameworks in the order in which they appear in Table 10.1.

Formative Evaluation (formal and informal) is an observational and empirical evaluation framework that assesses user interaction through an iterative process. Any formative evaluation can be placed on a formal to informal continuum, with informal evaluation providing mostly qualitative results such as critical incidents in which the software is not functioning well, user comments, and general reactions, and formal (and extensive) evaluation producing both qualitative and quantitative results (Bowman *et al.*, 2002), throughout the development life cycle.

This type of framework, with the use of iterative design and improvement has the advantage of being able to create tools that match the user requirements closely from the beginning of the product life cycle. This would be particularly useful in geospatial technologies that are developed within research settings – for example, geovisualization tools, where development teams are often small or consist only of individuals, developers and users are often the same

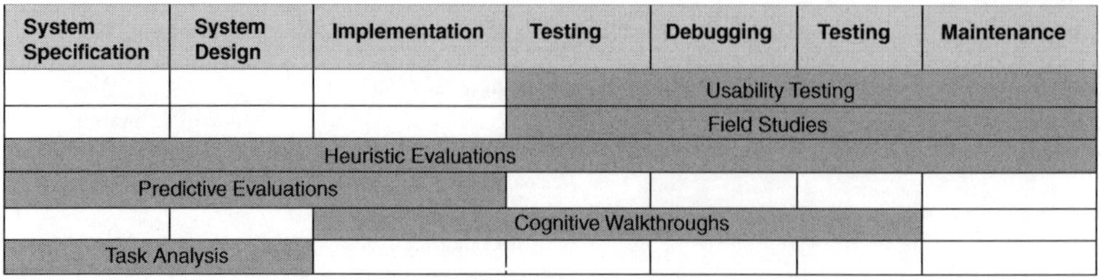

System Specification	System Design	Implementation	Testing	Debugging	Testing	Maintenance
			Usability Testing			
			Field Studies			
Heuristic Evaluations						
Predictive Evaluations						
			Cognitive Walkthroughs			
Task Analysis						

Figure 10.2   The role of methods in the application development process

person. Bugs or problems can be dealt with during product development, and there should be no major time consuming re-designs required in the final stages of the project. On the other hand, and especially in its formal incarnation, this process can be costly in terms of time, resources and monetary issues, especially and contradictorily, for small development teams.

Summative or Comparative Evaluation (formal and informal) is an evaluation and statistical comparison of two or more configurations of user interface designs, user interface components, and/or user information technologies. As with formative evaluation, evaluators collect both qualitative and quantitative data, and again can be either formally or informally applied (Bowman *et al.*, 2002). These evaluations are applied at the end of the development cycle. It should be noted that when dealing with successful websites, this framework can be very effective and successful. It is possible to create the two configurations and deliver them randomly to visitors, and therefore evaluate the performance of each alternative by checking the way that visitors perform tasks on the page.

The advantage of summative evaluations is that they are not as costly as formative evaluations and allow for the comparison of several potential designs or similar products. However, the end of the development cycle may be too late for certain usability problems to be rectified, with only major problems being dealt with, thus system usability may be sub-optimal.

Quick and Dirty Evaluations can be quick and dirty, where a designer acquires informal feedback to test if ideas are in line with user needs and requirements (Sharp *et al.*, 2007). This is not a real framework in the sense that it is one that was carefully thought through in any formal way. It is characterized by ad hoc implementation, based on whatever knowledge the developer holds.

The advantages of this method is that it can be done at any stage of development, and quick and dirty methods yield fast input rather than carefully documented findings (Sharp *et al.*, 2002). This has potential in geospatial technologies for rapid prototyping and for applications with a very small number of users. The disadvantage of this method is that the results are more subjective than more formal techniques.

Longitudinal Studies are undertaken over extended periods of time. A study by Bekker *et al.* (2004) of initial and extended use of children's computer games discovered that some initial usability problems encountered were overcome during longer-term use, though some of the problems remained throughout initial and extended use. Also, new problems were identified during longer-term use as experienced participants started to use the application in a different way. The longitudinal label can be applied to a variety of different studies. For example, a study that is carried out every day over several months is a longitudinal study. However, short studies can also be considered longitudinal, for example, a study that is carried out over two days is also longitudinal relative to usability testing, which usually lasts no more than an hour. There is no formal definition of granularity of longitudinal studies. It is possible to define two levels of longitudinal study. Longitudinal studies that take place over several days are defined as short-term longitudinal studies. Studies that occur over several years are defined as long-term longitudinal studies.

Although Bekker's study considers children's computer games, this pattern is applicable to geospatial technology and many other contexts. This is likely to be a good reflection especially of geovisualization software with many users investing much time and effort in learning how to use a new tool and building their knowledge from the information. As users become more knowledgeable about their tool of choice over time, the way they interact with it will change. Both Plaisant (2004) and Robinson *et al.* (2005) state that researchers may look at a dataset for months, examining different views and perspectives.

On the disadvantages, this kind of evaluation is expensive in terms of resources and time, not only for conducting the test, but also for processing and analyzing the data, especially for projects with one evaluator. As Plaisant (2004) states, longitudinal studies may be helpful but they are difficult to conduct. However, typically in usability evaluations, users are observed over a short period of time. These users are generally novice, still learning low-level syntax of the interface and are not able to focus on the meaning of the visualizations.

Convergent Method Paradigm is the use of multiple methods to provide converging evidence toward reaching a conclusion about the application's usability. This, in a way, is the 'all the above' option where both qualitative and quantitative methods are used and the results are integrated.

In the area of Web-based geographical information delivery, Buttenfield (1999) describes the use of convergent methods in the evaluation of the Alexandria Digital Library, allowing the assessment of the digital library and the evaluation methods themselves. Convergence allows the evaluator to have more confidence in the results gained. Another study by Marsh (2007) is also using multiple methods in the area of geovisualization. The downside to the use of multiple methods is the possibility of contradicting results for combining methods with varying advantages and disadvantages.

Case Study is an intensive study of a single unit with the aim of generalizing across a larger set of (similar) units. A 'population' is comprised of a 'sample' (studied cases), as well as unstudied cases. A sample is comprised of several 'units', and each unit is observed at discrete points in time, comprising 'cases'. A case consists of several relevant dimensions ('variables'), each of which is built upon an 'observation' or observations (Gerring, 2004).

The advantages of case studies include their exploratory nature, which makes them suitable for research applications. Case studies are most useful when propositional depth is prized over breadth and boundedness (Gerring, 2004). Such depth will be required for understanding the complexity of some geospatial technologies such as geovisualization, its usability and its support of ideation. It can also be effective in understanding the use of existing systems such as spatial decision support systems. However, although well considered case studies will be internally comparable, it can be difficult to generalize, although this is the stated aim of case studies. Much time and effort is required to produce case studies to gather information about them and carry them out.

Remote studies are situations where the user and the usability evaluator are not located in the same place. They can either be synchronous or asynchronous (Dray and Siegel, 2004). The way to implement the study is through computer mediated communication, or through methods that allow the evaluator to see the user's computer monitor – for example, through the use of add-ons to communication applications such as Skype.

Remote studies are especially helpful when the user population is small, specialized and dispersed, or difficult to get to geographically. With the high costs of traditional HCI evaluation, fewer iterations can be conducted than required. However, remote evaluations are cheaper and allow for a more intensive development cycle (McFadden *et al.*, 2002). However, in order to capture rich and dense data, a lot of equipment and software is required, which in certain situations can be problematic (McFadden *et al.*, 2002). In other cases, evaluators must rely on participants for data capture, which can vary (Castillo *et al.*, 1998). Little research has been conducted on remote usability in HCI research in general, thus how to conduct such evaluations effectively is not agreed upon. Haklay and Zafiri's (2008) screenshot study can be considered in this category.

Participatory design was discussed in some detail in Chapter 5, but one technique that is borrowed from it is the participatory design workshop in which developers and users work together to design a solution. It gives users a voice in the design process, enabling designers and developers to understand their users and provide a forum for identifying issues (Schuler and Namioka, 1993). Participatory design (PD) is an approach that focuses on the intended user of the service or product, and advocates the active involvement of users throughout the design process. User involvement is considered critical because users are the experts in the work practices supported by these technologies and because users ultimately will be the ones creating new practices in response to new technologies (Zaphiris *et al.*, 2004).

The advantage of participatory design as a framework is that this technique can be used at various points in the development life cycle. For example, it can be used at the beginning of the process to identify requirements and at various other points to assess if the product is suitable for intended users. A difficulty that may be encountered in participatory design is increased costs due to the need to explain the process to the users and changes in their understanding. Participatory design is especially effective in some areas of geospatial technology development, such as geovisualization and spatial decision support systems, it is difficult to define users as 'novice' or general users, rather than 'expert' users. Harvey (1997) provides an example of participatory design in GIS.

10.2.2   Methods

While frameworks provide the high level structures for considering the appropriate approach to the evaluation of user requirements and ensuring that these requirements have been met, methods provide a specific way of collecting and analyzing information about a situation. Methods can incorporate different data collection and data analysis techniques. Following Table 10.1, the methodologies that are suitable for geospatial technologies include the following.

Usability Testing tests user performance in terms of number of errors and task completion times. Such tests are based on a protocol for testing the interaction of the user with the system. Commonly, these studies will include a pre-test questionnaire to find out about their background, followed by a set of tasks that the user is asked to perform, and finally a post-test questionnaire or interview. User satisfaction questions and interviews are also used (Sharp *et al.*, 2007). Such testing is usually conducted in a laboratory setting, which is also known as *in vitro* testing (Dunbar and Blanchette, 2001). As users complete more tasks, they are exposed to more of the tools available in the interface and they will learn and remember where functions are placed and what they are used for. Consequently, they will perform tasks more quickly than they otherwise would. Thus, randomizing tasks is suggested to allow for the distribution of the efficiency effect over all tasks (Sauro, 2006).

The advantages of usability tests are that they are strongly controlled by the evaluator. Everything the user does is recorded and performance is quantified. This is the most common form of evaluation used. They are also useful for a wide range of purposes, from the understanding of how users use an existing system to the evaluation of new interaction techniques with research projects.

However, strongly controlled environments are not particularly ecologically valid in the sense that they often don't represent the real situation in which the system is used. Usability testing tends to focus on finding functional impediments, which is primarily what they are designed to do. However, such testing does not give any indication of task completion support, which is the aim of tool developers and analysts.

Field Studies (Participant Observation/Ethnography) is a technique for observing people by joining them in their activity, for example, working with people in their workplace to see how they use the computer software in their 'natural environment'. Techniques that can be used are asking users to fill in diaries, or shadowing the users while they are carrying out their task. Field studies are undertaken in natural settings aiming to increase understanding about users and technology. They are used to: identify opportunities for new technology, determine requirements for design, allow introduction to new technology, and evaluate technology. Field studies are also known as in vivo evaluations (Dunbar and Blanchette, 2001).

Participant observation is useful for gaining trust with the observed people and for developing detailed insights into behaviour and reasoning in a situation. This is particularly advantageous for understanding the diversity of user groups and the complex thinking and reasoning that takes place during the undertaking of tasks that rely on geographical information, such as geovisualization or spatial decision support systems. Field studies have the advantage of being more ecologically valid than in vitro, as the circumstances reflect real-world use of the system more closely.

On the other hand, field studies require a lot of organization and are expensive to run. Some situations, such as dealing with a specialized application that is used by a specific group of users who are diverse and disparate, may be very difficult to organize. It is possible that the observer will bias the behaviour of the observed. It is also possible that the data analysis will be biased if there is a single observer or analyst in the study, which is true of many data collection techniques. Plaisant (2004) also suggests that in the complex environments that are often observed in field studies, it is difficult to relate the user actions to a specific application of software, especially in settings where several tools are used.

Predictive Evaluation is a process where experts apply knowledge of what they assume will be typical users. These evaluations are often guided by heuristics, to predict usability problems, or use analytical based models (Sharp *et al.*, 2007). The experts might use a scenario – a carefully designed description of a certain way in which the system can be used by a user. The scenario can be enhanced by creating a persona who is a fictitious user, including a picture and background description, that provides the developers with a tangible user. By using several experts, it is possible to compare the subjective observations of each expert and 'triangulate' them into a common analysis.

Such evaluations are advantageous for evaluations with time and budgetary constraints, as there is no need to recruit users. The expert will be familiar with generic guidelines, so they are using wider knowledge than an average user. However, this is also the weakness of the methods – experts might fail to understand how a real user will think about the problem and they might be lacking specific domain knowledge.

Heuristic or Guidelines-Based Expert Evaluation is a method in which several usability experts separately evaluate a user interface design (probably a prototype) by applying a set of heuristics (for example, those available at Nielsen, 1993) or design guidelines that are relevant (Bowman *et al.*, 2002). Other expert review techniques include consistency inspections (Shneiderman, 1997).

Similar to the previous method, the advantage is that results from several experts are combined and ranked to prioritize iterative (re)design of each usability issue discovered (Bowman *et al.*, 2002). On the downside, the informality of this technique leaves much to the evaluator. Consequently, the evaluator's skills and expertise have a large bearing on the results (Hertzum and Jacobsen, 2001), as no representative users are involved (Bowman *et al.*, 2002).

Cognitive Walkthrough is an approach to evaluate a user interface based on stepping through common tasks that a user would perform, and evaluating the interface's ability to support each step (Bowman *et al.*, 2002). It has been recommended that cognitive walkthroughs should be performed by groups of cooperating evaluators, but the descriptions of the method maintain that it can also be performed by evaluators working individually (Hertzum and Jacobsen, 2001). This approach is intended especially to help understand the usability of a system for first-time or infrequent users; that is, for users in an exploratory learning mode (Bowman *et al.*, 2002). Cognitive walkthroughs are often considered to be an expert review technique (Bowman *et al.*, 2002 and Shneiderman, 1997) and do not include representative users. Richards and Egenhofer (1995) provide an example of a cognitive walkthrough in the context of GIS.

The advantage of such a technique is that it can be used to explore advanced tasks such as exploration and learning. However, as in the two other methods, it is potentially very difficult for HCI experts to simulate 'typical users' in cognitive walkthroughs, especially expert users who have very specialized knowledge.

Co-discovery is a method where two participants work together to perform a task. The main request from the participants is that they should articulate what they are doing with the system and how they are trying to complete the task. Talking to another person is more natural than individual think-aloud, and thus can yield more natural results (see Section 10.2.3 for further details on Verbal Protocol Analysis).

Co-discovery tests often yield more information about user thought-processes and interaction/problem solving strategies. However, it is likely to be more difficult to observe two people rather than one (Dumas and Redish, 1999).

Task Analysis is used to investigate existing situations, rather than envisioning new systems (as explained in Chapter 7), and is used to analyze the underlying rationale and purpose of what people are doing, what they are trying to achieve, and why. This establishes foundations on which new requirements and tasks can be designed (Sharp *et al.*, 2007).

The advantage of using task analysis is that it provides knowledge of the tasks that the user wishes to perform; thus it is a reference against which the value of the system functions and features can be tested. But as was noted in Chapter 7, task analysis faces difficulties when the way in which people will use the system is ill-defined (such as spatial decision support systems or geovisualization systems) and also the level of the task and the identification of the task itself can be challenging, so a decision about what constitutes a task is highly important.

Case Study 10.1: Usability evaluation of GeoCommons thematic mapping wizard

GeoCommons is a website that was developed by FortiusOne, to allow users to integrate large geographical datasets and then visualize them through a Web-based interface. The visualization allows the user to understand the patterns in their data, so the application belongs to the class of geovisualization applications. As we have seen in Chapters 3 and 8, the production of a meaningful thematic map can be a challenging task even for experienced cartographers. Because GeoCommons Maker is intended to be used by end-users who are unlikely to have experience in cartography or geographical visualization, there was a need to develop a specialized application that will help them to progress through the creation of a map, and help the users to make the correct decision.

To deal with this challenge, FortiusOne approached Axis Maps, which specializes in Web cartography and visualization. The challenge that was put to Axis Maps was to allow the user to produce a meaningful map in 5 minutes, and for that purpose they have created Map Brewer. To this end, they have designed a structured five step process (Figures 10.3 to 10.6 show the first four steps for a point data).

The process is based on a 'wizard' based approach, in which the user makes a selection at this stage through a small window that provides the options. With each selection, the map immediately changes to provide feedback to the user.

To evaluate the development, Mark Harrower from the University of Wisconsin carried out the usability evaluation. The evaluation was carried out through a mix of different techniques, many of them fairly informal (yet structured). The goal of the evaluation was to make sure that Map Brewer made sense to lay users, that users don't get lost or confused, and if they did, they could find their way back. The evaluation also looked for places the user interface was broken, i.e. didn't function properly or didn't offer enough choice (e.g. one participant was looking for a cancel button and couldn't find one). In other words, the emphasis was not on quantifiable variables like the speed of performance from start to finish, but rather a more grounded usability approach focusing on aspects such as confusion, workload, and how the application is meeting users' needs.

The evaluation was carried out over a four-month period with users of differing skill levels. During this period, three evaluation stages were carried out. The first was based on non-interactive PDF mock-ups of the user interface, which were used to ask participants to actually click, as with a mouse, on the mock up and talk about what they were doing and why. Later on, as the real application was evolving, another two rounds of evaluation used the working-but-incomplete versions of the application.

The studies were based on one-on-one testing, with a script to guide the process, taking roughly 15-30 minutes per participant and without monetary compensation. The techniques that were used include a mix of think-aloud and task-based activities (such as 'change the colour of the map'). In terms of demographics, the evaluation team recruited a broad group of users from people with little or no Web-mapping experience through to power users, researchers and developers. At each round, a sample of four to seven subjects were included, based on recent and growing evidence that 90% of the problems with a user interface will be exposed by the first three to four users.

The results of the usability evaluation led to a range of changes to the interface, for example making the text on the 'learn more' button clearer, or changing the text at the last stage from 'make my map' to 'finish' which is more familiar to users from other applications.

The results from these detailed user studies is a truly easy-to-use application that can help users to produce useful and effective maps.

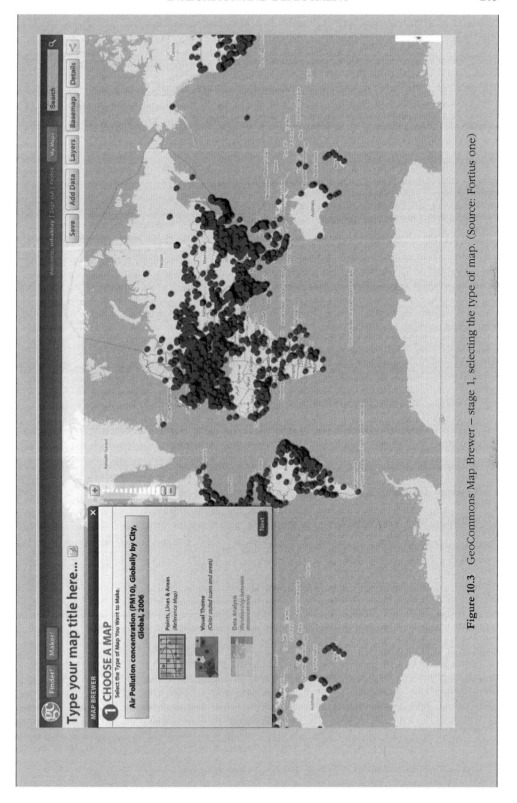

Figure 10.3 GeoCommons Map Brewer – stage 1, selecting the type of map. (Source: Fortius one)

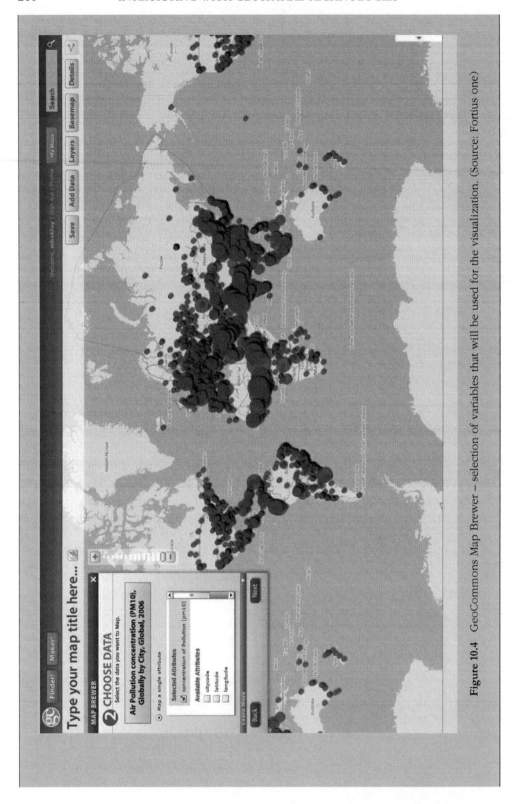

Figure 10.4 GeoCommons Map Brewer – selection of variables that will be used for the visualization. (Source: Fortius one)

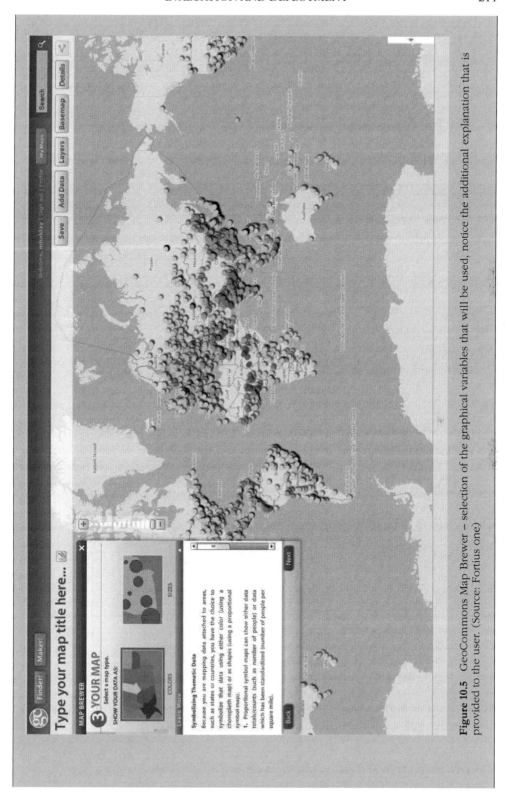

Figure 10.5 GeoCommons Map Brewer – selection of the graphical variables that will be used, notice the additional explanation that is provided to the user. (Source: Fortius one)

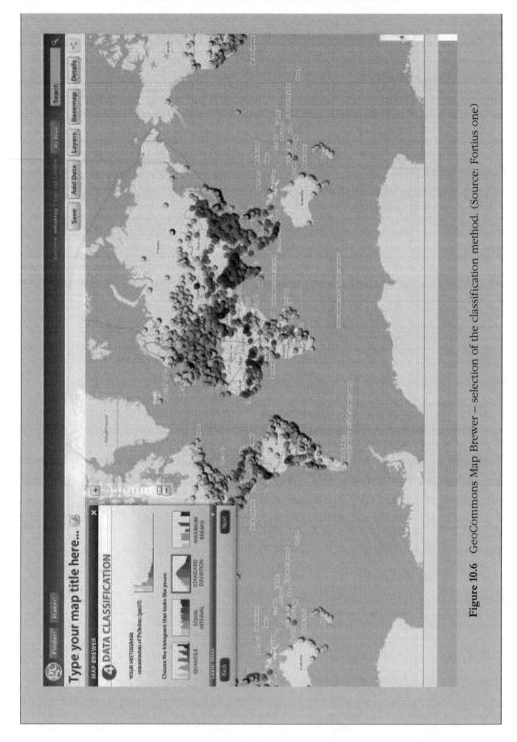

Figure 10.6 GeoCommons Map Brewer – selection of the classification method. (Source: Fortius one)

10.2.3   Data collection techniques

Once the framework and methodology for studying user needs or performances have been established, there is a need to move to detailed data collection techniques. In this section, we describe data collection techniques which are applied in HCI, and can be used in combination depending on the HCI methodology being used. The order, as in previous sections, follows Table 10.1.

Questionnaire is a written set of questions used to obtain demographic information, views and interests of users before or after they have participated in a (typically formative) usability evaluation session (Bowman *et al.*, 2002). Questions can be open-ended with users inputting free text, or closed questions, which can be administered with tick boxes. One specific example is that of Likert scaled questions where respondents are asked to rate the level at which they agree or disagree with a given statement. This can allow the evaluator to explore certain perceptions and views of the participants by using different statements.

Questionnaires are good for collecting subjective data and are often more convenient and more consistent than personal interviews (Bowman *et al.*, 2002). Questionnaires allow for qualitative and quantitative results. However, while questionnaires may be more convenient to administer than interviews, the data they yield is likely to be more superficial. In addition, special attention should be put on the exact phrasing of the questions and their order, as these can bias the answers.

Interviewing is a technique for gathering information about users by talking to them directly. Interviews are good for getting subjective reactions, opinions, and insights into how people reason about issues. 'Structured interviews' have a pre-defined set of questions and responses. 'Open-ended interviews' permit the interviewees to provide additional information; this type of interview asks broad questions without a fixed set of answers, and explores paths of questioning which may occur to the interviewer spontaneously during the interview (Bowman *et al.*, 2002; Krueger, 1994). Another technique is that of contextual inquiry, a structured technique to gather field data through interviewing (in a participatory design sense) and collecting information about the environment (the context).

An interview is an excellent technique to gather more information than a questionnaire and may go into a deeper level of detail on subjective reactions, opinions, and insights into reasoning (Bowman *et al.*, 2002), on the other hand, becoming an impartial and effective interviewer requires a lot of skill and practice. Interviews also require more time and organization than questionnaires.

Focus Groups are a way of gathering data through group participants' interaction (Krueger, 1994). They have emerged from market research and are based on gathering a group of people with similar characteristics – for example, potential users of a system with limited cartographic and geographic knowledge – and organizing a group interview. This is done through the form of a structured discussion where the facilitator presents different issues for discussion and the different participants respond. The main aspect of a focus group is that they expose social interactions and issues that emerge from inter-personal interactions.

Focus groups can be a more natural way of yielding information than individual interviews as participants ask each other questions, seek clarification, comment on what they have heard, and prompt others to reveal more (Finch and Lewis, 2003). However, there is a need to control and balance individual contributions during the discussion as dominant individuals might divert the discussion (Finch and Lewis, 2003). This may be particularly true when a number of diverse users are taking part in a focus group, including general users and expert users.

Verbal Protocol Analysis (VPA), which is also known as think-aloud protocol, evaluates cognitive processes. It analyzes what the participants think, ascertaining the internal processing conducted by the user while carrying out the task. This facilitates the understanding of human performance levels (Haniff and Baber, 2003). A video camera can be used to capture what is happening on the users' screen, their interaction and navigation of the resource, and the users' 'think aloud' comments as they interact with the system. In usability observation, a representative user is asked to do tasks using the prototype or instructional resource (Haniff and Baber, 2003). VPA can be used concurrently (CVPA) whilst users are interacting with a tool, or retrospectively (RVPA) when users describe actions and thoughts after the event. For a discussion on the use of VPA in cartographic research, see Suchan and Brewer (2000).

The advantage of VPA is that it acquires information that would otherwise be unobtainable. VPA is also an unobtrusive method for extracting cognitive and physical actions (Haniff and Baber, 2003). VPA will have the advantage in geospatial technologies of not adding to the cognitive load of an evaluation, as simultaneous interacting and talking is relatively natural to most. The disadvantage of VPA is the possible inability of the subject to express what they are thinking about and view thinking and verbalization as two separate processes (Haniff and Baber, 2003). It is also hypothesized that VPA will be difficult to sustain over long periods of time.

Onscreen Interaction Capture In addition to the use of video cameras to capture the interaction of the user with the system, it is possible to install specialized software that captures participants' interaction through recording the computer screen. This is often captured audio and video data (i.e. VPA) for the context of user. If a Web camera is installed on the screen, it is possible to capture facial expressions. The software can provide a full record of what the user was looking at on the screen and possibly keystroke and mouse clicks. For the analysis of the recording, either the video can be viewed using media player software, or through the use of specialized software such as Mangold Interact or Noldus the Observer package.

Onscreen interaction capture gives the evaluator the ability to record how the application is being interacted with, and a visual record of usability problems experienced. The recording is unobtrusive. However, it is difficult to record long sessions of software use because most of the software is writing the video to memory. Therefore, long periods of recording slows the computer processor down. This may make longitudinal evaluations difficult to capture in detail. For an example of a study that uses this method, see Jones *et al.* (2009).

Diary Keeping/Note-Taking are often used as activity diaries and require the informant to record activities they are engaged in throughout a normal day. Diaries may vary from open-ended, where the informant writes in their own words, to highly structured tick-box forms, where the respondent gives simple multiple choice or yes/no answers to questions (Maguire, 1998).

Diary keeping is useful for capturing user behaviour over a period of time and it allows data to be captured about everyday tasks, without researcher intrusion (Maguire, 1998). One example has been found of the use of note-taking to assess knowledge construction of students as reported by Lowe (1999). Lowe assessed the effectiveness of interactive animated sequences of weather maps in promoting knowledge construction among novice meteorologists (students). Lowe had his students use notes prepared while they were interacting with the maps to later predict the weather from previously unseen maps. Lowe analyzed the content of these notes to determine what information students were extracting from the visualization to create their mental schemata of meteorological processes.

There are also challenges that are associated with the use of diaries. Sillence *et al.* (2004) used logbook entries and data logging as a triangulation method for understanding selection and rejection of health websites. Participants were asked to record their perspectives of each and every website they visited. A logbook consisted of tick-box questions and space for a short description. Sites visited and the duration of the visits were recorded automatically. It was discovered that participants failed to record each and every site they looked at; often sites that were looked at and rejected in less than 2 minutes were not logged. In one session, data logging showed that 80 sites were visited, but only 34 were recorded by participants. Certain findings of this study call into question the amount and variation of detail that could be recorded from note taking (not only between participants but also within one participant's notes). Steves-Potts *et al.* (2001) used diary keeping to track tasks and actions participants were undertaking in the groupware tool they were evaluating, but the authors make no comment on the effectiveness of this technique. This technique is likely to suffer from some of the problems of task analysis such as the user omitting evidence from the description of something routine (Mills, 2005), something may be overlooked as it is considered too simple from the point of view of the user, although they are critical for the task at hand.

Scenarios are characterizations of users and their tasks in a specified context. They offer concrete representations of a user working with a product in order to achieve a particular goal. The objective of user scenarios in the early phases of development is to improve the accessibility of end-user requirements and usability goals for the design team. The scenarios can be written to describe typical use situations and the way in which the user performs daily tasks. Later scenarios can be used during design and evaluation activities. In the area of Web-based geospatial technologies Buttenfield (1999) collected scenarios of use to inform the Alexandria Digital Library Project and provides an example for their use.

Scenarios make sure that developers and designers have a specific idea of who the product is targeted at and that consideration has been given to the different types of users and how their needs and goals may be different. However, they can also suffer from omitted detail, like many other techniques such as task analysis. For example, simple but important tasks which are easily overlooked are not identified and included in a scenario.

Affinity Diagramming is a technique that can be employed in participatory design workshops (Zaphiris *et al.*, 2004). Affinity diagramming consists of placing related items together in graphic form. For example, individuals will write system requirements on post-it notes and then as a group. These requirements are then placed in related clusters, as a way of gathering requirements for a prototype. Marsh (2007) provides an example for the use of affinity diagramming in a study of the usability of geovisualization application.

Affinity diagramming is a simple but powerful technique for grouping and understanding information that is presented by participants in a discussion and providing a way to identify and analyze issues. On the other hand, if different levels of user are participating in the exercise, the end result may be greatly influenced by dominant and more accomplished users.

Card Sorting is a technique for exploring how people group items, to develop structures that maximize the probability of users being able to find these items, so it can influence the structuring of menus and general information architecture. Names of items to be categorized are put on individual cards. Participants are asked to group items in a way that makes sense to them and they may be asked to name the resulting groups. Such techniques are used for requirements gathering and guideline generation (Gaffney, 2000).

The advantage of this method is that once all participants have completed the exercise, the groupings can be examined and cluster analysis can be used for a graphical representation of the resultant groupings. However, there will be groups with consensus and groups without. The evaluator should examine the semantics of group names without consensus, and evaluate if they were to be renamed could they be included in another category? The final process of grouping is open for an evaluator bias.

User-Defined Tasks and Product-Defined Tasks User defined tasks (UDTs) are tasks that participants bring into the usability evaluations as opposed to product-supported tasks (PDTs) that make up most usability tests (Cordes, 2001). In both cases a usability evaluation study will follow these tasks and the way that they are performed with the specific system in use.

By having users bring tasks they want to perform, and believe they should be able to perform with a tool, a truer picture of usability can emerge. However, users might self-limit what task they choose, which is likely if they know the system well. This is not necessarily a problematic bias as it accurately reflects the user requirements. One stated problem of using UDTs is the variation in the quality of tasks (Cordes, 2001), but if expert users are involved, this is likely to be less of a concern.

Paper-Based Prototyping is a paper sketch of a user interface with detail of layout, function and design. These can be used to make design decisions or be used in usability inspections, focus groups, or simple user tests (Shneiderman and Plaisant, 2004). They can be produced by groups or individuals.

Paper-based prototyping has the advantage of being produced quickly and can be used to gain ideas and opinions without having the preconceptions that 'functional' prototypes face, i.e. being perceived as working software. It has been successfully used in information visualization (Zaphiris *et al.*, 2004) and in the case study of GeoCommons, above. Such prototyping may lead to a focus on layout and design, rather than the underlying processes.

Eye tracking uses specialized equipment that allows the capture of the user's gaze at the computer's screen (Figure 10.7). The tracking device is usually located below the monitor and projects infrared light to the participant's eye, and a specialized software records the size of the pupil and the movement of the gaze across the screen. Special attention is given to episodes of fixation when the user focuses on specific areas in the interface (Dix *et al.*, 2004).

Figure 10.7   Eye tracking experiment, the eye tracker is just below the computer's screen

The main advantage of eye tracking is that it provides physiological evidence of what the user was actually looking at, and if this is combined with think-aloud protocols it can provide useful insights to the part of the interface that the user responds to. On the other hand, the equipment remains fairly expensive, and it can be tricky to calibrate to a specific participant. Most eye trackers cannot deal with users who are wearing glasses.

10.2.4 Data analysis techniques

In this section we provide an overview of some of the techniques that are used to analyze HCI data. Though this is not necessarily a comprehensive list, we try to cover the most commonly cited techniques in the HCI literature. There is no simple connection between data analysis technique and each data collection technique. How the data is analyzed depends on the purpose of data collection, apart from the use of quantitative analysis for quantitative data, and qualitative analysis for qualitative data. Notice that in most cases, the use of both qualitative and quantitative data is necessary within most usability studies.

Content Analysis is suitable for qualitative data (video and audio data, for example), such as that obtained from focus groups and interviews. This data can be transcribed and then content analysis applied to elicit significant characteristics (Patton, 1990; Silverman, 2001). This involves categorizing data through coding and establishing linkages between and within categories. Analytical coding schemes are developed before coding begins (Griffin, 2004; Silverman, 2001). Organic schemes are developed whilst coding takes place (Silverman, 2001).

Analytical coding allows for a higher degree of coding consistency, which is important when a single coder is analyzing the data. Yet, it is possible that certain information will be missed if it does not fit into the existing coding schema and it is likely that the coding will be biased by the coder's perspective (Griffin, 2004). It is advisable, therefore, that at while the codes are being established, several coders are used and a coding scheme is agreed upon.

Statistical analysis and ANOVA Analysis of Variance (ANOVA) is a test of statistical significance of the differences among mean scores of two or more groups of one or more variables. For an example of its use in geovisualization usability testing, see Tobón (2005). More generally, statistical techniques can be used to evaluate the results of questionnaire and other quantitative outcomes from a study, such as the time that it takes to complete a task, or the average fixation in eye tracking studies.

The advantage of these techniques is that it can provide statistically significant results of trends found in quantitative data. However, results are not statistically significant if evaluations use small numbers of users, thus results may not be widely accepted. Another issue is that statistical analysis provides information about what is happening, but rarely provides strong enough evidence for causality.

Severity Rating is the rating of usability issues and assesses how severe the problem is, for example, from minor irritation to critical, such as system failure (Hartson *et al.*, 2001) usually on a scale of 0 to 5.

Severity rating allows the analyst to distinguish between trivial and important usability issues. It is often used to prioritize problem fixing, though there is a lot of subjectivity in the rating of severity which may differ between evaluators (Hartson *et al.*, 2001).

Problem Frequency Counting This method counts the number of times a usability problem is experienced during a usability evaluation, and therefore provides a guideline for the developers about which problems should be fixed.

It can identify which problems are being experienced most frequently, but this method does not indicate whether the problems are severe or trivial.

Performance measures are quantitative and can be taken from data recorded in usability tests, such as task completion times (measured against a predetermined benchmark), task completion success rates, how many errors are made, how many times the error is repeated, or counting keystrokes (Dumas and Redish, 1999).

Collecting and measuring these requires careful observation but no judgemental decisions, thus there is no subjectivity in the results. Some measures that do require judgement include: observations of frustration, observations of confusion, expressions of satisfaction (Dumas and Redish, 1999). However, although these measures quantify aspects of usability of the system, they cannot suggest what was going on in the user's mind. In addition, certain measures may be very difficult to apply to ill-defined tasks.

Subjective Measures can be qualitative or quantitative and measure users' responses to the system, including rating ease of learning, ease of use, or preferences, for example (Dumas and Redish, 1999).

User response is an important aspect to the uptake of tools, and captures information that cannot be gained from performance measures. However, as their name implies, they are highly subjective and require a lot of interpretation by the evaluator.

Quality assessment of usability issues is an analysis of how well specified the problems are and the accuracy of their description. Quality assessment is closely connected to the potential of fixing the problem through more accurate descriptions of problems, how well have they been reported and how thoroughly the situation that needs to be fixed has been understood can speed up the process of fixing these problems (Doubleday *et al.*, 1997).

Quality assessment (in addition to problem frequency counting) sheds light on not only the quality of the interface but also the quality of the evaluation techniques. If the definition of error is imprecise and boundaries between errors are fuzzy, one stated error could be made up of several separate errors (Doubleday *et al.*, 1997). Therefore, quality assessment allows for rectification of evaluation techniques and processes themselves. However, quality assessment itself is problematic. No standard definition or technique has been found for implementing quality assessment, thus it is difficult to generalize and compare results across studies by different evaluators.

Discourse Analysis is a method that has been borrowed from social science, where it is commonly geared towards identification of power relationships that are described within a discourse. In its usability form, it is understood as the study of the rhetorical and argumentative organization of talk and text (Silverman, 2001). Discourse analysis can be used as an alternative approach to video and audio data, focusing on the dialogue i.e. the meaning of what is said, rather than the content (Sharp *et al.*, 2007). This technique is particularly relevant to ethnographic studies.

While it holds the potential for a deep understanding of users' needs, it is highly subjective and requires a lot of interpretation by the evaluator.

10.3   Methodological consideration of usability techniques

The HCI literature includes discussions on the limitations and methodological considerations of HCI techniques. This includes literature on the theoretical methodologies of evaluation and comparative studies reporting on the effectiveness of varying methodologies in specific contexts. However, researchers are far from being in agreement on a standard means for evaluating and comparing usability evaluation methods (Hartson *et al.*, 2001). The reasons that it is difficult to reliably compare them is that there is no standard criteria for comparison; no standard definition, measures, and metrics, on which to base criteria as well as stable, standard processes for usability evaluation methods comparison.

Another aspect of usability evaluation methods and techniques, is their various levels of validity (Gray and Salzman, 1998). Gray and Salzman identify four types of validity:

- Statistical conclusions validity – Establishes real differences between groups of participants, as without it noticeable differences between participants might not be true, or real differences in performances are missed.
- Internal validity – Concerns whether these results of an experiment are causal as opposed to correlated but another variable is the real explanation of the observed results. Comparing the results of different methods is valid only if they do not favour a specific condition over another.
- Construct validity – Concerned with the definition and scope of methods used. Is the evaluation manipulated by what they claim to manipulate and are they measuring what they claim to measure?
- External validity – Concerns generalizing to particular target persons, settings and times. With few people per group, small variation in individual performance could have a large influence on the stability of measures of group difference.

The factor of setting included in internal validity is discussed more widely in the literature as 'ecological validity'. Ecological validity can be defined as the extent to which the conditions simulated in vitro reflect real life conditions. The phrase 'ecological validity' was first used by Brunswik (1955), who focused on analyses of the impact of the environment on human judgement. Hook (undated) states that in HCI, we are turning more and more towards ecological validity. Systems need to be brainstormed, developed and tested in settings that are very close to the 'real' settings they will be used in. Hook also notes that when people are brought into the laboratory, a number of unwanted consequences arise. The participants in the experiment will feel inclined to be nice to the person who designed the system being tested or who set up the study. Therefore, differences between what people say in questionnaires or interviews and how they actually behave with the system are found. What endangers the ecological validity of usability test results is the fact that the test situation removes both the surroundings and the dynamics of the activity in the real world. It may be that a usability problem detected in a usability test is not a usability problem at all if the individual is allowed to work in the natural setting. Plaisant (2004) suggests that to be reliable usability assessment needs to be in a real setting. Hook (undated) found when comparing controlled lab studies and open-ended interpretative studies, the ethnographically oriented studies were far more useful to the design process. Although such studies are not statistically significant, they are no less scientifically significant.

The ability to generalize from a study is considered by Gray and Salzman (1998) within external validity. Many evaluations also use a few low-level tasks in circumstances unrelated to actual work goals or the conditions of the workplace. New technological devices no longer fit the traditional 'one user, one computer' scenario that was once central to HCI research and practice, thus the problem of ecological validity is particularly acute in experimentation with new or emerging applications. Each situation has unique characteristics that affect the task demands and their interaction with the technology. It is difficult, if not impossible, to re-create these kinds of situations in a laboratory with sufficient validity to yield the kinds of user behaviours one may expect in 'real life' (Lindgaard *et al.*, 2005).

There are several potential ways to improve HCI evaluations. For example, in vitro-based tasks need to be simple enough to be completed in a predictable amount of time and specific enough to measure. One possibility is to allow users before or after controlled tasks to explore the application freely on their own and report findings and understanding. This could lead to informative results when subjects are motivated, but it is likely to fail with subjects of typical studies using typical subject pools, advocating the use of expertusers. Using domain experts will lead to more realistic results but individual differences between subjects should be controlled (Plaisant, 2004).

Summary

In this chapter we have reviewed the main usability evaluation frameworks, methods, data collection and data analysis techniques. As noted above, there is no method that will discover all usability problems or help in developing a perfect interface. The methods should be chosen according to their suitability to the application at hand, the budget and time available and the knowledge of the team that implements them.

Most importantly, when implementing these various tools within a usability study of geospatial technologies, take into account the specific issues that are pertinent to this class of technologies. Of particular importance is understanding the existing spatial knowledge of the expected users and those who are recruited for the evaluation. Ask specific questions that are about the geographic or cartographic nature of the information. Carefully consider how the way in which the application is used will be influenced by the fact that the output is relevant for geographic tasks. By taking these aspects into consideration, you will ensure that your usability evaluation is fit for your application.

Further reading

Dix *et al.* (2004) and Sharp *et al.* (2007) provide detailed introduction to the methodologies and the techniques that can be used for usability evaluation in general.

However, for the specific area of geospatial technologies, some of the chapters in Medyckyj-Scott and Hearnshaw (1993) *Human Factors in Geographical Information Systems* are relevant for the design and implementation of usability frameworks. Davies and Medyckyj-Scott's (1996) paper 'GIS users observed' provides an example for an *in vitro* study, of which not many have been carried out since.

Stephanie L. Marsh's (2007) PhD 'Using and Evaluating HCI Techniques in Geovisual-isation' (supervised by Dr Jason Dykes at City University, London) on which this chapter is based is a valuable resource detailing how usability evaluation can be used, especially in the context of geovisualization applications, which are some of the most cognitively complex geospatial technologies.

Revision questions

1. What are the main frameworks and methodologies for usability evaluation and at which stage of the design process can they be applied?
2. What should be the mapping between usability evaluation methodologies and data collection techniques? Try to develop the linkage between the two.
3. You are asked to design the usability evaluation for an application that will allow local citizens to find out about their nearest recycling centre. Which methods and techniques will you use and how?
4. There are now eye trackers that can be used in mobile situations and follow what the user is observing while they are on the move. How will you integrate such a device into the design process of a new Portable Navigation Device (PND)? Consider in detail at which stage of the development you will use it, which other techniques you will also use and how will you analyze the results.

11 Single user environments: desktop to mobile

Mordechai (Muki) Haklay and (Lily) Chao Li

The most common situation for the use of computers is when a single user is using a specific computer terminal to enter commands or information and receive feedback. Some devices, such as the Luminous Planning Table, which was discussed in Chapter 4, allow many people to interact with the technology simultaneously. These interactions are still very rare, and even in highly collaborative applications the interaction remains at a personal level with a device that the user uses to enter information, usually with a keyboard, and get a response, usually through a screen.

The requirements of geospatial technologies in terms of storage, processing and display mean that until the early 2000s, the stationary computer was the most suitable platform for geographical information processing. There are still many applications that require significant computing power, and there are many millions of users who use desktop GIS on a daily basis. ESRI, the biggest vendor of GIS with about a third of the market, estimated in 2005 that there were over a million users of their software (Longley *et al.*, 2005). In addition to these specialist users, there are many millions who are using bespoke products that rely on geographical information, such as dispatching applications. These two types of applications are at the centre of the discussion of this chapter. We will look at their context and some of the issues that are central to the design of usable applications.

With the growth in the use of mobile devices and the growing availability of fast wireless Internet connectivity wirelessly, there has been a rapid increase in mobile geographical applications. Towards the end of this chapter we look at these applications.

Importantly, we should not forget that the largest group of users is for the most basic applications, which are aimed at the general public and are nowadays mostly provided over the Web. These applications are the focus of Chapter 12.

11.1 Technological considerations

As with every computer system, the physical parameters of the working environment and the technical capabilities of the computing machinery have a critical influence on the usability of the application and the way that the user interacts with it. In the case of geospatial technologies, there are special demands on computing that mean many of these applications benefit significantly from fast processors, large operational memory, large storage space, and fast graphical processing. These all stem from the computational characteristics of geographical and spatial processing. The data sets are large and continue to increase in size, with files that

Interacting with Geospatial Technologies   Mordechai (Muki) Haklay
© 2010 John Wiley & Sons, Ltd

are commonly of several gigabytes in size. Because of the size of the data, and the nature of geospatial algorithms, which on many occasions rely on complex scientific calculations, they require fast processors and because of the way that these algorithms are implemented, need large operational memory (RAM). The need to render large amounts of data and provide the user with maps and visualizations that respond within a short time frame of several seconds, combined with the trend towards larger monitors, means that fast graphical processing is needed too. Thus, it is not surprising that together with Computer Aided Design (CAD) applications, GIS has remained resource hungry since the early days of the personal computer in the 1980s, and they are used as demonstrator applications for the latest hardware and software.

Yet, there are often significant disparities between a few high-end workstations that can be used by developers of geospatial technologies and some of its users, and the vast majority of computers in which it is used in an organization. This more typical configuration must be taken into account when designing an application. The careful analysis of the common hardware and software that will be used for the application must include details of the computer ability, typical screen resolution and size and other critical parameters that can influence the application that is being designed. Ideally, a typical machine should be used by the development team to ensure that performances are satisfactory for the vast majority of users.

Another aspect that should be considered in specialized applications, where specific hardware devices are necessary, such as when a system is used to capture data from aerial photographs in a photogrammetry workstation (Figure 11.1), special attention should be paid to ergonomic aspects of the work environment. Davies and Medyckyj-Scott (1996) have noted that GIS users are doing many repetitive activities. This is still true and even at the most rudimentary level GIS operation usually requires many more switches between the keyboard and the mouse, in comparison to other common applications. Thus, consideration on the way that the user manipulates the mouse and the space that is available for it should be central to the design of the physical environment of the user.

Of course, consideration of the immediate work environment of the user in isolation is not enough. Most modern geographical applications will rely on computer networks to deliver data to and from the user's computer. Before developing an application the evaluation should include the existing computer network and its capacity. While the bandwidth that is required to deliver geographical information to the end-user's computer is not as significant as streaming high quality video, it will consume significant bandwidth – more than most other office applications. The reason that the evaluation is required is that users consume geographical information by zooming in and out, with every operation triggering a query to retrieve more information over the network. Because of this, network latency can have a significant impact on the performance of the application. If remote servers are used over the Internet to provide background mapping (which is the case with OGC Compliant Web Mapping Services), then this connection should be evaluated, too.

The final aspect of the physical environment to consider is how the system will be used outside an office environment. The range of devices available is staggering, The most basic devices can be standard mobile phones, with display area of 2.7×2.7 cm (Figure 11.2), while at the other end of the scale there are laptops with screen size of up to 19 inches, and full size keyboard. Because of this wide range of hardware, the analysis of the user's context should take into account what are the physical constraints that influence the user during the use of the application. For example, while high-end laptops are very powerful and this might be necessary in some applications, they are also heavy and require some support during

Figure 11.1   A photogrammetry workstation – here consideration of the way the operator interacts with the hardware must be part of the design process

operation. Other devices that require dual handed operation, such as holding a handheld device and using a stylus for input, might be unsuitable for a situation where the operator can release only one hand. Although these considerations might seem trivial, in many cases they are ignored by designers who are not aware of the physical operating environment of the application. It is highly recommended that in mobile situations a paper prototype or similar mockup should be used to evaluate how the device itself will perform before developing the application itself.

When considering mobile devices, there are also severe limitations on bandwidth and network reliability. The design process should evaluate realistically the network characteristics during operation. The analysis must not be carried out in isolation of other applications and activities that are likely to be in use at the same time and in the same place. This is especially critical for applications that should be used in emergency situations, where disruptions to network, or very high demand for bandwidth are likely to occur.

Figure 11.2   Some devices pose severe limitations – for example, the Nokia 6230 has a screen of 2.7 cm, and resolution of 128 × 128 pixels (Data source: Google maps)

Case Study 11.1: The impact of a large monitor on user productivity in GIS

One of the often overlooked aspects of computer maps is the impact of screen resolution on the display of information. Current pixel sizes of standard computer monitors mean that the physical resolution of the information displayed on them is equivalent to about 90 pixels per inch on a desktop computer and about 120 pixels per inch on a laptop. In comparison, printed maps (such as Ordnance Survey or tourist maps) are printed at printing resolutions of about 1200 dots per inch. Although the relationship between screen pixels and printing should not be viewed and calculated simplistically, the significance of this is that a paper map can display 6 to 10 times more information per square inch than a standard computer monitor.

To make this issue less theoretical, consider Figure 11.3. This is a scan of Mini London A–Z, measuring (when open) about 153 × 180 mm. The area marked on the map shows the coverage area of Google Maps, on a 1024 × 768 monitor, measuring

183 × 245 mm. Even so, the Google Map does not include all the street names and features that can be found on the A–Z, simply because of the relatively low resolution. Both maps are at about the same practical resolution – you can read all the street names and see all the features – yet you can cram over 12 computer maps into an object that is two-thirds of the physical size of the computer monitor and still read the text clearly.

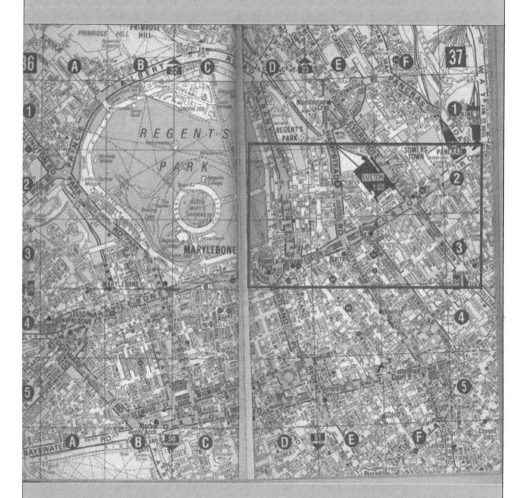

Figure 11.3 A mini A–Z for London, showing the area covered by a digital map (Data source: Geographers' A–Z map company) (a full-colour version of this figure appears in the colour plate section of this book)

With maps, the amount of information that can be viewed at one time is critical, since maps are about the context as much as about the specific location that the user is focusing on. This difference in resolution and the need to view information within the context are the reasons that some of the most common operations in any

GIS are zooming and panning. However, zooming and panning are not actually part of the user's task, which is about navigation, designing a new road or performing socio-economic analysis. Often time is spent interacting with the map rather than completing the task in hand, and all this interaction means that the information we can retain about the task in the short-term memory is reduced as data patterns are lost whilst the map takes time to render as it pans/zooms. This has a direct impact on the user's productivity.

During a snapshot study that was carried out to evaluate how users use their GIS (Haklay and Zafiri, 2008), the fact that the effective map area in GIS applications is limited came to light. In this study, Haklay and Zafiri identified cases where the screen area was a mere 25%. On average, ArcGIS users have 56% of their screen dedicated to the map. In other words, the software presents the user with an interface whereby, in the worst case, almost three-quarters of it is taken up by functions, buttons, and commands not used frequently and, because of the small map area, the user is forced to use many more pan and zoom operations.

Importantly, pan and zoom operations are not free. With each operation, the GIS retrieves data and render it: an operation that can take from a fraction of a second if the map is simple to 20 or 30 seconds in more complex datasets. According to common usability guidelines, an application is considered interactive if it responds within 1 to 2 seconds, as after this period the user's attention can be diverted and the previous operation is forgotten.

To investigate the impact of screen size and the time spent on the non-task specific interactions, we devised a simple pilot experiment that was carried out by two intermediate GIS users. The users created two simple maps of car ownership across London and its suburbs using census data. In the first experiment the users created a map of households with a car, for a town centre in the east of London, using MapInfo and a large 24-inch monitor and resolution of 1920×1200. In the second experiment the users created a map of households with two cars, for a town centre in the west of London. This time MapInfo was used but with a much smaller monitor of about 17-inch and resolution of 1024×768. Both participants conducted the experiment on the larger screen first.

For each experiment a desktop recorder was used to capture the on-screen activity and interactions of the users within the GIS. This enabled the interactions linked to zooming and panning to be coded using video analysis software. The majority of interactions occur in the later part of the experiment when the users were trying to find the correct scale to display the map. In the experiment with the large monitor, participant 1 created the map using less than 10 interactions, whereas participant 2 used 36.

By comparison, when the participants work on the small monitor, both used a different technique to create the maps. In the second experiment 71% more panning interactions were used than when the experiment was conducted on the large monitor. An increase was also observed in the amount of time the users spent zooming in and out of the map, where 67% more interactions occurred when using the smaller screen.

In total, the participants used 69% more interactions to complete their task using the smaller screen, all contributing to a reduction in productivity. Each participant stated that it was much easier to work on the larger screen, because there was room for a large map and space to have the floating toolbars. So whilst this was a pilot experiment and the number of participants prevents any detailed statistical analysis of the results, there is certainly enough initial evidence to imply that larger screens increase the efficiency and effectiveness of the GIS. The evidence from other areas of research show the same link between productivity and screen size (Nielsen, 2006), though in the case of GIS the productivity gains are more significant.

Until Usability Engineering becomes more ingrained in GIS systems development and the applications themselves can be considered more usable, there are ameliorating actions that can be taken to reduce the impact of the poorly designed interfaces. Increasing the size of the monitor on which GIS users have to work will simply enhance their work experience and increase the efficiency and effectiveness of their work. Considering that the cost of a modern, good-quality 32-inch monitor is not very high, it seems ridiculous that, with such an impact on productivity, GIS users are not equipped with these monitors as standard.

11.2   Understanding the user context

In the previous section we started to touch upon the issue of the user context, in as much as it influences the physical environment in which they operate the application. However, understanding the user context goes beyond these physical aspects, and should take into account why they are going to use the application, which other applications will be used in conjunction with the one that is being developed and other aspects of the cognitive and physical state of the user.

Most geospatial technologies applications are being developed to be used in the workplace. Therefore, the user will not have an option of selecting among a range of applications to complete a task, but will be required to use the system that the specific organization selected. Most users will find themselves using a series of organizational systems that are required to accomplish their work processes, and therefore the tasks that they need to carry out. For some users, the GIS will be central to their work – they might be using it for designing new power lines, or maybe updating existing maps. In other cases, the user's main work is to handle land ownership records, and within this work process there is an element of capturing the location of the property on the map, although most of the work is not linked to mapping and use of geospatial technologies. Finally, there are users who just need to use a system infrequently, maybe a few times during the week to ascertain the location of some addresses in their database. The level of expertise that can be expected from the user is different in every case. When a user uses an application for most of his working time, as the main tool of work, the application should provide scope for a user to become an expert in its operation, including providing short-cuts, end-user programming, i.e. writing short programs or scripts to automate repetitive activities and potentially some form of command line interface. Users who will spend a lot of time using a specific application are also more likely to be able to customize their physical environment to suit their specific needs. Finally, while not ideal, the application can rely on using non-conventional interaction metaphors if they significantly improve productivity and efficiency. For users who are using an application infrequently, the application developer needs to take into consideration not just the reduced capacity for training and the need to include cues that will remind the user about the functionality of the application, but also the wider context of the user's task. The most important rule for the design of infrequent application may seem obvious – 'most of the time, the user is NOT using your application'. The reason that this is a very powerful rule is to remind the designer that because of this, the application must be based on interaction practices that the user uses most of the time. For example, if for most of the other applications in use the user is expected to rely on keyboard entry of information through forms, then the geographical application should ensure that the 'tab' button operates properly and allows the user to move between fields, and the amount of reliance on the mouse for operations should not be as high as in full GIS applications. Finally, if the user is using the application only occasionally, special attention should be paid to the ability to run it with minimal training, if any, and to provide the information quickly.

In all these work cases, the designers need to take into account the most typical tasks in which their system will be put into use. Naturally, developers like to deal with challenging aspects of geospatial technologies, such as clever optimization of algorithms for routing. However, as Davies (1998, 1995) found in the early 1990s, the vast majority of tasks that are being carried out with GIS are related to producing maps by combining data sources and with little analysis and manipulation of the data. Interviews with practitioners and anecdotal evidence show that this is still the case. Thus, in a GIS application that is being developed for the workplace, the most significant productivity gain will come from ensuring that the most basic process of map production is efficient and effective – like the structured process that was described in Case Study 10.1, within the GeoCommons application.

Another type of geospatial technology is applications that are designed to be used for leisure activity, in games or for educational purposes. In these cases considerations of the user context should include where the application will be used – for example, using an application in the classroom should allow the teacher to maintain control over the class and structure the progress of the students, compared to using an application at home, where the students are working by themselves. Other issues that are required for this application is to consider the balance between entertainment and learning, complexity of interface (which in some games provides additional enjoyment to the players) and the way in which the information will be updated and provided to the user. Generally speaking, this area of application is the one that is experiencing the fastest transition to the Web, so further discussion is provided in Chapter 12.

The final issue that requires attention when analyzing the context in which the application will be used is the needs of disabled users, as well as older users. While the concept of universal usability which Shneiderman developed (Shneiderman and Plaisant, 2005) encompasses many aspects, the issues that are linked to human ability and limitations are worth special attention. Of course, geospatial technologies can empower disabled users in their interaction with the world (Golledge, 2005), but the more general design of systems usually ignores disabled people. For example, many designers ignore accessibility guidelines, and the following statement (taken from a UK local government website) is typical: 'Unfortunately, due to the graphics-intensive nature of GIS, this information cannot all be presented in an accessible format'. This is, however, untrue. If the application designer had considered the user's task, they would probably have discovered that most of the tasks can be implemented and provided in a way that enables all users to benefit from the system. As we have seen in Chapter 3, taking into consideration colour blind users in map design can help in making maps more accessible and useful for more users – for example, elderly users who have difficulties in differentiating between colours. Thus, consideration of accessibility and universal usability can improve the performance of an application to all users, and should be included in the design process from the start and not as an afterthought.

11.3   Designing desktop applications

11.3.1   Maps are not the only way to represent geographic information

The first, and most important consideration for any geospatial technology is the use of maps as the main interaction and output device. As noted in Chapter 2, it is wrong to assume that map reading skills are universal and that maps are readable by all. The process of learning how to read a map and relate it to reality is not a trivial one, and requires a process of learning (Lobben, 2004). It is therefore important not to confuse aspects of spatial cognition, which are universal, with map reading, which is not. In many cases, the application can provide an excellent service to the user without showing any map. Even when a map is the appropriate output, it is highly important that it contains enough information that will allow the users to orientate themselves quickly and easily such as a clear mark of the place that was queried by the user, ensuring that thematic maps are using base maps, labels and landmarks that are familiar to the user etc.

Having decided that maps are the appropriate output, the issue of using maps as an input metaphor should be examined carefully. For infrequent and occasional users the manipulation of graphic objects on the computer screen is not necessarily familiar or natural. While the increased use of digital cameras means that more people are familiar with some level of graphical manipulation, for the vast majority of users operations such as selection of several objects on the screen may present a challenge, as they are more used to operating computers by typing search phrases, selection from drop-down boxes or by using check boxes. Thus, the map should be used as the input device only when it is necessary and appropriate.

For many users and applications, the appropriate output and input will be through maps. This is the case with most GIS applications, and the user is expected to learn how to control the interface and manipulate both geometrical shapes and attribute information. As noted in the previous section, the main tasks that these applications are used for are to produce maps, and this task should receive special attention during the design phase. Of course, GIS is not used solely for the production of maps. As Davies (1995) found, editing, updating and integrating data is also a central operation within everyday use of GIS, as well as basic querying of the information that is presented on the screen. Analysis functions are far less frequently used, so even within high-end applications analysis should be treated as an infrequent part of the application and include clear cues to assist the users when they require them.

11.3.2   Geographic Information usability

The geospatial technologies market is divided, for many good reasons, into data producers who specialize in the creation of datasets such as road networks or aerial imagery, and software vendors, who are developing the software that facilitates the manipulation of these data sets. What is usually forgotten is that from the user's perspective, the application is a unity of information and software. Unlike word processing, where the application is purchased from a vendor and all input can be provided by the user, in geospatial technologies the starting point for any task is an integration of data sets and no activity can start before a base map and other layers are uploaded to the application. To date, very little research has been carried out on the usability of geographical information, though consideration of the user needs by data producers are starting to emerge (Davies *et al.*, 2005). While vendors of generic GIS packages can try to make their software applicable to a wide range of data sources of different types, in any given application – especially within an organization – special attention should be given to the usability of the information itself. This should go beyond making sure that the metadata fields, which provide details on data sets, are filled. The naming of layers, level of granularity of base map data and even the conventions for naming columns in a table can have a significant impact on the productivity of the users. For example, consideration of delivery of the base map as pre-rendered image files or as a vector layer will have an impact on each user in terms of the rendering time of the application for every zoom in, zoom out and pan.

Rendering and the speed of data retrieval should be considered and tested during the application design. Even if this means that a special effort is required to create synthetic data to allow for testing, this is an aspect that should receive attention during software development and testing. Today, most GIS applications just assume that the user will wait until the application retrieves and renders the information that should be displayed on the screen. Almost no application is truly interactive and provides a responsive application to the user within two seconds of an operation. Applications such as Google Earth (see Case Study 9.2) has shown that it is possible to provide intelligent progressive rendering that starts from the centre of the screen with a rough resolution which progressively improves. This enables the user to start with the next operation – be it pan or further zoom – without waiting for all the data to be downloaded. Of course, rendering vector data is a much more challenging task, but the current situation in which users find themselves waiting for sometimes over a minute after a simple zoom or pan operation is clearly unsatisfactory.

11.3.3   Interface design and metaphors

When geospatial technologies rely on maps as the main interface, it is critical to consider how the screen resources are allocated between the elements of the interface such as the menu, toolbars and other elements of the interface such as the 'table of content' that informs the user which data sets are currently open. Because GIS applications include many thousands of commands and deal with both geometric and attribute information, there are always multiple demands on the interface, and the designer needs to carefully use the space in a way that will allow the user to work productively with the specific task at hand. Figure 11.4 is an illustration of these challenges, and was taken from a snapshot study (Haklay and Zafiri, 2008). The active area of the application (the map) is surrounded by a toolbar, large table of content and other interface elements. As a result the active area is rather small and will, by necessity, require the user to zoom in and out to perform the map design task.

Current eye tracking studies show that people scan the interface from the top left to the bottom right. On Web pages, an F-shaped pattern was identified with a detailed scan of the top from left to right, and then followed by looking down and making a smaller scan to the right as the gaze moves down. This means that the most important information should be located towards the left, and it is valuable to put more frequent operations on the left, so the user can locate them quickly.

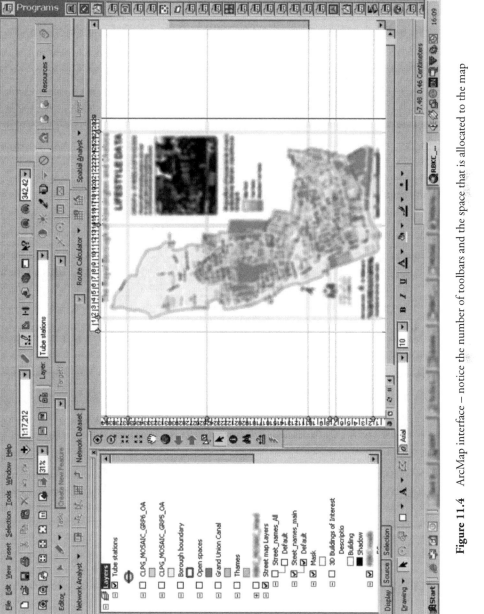

Figure 11.4 ArcMap interface – notice the number of toolbars and the space that is allocated to the map

Figure 11.4 shows another challenge in geospatial technologies – the number of buttons and options on the interface. This is a result of the many operations that are available to the user and the need to expose them on the interface. When there are many operations, sometimes the designers try to manage them by splitting each set of operations to a specific toolbar. However, the proliferation of toolbars can confuse the users who might feel that unless they dock the toolbar to the visible area of the interface, instead of switching them on and off, they will have to make a significant effort to identify it again. By doing so, they reduce the effective area of the interface and reduce the overall productivity of the application. In such cases, a command line interface might be more effective. The command line interface should be accompanied with a well designed, consistent and meaningful command structure. It should also allow the users to write a short set of instructions, or scripts to automate some routine operations (we will discuss end-user programming later in the chapter). AutoCAD provides an example for a well integrated and structured command line interface which is included in a graphical environment (Figure 11.5).

Some operations can be implemented in both graphical interfaces and command line form. For example, Bruns and Egenhofer (1997) provide a detailed comparison of map algebra operation in a GIS. As the example that we have seen in Chapter 7 shows (Figure 7.1) some models can include many data sets and complex interactions. In such cases there is a special need to consider how the user can adjust the interaction and edit it in a way that supports the user – so if a data set is used in several places in the model, and the user changes the data set in one part of the model, the user should be alerted to other occurrences of the same data set. All the interfaces that are reviewed by Bruns and Egenhofer can support this type of editing and updating of models.

Graphical user interfaces and command line or text-based interactions are also relevant to another common task in geospatial technologies – querying. The most basic queries should be supported by direct manipulation – the process of pointing on an object that is visible on the screen and executing some action with the mouse, such as double click, or possibly pressing a key on the keyboard. In some interfaces, the information about the object can appear in a separate part of the interface when the user clicks on it. However, this type of query is very basic, and in most cases there is a need to support the user in selecting a group of objects or selecting objects on the map on the basis of criteria that applies to their attributes. To this end, the application should support the user in issuing and executing queries. The popularity and likelihood of familiarity of the Structured Query Language (SQL) make it a natural candidate for this role. However, special attention should be paid to the integration of spatial extensions into the language, and the support mechanism to assist the users in constructing a query – for example, by highlighting reserved words and providing lists of fields and tables. For less advanced users, some visual support for query construction can be helpful. This is especially true if the application is implemented within an organization that is using applications such as SAP Business Objects, where certain ways of constructing a query and executing it are familiar to the user. Within stand alone geographic applications, it is possible to implement effective mechanisms for running visual queries such as the one proposed by Guo (2003), and even to implement the whole query through graphic interaction, as proposed by Egenhofer (1997).

As noted, in graphical user interfaces the topic of direct manipulation, which was noted in Chapter 9, is quite central to the way the user interacts with the application. In direct manipulation, the user points to visual representations of the object that they want to interact with, and by using the mouse and moving the object they can perform a task. 'Drag and Drop', for example, is a common approach across graphical user interfaces, in which the

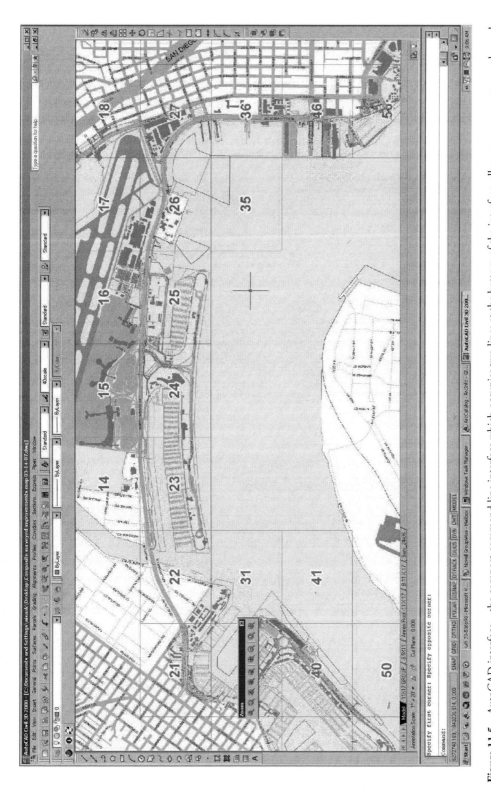

Figure 11.5 AutoCAD interface – the compact command line interface which occupies two lines at the bottom of the interface, allows power users to use the package effectively (Image courtesy of Ari Isaak, Unified Port of San Diego)

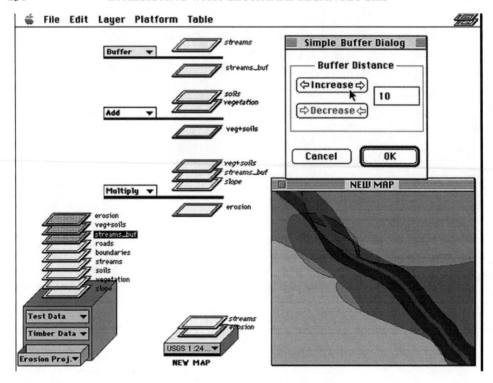

Figure 11.6 The Geographer's Desktop (Image courtesy of Max Egenhofer) (a full-colour version of this figure appears in the colour plate section of this book)

user points at an object, clicks on the mouse and keeps the button pressed, then moves the mouse and therefore the object to a target area on the screen, and when the mouse button is released, an operation is executed. The most familiar action of this type is to drag a file to the trash bin to delete it. In the area of geospatial technologies, Richards and Egenhofer (1995) demonstrated how direct manipulation can be an effective metaphor for organizing layers in a GIS and even for executing map algebra operations. Their application, The Geographer's Desktop (Figure 11.6), allowed users to view data layers as maps that are taken from the mapping cabinet, and visually organize map algebra operations.

In the area of geovisualization, in which the application is supporting the process of knowledge construction and ideation, direct manipulation is especially important (MacEachren and Kraak, 2001). This is because of its ability to imitate 'natural' operations which allows the user to concentrate on the more cognitively demanding tasks.

One interesting aspect that explains some difficulties of users with conventional GIS packages, and also with some other geospatial technologies such as hand held data capture, relates to modes. As Johnson *et al.* (1989) noted, modes – the situation in which the software is under specific constraints that allow just specific operations and interactions – while they may be helpful to novice users, are generally adding to the cognitive load of the user and should be used sparingly. In contrast, most GIS packages do include many mode-based situations, in which the user presses a button and the following interactions with the package are dictated by this mode until the user actively changes it. For example, most packages will have zoom in, zoom out and pan buttons, and if the user clicks on zoom in, every mouse click on the map is interpreted as another request for zoom. Because the design of most other packages is strongly geared towards avoidance of modes, this aspect of geospatial technologies can be very confusing to new users.

An example that clarifies the difference between the use of modes in GIS and other software packages is the use of noun-verb command syntax in general computing, compared to verb-noun command syntax in GIS. A noun-verb command syntax means that first the objects on which the operation will be carried out are selected, and then the user selects from the set of operations that are relevant to the selection (Johnson *et al.*, 1989; Buxton, 1993). For example, in word processors, the user first selects a block of text and then can decide if to delete it, move it to another area of the document, change the font or italicize it. If the user decided not to do any operation, than she can move to the next selection or operation without any implications. The noun in this case is the text that was selected, and the verb is the set of commands that are available to the user. In contrast, most GIS packages follow the opposite syntax. First, the user is required to select a button or operation, and then select the objects or layers on which the operation will be carried out. When it comes to simple operations such as zoom, pan and selection of objects, the use of the verb-noun structure adds to the cognitive load of the user during interaction with the system, especially when every other system is using the opposite syntax.

11.3.4 Advanced applications

In many advanced applications of geographical information handling, the user is required to carry out a repetitive task that is made of several steps. For example, when editing a geographical data set, it is common to select a set of objects and change one of their attributes, such as updating the road type from 'generic' to 'residential'. When this operation is done interactively, the user needs to repetitively select an object, open its attribute information, change the type and save the object with the changes.

To support the handling of such tasks, applications usually include some support for end-user programming. The ability to adapt an application to user needs depends on two main aspects – the characteristics of the language that is being used and the level of detail in which the objects in the GIS are exposed through the programming interface. The language that is being used can be a specially designed language, an integration of an existing language or a graphic language that does not require any written code. The latter is especially suitable for novice users who want to concatenate a series of actions and has been proven useful for GIS application by Traynor and Williams (1997). ArcGIS model builder (Figure 7.1) is another example of visual programming. Of the two other options, the creation of a consistent and specific language for the GIS application is a challenging task. First, there is a need to ensure that it is a consistent and useful computer language, which is not a trivial task. Secondly, from the user's perspective, there is a need to spend time and effort learning the language and how to perform operations with it. MapInfo Professional's MapBasic is an example of such a language. Although it is based on Basic, and it is fairly easy to learn, its ability to express complex structures and actions is limited and the user is required to consider work-around solutions in order to complete a task – in other words, they are required to focus on the task of programming and not on what they are trying to accomplish. Yet, MapBasic does have a very useful feature that assists users in learning how to program it. In Mapinfo Professional, the user can choose to open a code window, in which the system is generating the code from interface driven operations such as queries, opening and closing files etc (see Figure 11.7).

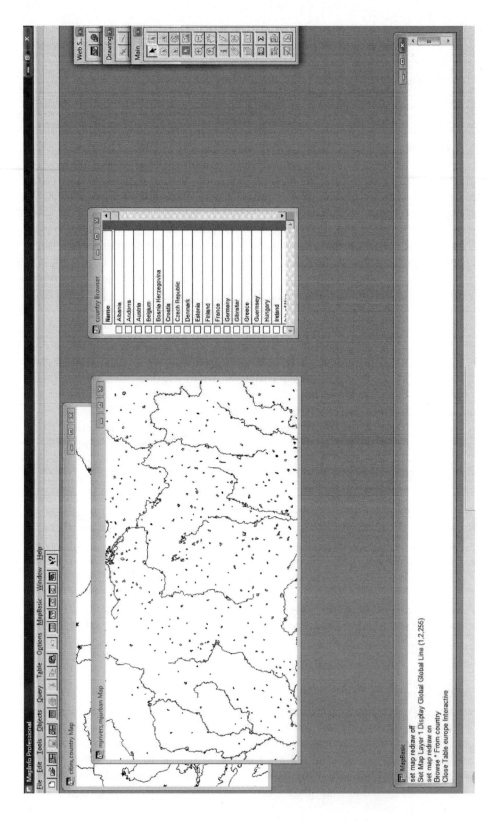

Figure 11.7 Mapinfo Professional – notice the code window at the bottom of the screen (Data source: DCW)

Because many of the repetitive operations can be executed by using the menus and toolbars, the ability to produce the code, and then copy it to the MapBasic environment in order to adapt it is assisting the user to learn this complex task.

The final option is to rely on existing computer languages, ideally popular ones, and integrating them into the application. At the level of the end-user, ArcGIS integrates Visual Basic for Application, Python and VBScript. Another example for handling programming by end-users is provided by Manifold GIS. In this case, the application is open to any ActiveX scripting language, which includes VBScript, JavaScript, Python, Perl and several other languages. The application also includes some support for programming by including some features of a software development environment such as keyword highlighting and a debugger. In both packages, the use of programming languages that are popular and might be familiar to users from other activities, such as constructing a Web page, reduces the learning effort of the user.

The second element that was noted with respect to end-user programming is the level of exposure of the inner software structure to end-users. While there will be always developers who would like to be able to manipulate every aspect of the software, exposing the complete inner structure of the software might cause difficulties to novice or occasional programmers. As was noted in Chapter 7, most users' conceptual models of the software are very different from the technical architecture of the specific implementation. From the users' perspective, they are likely to want to manipulate a selected geometry, or a layer. Thus, the objects that are most likely to be manipulated by end-users should be provided through the programming interface in a way that is easy to understand and handle.

Finally, there is a need to pay special attention to some classes of applications where the use of the system is more complex both at the cognitive and the operational level. These types of applications include geovisualization software and Spatial Decision Support Systems (SDSS). In geovisualization applications, a large amount of data is represented through different visual representations. The aim of geovisualization is to allow the user to explore the data, generate ideas and hypotheses, construct knowledge and consider solutions to problems through the use of interactive geographical and non-geographical representations which are synchronized (Koua and Kraak, 2004). Koua and Kraak (2004) have proposed a usability framework for such systems, and it is worth highlighting some of their recommendations. They notice that the requirements from the application include accuracy of results, flexibility in use, adaptability, support of exploratory activities, ability to provide multiple views of the same data and detection of irregularities and knowledge discovery. In terms of task analysis, there are three key exploratory tasks – categorizing and classifying the data, comparing different groups of data for commonalities and differences, and reflections – evaluation, integration and generalization of the knowledge that has been constructed. In terms of evaluating geovisualization applications, Marsh (2007) provides a detailed analysis of the different methods and their efficacy for geovisualization applications. She has noted that because of the focus on knowledge construction and ideation, the questions that are asked and the data that is collected through such a study should be constructed specifically for each application.

In most cases, SDSS are used in a group decision making setting (Jankowski and Nyerges, 2001), and have been discussed in Chapter 4. In terms of the application design, they have some commonalities with geovisualization, as they should allow users to explore different options in the process of making a decision. The difference is that the system is geared towards problem solving more than towards data exploration and hypothesis construction. Densham and Armstrong (1995) provide a detailed description of the design process of an

SDSS for location analysis. Another aspect that is typical of SDSS is the use of optimization algorithms such as multi criteria evaluations or route finding. Some of these algorithms are quite computationally intensive and the consideration of the level of interactivity and the speed in which the user can receive the output for a scenario should be central to the design process. In some cases, it is valuable to provide simplified algorithms at the early stages, to ensure fast computations and response time, with more detailed computations provided at later stages when the number of options is reduced.

Another major difference between geovisualization applications and SDSS is the level of knowledge that is expected from the user. Many geovisualization applications are designed to be used by domain experts (Marsh, 2007) and therefore the designer can assume a high level of familiarity with the data and the meaning of the information. In SDSS, the reverse is true. The user is expected to learn about the problem and select between possible solutions without significant knowledge about the specific domain, or to deal with multiple domains of knowledge in planning decisions. This aspect means that providing hints and information about the specific issues that are being evaluated and constructing an information architecture that promotes learning in a way that will support the decision-making tasks is necessary.

11.4   Mobile devices

With the wide availability and fast development of mobile devices, growing attention is paid to human-computer interaction aspects of mobile devices. This field of research is often referred to as mobile human-computer interaction (mobile HCI). With an increasing number of applications using mobile devices (e.g. location-based services), mobile HCI research gives more emphasis to human-device interaction in terms of developing mobile context-aware applications (Borntrager et al., 2003, Brimicombe and Li, 2009).

The interaction between users and mobile devices, in general, can be regarded as a new kind of human-computer interaction because of the differences in the physical and psychological contexts. The conventional desktop-based human-computer interaction has traditionally focused less on the surrounding environment of the user. This is because the environment in such systems is either so static that it is not significant or it cannot be locally measured. For mobile human-computer interaction, the surrounding environment has started to be brought into consideration, for example by directly observing the phenomena and people (Oulasvirta et al., 2003; Paay, 2003; Kaasinen, 2003). However, in the early phases of mobile HCI research, the majority of the research and approaches have tended mainly to focus on the interaction between human and mobile devices whereby the dynamics of the surrounding environment do not have a significant role. In addition, less emphasis is placed on addressing the question of what is useful from a user perspective within a dynamic environment. As the review carried out by Kjeldskov and Graham (2003) indicated, the main emphasis in mobile HCI research methods is towards building systems and tends not to focus on understanding design and usage, which is limiting the knowledge development in this area. Approaches have tended to be laboratory-based, and whilst interaction between human and mobile devices can be studied in detail, the surrounding environment has not had a significant role in such experiments. In mobile HCI, the surrounding environment, particularly geographical location, should be regarded as a mutable information source with which people are interacting. Therefore, an explicit focus has to be put on the dynamic interactions between individuals, mobile devices and environments (Li, 2006; Li and Longley, 2006; Li and Willis, 2006).

One of the main goals in mobile HCI is in developing mobile context-aware applications (Bornträger *et al.*, 2003). In order to provide timely services to users in supporting their activities in mobile situations, context-awareness of users and their environment is important (Beale and Lonsdale, 2004). Mobile HCI is strongly influenced by and linked to relevant contexts. Context-awareness can be used to improve system design, identify relevant content, enhance communication and deliver services (Brimicombe and Li, 2009). Therefore, understanding context is central to mobile HCI. Important context information can include user location and the surrounding environment, user situations during activities, and the technologies involved (devices, networks and systems). Context can be linked to a specific place and a specific time (spatio-temporal), and can also be related to user characteristics, personal preferences and behaviours.

Research into context awareness, in general, focuses on location, mobility and time (Markopoulos *et al.*, 2007). Location can sometimes be a position on the face of the Earth without any unique meaning, a surrounding environment or a meaningful place such as home or office. Mobility can reflect users in situations where they are continuously on the move or users at a certain place. Time can add another dimension to context, either as a momentary instant or as a fixed or floating period; time can be absolute or relative. Context can also relate to user ability and preferences, and to the nature of activities being undertaken. Furthermore, the mobile technologies being used contribute context as well. Broadly speaking, using mobile devices (and mobile computing in general) can benefit from awareness of context in two ways: firstly, through adaptation of content to changes in the environment; and, secondly, through improvement of the interaction with users. Context information provides a means to filter the flow of information from service provider to user, can help address the problem of information overload and can be used to give additional meaning to the information request made by the user. Three main strands of context in developing mobile context-aware applications are discussed in detail by Brimicombe and Li (2009) and are briefly summarized here:

- Environment, as context, can be viewed as comprising the physical, socio-economic and cultural aspects of our surroundings. The environment context should also consider the context(s) of the various situations encountered by users.
- Context related to individuals (as users) can have a direct influence on how data and information services might be requested, accessed and used via a mobile device. User context reflects personal differences, preferences and user situations such as emotional state and physiological condition.
- Technology context concerns the technological elements involved in mobile HCI. Such context can relate to the nature of mobile devices, wireless telecommunication networks, and positioning technologies.

Furthermore, context in mobile HCI should be understood from a dynamic perspective (Li and Willis, 2006). Dourish (2004), for example, has argued that context be defined as a dynamic interaction issue, not just a presentational issue. The diagram shown in Figure 11.8 demonstrates such dynamic characteristics of context that should be considered in mobile HCI. As an example illustrating the dynamic nature of context in applications, Brimicombe and Li (2006) use a mobile space-time envelope that is constructed based on an awareness of the changing context of user location, direction of movement and speed of travel, and then adapts its shape accordingly to provide more immediate and geographically more wide ranging (and yet pertinent) information than just the proposed itinerary itself. Where a user

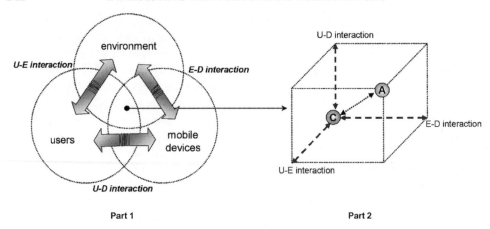

Part 1   Part 2

Figure 11.8 An interaction concept of dynamic contexts (from Brimicombe and Li, 2009). C symbolizes situations with greatest certainty; A symbolizes those with greatest ambiguity.

is stationary, the shape of the envelope can be clipped into a circular zone, implying that information and queries could relate to any direction of movement and pushed alerts for any event or location within this zone would be welcome. Where the user is moving, the envelope becomes proportionally elongated with velocity ahead of the user reflecting that there will be greater interest in what is ahead and less willingness to stop and turn around. Thus context in mobile HCI is relevant to particular settings (technological and environmental), the actions that take place and the parties involved in the action along with their particular goals and preferences. This is a much broader focus and imposes more complex demands than for many other types of IT services.

Summary

In this chapter we have looked at single user environment – applications where a user is interacting with a geospatial technology to complete a task. We have noted that many of these applications are based on a desktop computer environments and within an organizational environments. When designing an application for such conditions the physical, technical and organizational aspects should be taken into account. We have examined the development of desktop applications in terms of their interface design and the allocation of screen resources to different elements of the graphical user interface.

The final part of the chapter provided a short introduction to mobile HCI, highlighting the need to consider the user's context and the changes in the context under different geographical and temporal conditions.

Further reading

Both Shneiderman and Plaisant (2005) and Sharp *et al*. (2007) discuss the design of user interfaces in detail. Shneiderman and Plaisant are especially detailed in their discussion of universal usability, direct manipulation and information visualization – all are areas that they have carried out a lot of research on.

In the area of GIS, the work of Max Egenhofer and his students and researchers should be noted for the provision of new interface metaphors such as The Geographer's Desktop which we have discussed earlier (Bruns and Egenhofer, 1997; Richards and Egenhofer, 1995), as well as query-by-sketch (Egenhofer, 1997).

As noted above, Koua and Kraak (2004) provide a framework for designing geovisualization applications, and Marsh's (2007) PhD provides further information on this application. Dykes *et al.* (2005) is also a useful compendium of papers on geovisualization and includes many ideas that can be integrated into the design of these system, and finally, interested readers should explore the work that has been produced by the GeoVista centre in Pennsylvania State University under the guidance of Alan MacEachren.

In the area of spatial decision support systems, Jankowski and Nyerges (2001) provide information and guidance on the design of such systems, and interested readers should explore other writing by these authors in the area of usability of these systems.

Revision questions

1. What are the technological considerations that should be included in the development of a geospatial technology that will be used by geographical information specialists in an organization?
2. How do the specific technological constraints influence and impact the development of geovisualization application?
3. You are asked to redesign the interface of MapInfo Professional (Figure 11.7). What changes will you make to the interface? How will the various considerations that have been discussed in this chapter influence the design?
4. You are asked to develop an application that will help transport planners in the design of new bus routes. Consider which type of operation should be provided directly on the map interface of the application through direct manipulation, which operation should be provided in buttons on the tool bars and which should be supported in the command-line interface that will support advance analysis.

12 Web-mapping applications and HCI considerations for their design

Artemis Skarlatidou

The technological developments and the creation of sophisticated computer software reduced not only the costs of map production, but also improved the ways in which we can deliver them. The emergence of the Internet, and in particular the World Wide Web (WWW) is a major milestone in the evolution of maps, as it has changed the way people interact with digital maps and made them accessible to the wider public. Since the introduction of Xerox PARC Map Viewer in 1993 (Putz, 1994) a plethora of Web-mapping applications was developed, with some of them allowing the end-users to interact in new and innovative ways.

The rapid development of Internet applications and websites resulted in an increased number of Web usability studies in the early 1990s (Nielsen, 2003) which provided an in-depth analysis of the human aspects that should be taken into consideration when designing such systems. However, this is not the case with the area of Web-mapping, where studies of Web-mapping usability and their HCI aspects are still limited.

This is particularly problematic because a fundamental characteristic of these applications is that they are open to everyone with an Internet connection and the vast majority of the cases are people without any GIS knowledge or expertise. Therefore the overall design and implementation of Web-mapping applications is fairly challenging.

This chapter aims to discuss the elements which influence the design and use of Web-mapping applications and to present some of the principles which should be incorporated into the development process to improve the usability and user experience of Web-mapping systems. The first part of this chapter provides an overview of Web-mapping and discusses the basic architectures, Web-mapping system types and their advantages and concerns. In the second part a discussion of HCI principles, accompanied by examples of existing applications, is provided. This provides a framework of all the elements that should inform and guide the design and evaluation of Web-mapping applications.

12.1   Overview of Web-mapping

12.1.1   The Internet and the web

To understand what Web-mapping involves it is necessary to first explain the concepts of the Internet and the WWW. The Internet is a global network which consists of thousands of smaller networks. For example a university network is only one of the networks that belong to the Internet and it links to other networks, with an ability to send data across all these networks, if the address of the recipient is known. Bidgoli (2004) describes the Internet as an *'information superhighway'* because its purpose is to distribute information (data) to its connected computers.

The Internet data transfer is based on the packet switching technology, where the data is broken into small pieces (packets) of information and sent over the network, using the Transmission Control Protocol/Internet Protocol (TCP/IP). The Internet provides different services, for example, electronic mail (e-mail) and the World Wide Web (WWW).

The WWW was proposed in 1989 by Tim Berners-Lee. Although the terms Internet and WWW (or simply web) are used interchangeably, technologically they are different. The web runs on top of the TCP/IP protocol and thus on top of the Internet, using the Hyper-Text Transfer Protocol (HTTP). The hypertext allows users to access files and applications in a non-sequential manner and also to combine textual information with images, videos and sounds (hypermedia). HTTP is based on client/server architecture. The client is the user's system which through an application, usually a Web browser, sends requests to the server for accessing specific resources. The server is a computer which hosts services or applications such as websites and sends the resources to the Web client. It should be mentioned that a resource is a piece of information, such as a file or a query result and which can be identified by a Uniform Resource Locator (URL). Having realized the differences between the Internet and the Web, it can be concluded that Web-mapping involves the process of designing, creating and distributing maps over the web. Prior to the web, and bypassing it, maps can be distributed over the Internet using the File Transfer Protocol (FTP), although this requires technical skills to transfer and use the data.

While the purpose of this chapter is not to discuss the technological aspects of Web-mapping in depth, it is necessary to briefly introduce Web-mapping technologies and the types of Web maps.

12.1.2   Web-mapping architectures

There are two basic architectures for the implementation of Web-mapping applications. These are the server-side and the client-side architectures or hybrid approaches which combine the two. In this section the server-side and client-side architectures are briefly presented.

Server-side architecture

A client is the user's Web browser while the server is the computer which hosts the service or the application. In server-side Web-mapping the user/client requests a map from the server where all the data and data tasks are located. The server processes the request and sends to the user the resulting map. For example, in the case of Web static maps, the resulting resource that will be sent to the user will be a static map in a JPEG, PDF or other format. If the application requires an interaction between the user and the map, the server will consist of: web server (i.e. Apache), which is responsible for the communication with the client; a geographical information database, which handles the spatial and non-spatial data, usually in an object relational database; and a web map Server (WMS), which performs the Web-mapping tasks – combining the information with cartographic rules and creating the maps.

Apart from these main parts there are usually several other programs or components to support the communication between client and server. For example, Common Gateway Interfaces (CGI) applications which run on the Web server are responsible for processing the client's requests and also for ensuring that the resulting resource is in appropriate form. Web application servers are also used to connect the various software components with the Web server.

The server-side architecture, while useful for managing effectively the use and distribution of centralized data, can be highly influenced by the bandwidth and the size of the files that are requested by the users. For example, if the speed is too low and/or there are too many concurrent requests from the server, this can be frustrating for a user.

Client-side architecture

While the server-side architecture allows the client mainly to display maps, the client-side architecture can provide interactivity (i.e. draw, analyze, query). This means that the server does not process all the data and sends back the requested files, but in contrast some of the data is processed locally on the client's computer.

Client-side architecture can reside in applications (or applets) downloaded and installed on the user's computer (for example Google Earth which was discussed in Chapter 9) which then interact with the server only for requesting data. Alternatively, the Web-mapping application can be provided to the user through the client's Web browser. In the latter the increased interactivity of a Web-mapping application can be supported by the Web browser with plug-ins or applets, which in most cases are built specifically for certain types of Web browser. For example, Java applets can now be used in Web browsers and are commonly used in Web-mapping applications. Similarly Web browser plug-ins such as Adobe Acrobat or Adobe Flash can support client-side Web-mapping.

With the client-side architecture tasks can be completed locally at the user's computer and performance can influence how fast the tasks are completed. Furthermore, if the applets or application does not reside permanently on the user's side, then for each request the server must send both the resulting applet and the data and thus it can influence the waiting time for starting the interaction, although after the download the plug-in can support sophisticated applications.

12.1.3   Types of Web maps

There are now hundreds of Web-mapping applications available, based on different architectures or hybrid approaches, each of them providing maps with diverse functionality. One of the first classifications of Web maps was made by Kraak and Brown (2000), who distinguished Web maps into two categories: static and dynamic maps. Each category is further subdivided into 'view only' and interactive maps. Neumann (2008) provides a revised categorization of Web maps in an attempt to differentiate and describe the properties of the currently available Web maps. He classified maps in the following categories:

- *Static Web maps:* These are static maps which do not support any interactivity or animation (i.e. a scanned paper map).
- *Dynamic Web maps:* These maps are generated in the server side so each time a user requests a new map, the map is created at the server and sent back to the client.
- *Distributed Web maps:* As their name reveals they are created after combining distributed data located at different servers.
- *Animated Web maps:* These maps represent not only spatial elements but also changes in temporal variables and most commonly changes over time (i.e. weather maps, climate maps).

- *Real-time Web maps:* These maps are based on real-time data which is gathered continuously through sensors and used to update the map either immediately or by request (i.e. traffic, weather maps).
- *Personalized Web maps:* These maps are created according to individual preferences. For example the users can define the map's content (i.e. select or deselect data layers) and how the symbology will appear on the map.
- *Interactive Web maps:* These can be either static maps (clickable maps) which include hyperlinks and other information, or dynamic maps. The interactivity of dynamic maps supports more complex tasks, such as for example zooming in and out from specific areas, calculating distances and routes and so on. Java, JavaScript and other technologies can help in the creation of interactive Web maps.
- *Open Web maps:* These maps are open to the public and can be reused by people for their services or websites. Application Programming Interfaces (APIs) can be used to load existing Web maps, as for example the Google Maps API. Such APIs can also be integrated with additional geographical data on top of the API, creating the so-called mash-ups.
- *Analytic Web maps:* These are Web maps which provide increased functionality to support further analysis similar to that provided by GIS.
- *Collaborative Web maps:* The idea of creating collaborative projects is not new for the web. Wikipedia is an example of such a collaborative project that is created by individuals. More specifically, collaborative Web maps are created and edited collaboratively by a group of people who can add and share information based on their local knowledge. For example OpenStreetMap consists of such a collaborative Web-mapping project (see Chapter 4).

Although this categorization describes the basic properties of Web maps it should be noted that a Web-mapping application can combine the properties of more than one of these categories. For example, an interactive map can also be real-time or can support personalization options.

12.1.4   Web-mapping advantages and concerns

The web created new opportunities and changed the way people build, view, use and access maps. The popularity of Web maps and the rapid technological developments in this area resulted in the creation of several open source tools, which can be used to produce Web maps at low cost and with limited expertise. Also, Application Programming Interfaces (APIs) can help developers to easily combine (mash up) distributed data from other sources. Furthermore, the web helped to reduce the costs of map distribution as Web maps do not have to be printed on paper and then delivered. The only consideration is to follow open standards so that maps can be distributed to any operating system or Web browser.

In more advanced Web maps, users can personalize the maps according to their needs and interests. For example, layers can be selected or deselected, increasing or reducing the amount of information that is viewed at any one time.

Text, video, 3D visualizations and hyperlinks can now be easily combined with Web maps, giving immediate access to additional information which would otherwise be difficult to link – especially with paper maps. For example, as the user hovers with her mouse over certain areas, pop up windows can be displayed to provide additional information. Animated maps and real-time maps were also grown with the web, as now updates are easier to implement and cheaper.

Although Web maps have significant advantages over the traditional paper maps there are still some important concerns and in some cases limitations. For example, most cartographers and map designers know that copyright issues of geospatial data are an important concern for both paper and Web maps. Especially for Web maps copyright issues can be complicated as is the case with any other program, service or application provided through the web. Very often permission is necessary, and in other cases data encryption or other means for data protection are required. This means that while technological advances allow the design and implementation of Web-mapping systems with increased interactivity, the data (especially in vector format) should remain at the server's side for copyright issues. Privacy, reliability and trust issues also add an ethical dimension to Web-mapping. Monmonier (1996) in his book *How to Lie With Maps* explains that apart from the in-purpose lies that can be told with a map (i.e. political propaganda), map attributes such as scale, symbolization and projection can also misinform the users. Especially for Web-mapping, where the screen's size and the resolution is not the same for all the users and can be poor, issues such as generalization or symbology can be very important. Below is an example from the What's In Your Back Yard website provided by the Environment Agency (see a longer discussion in Chapter 9), where the symbology is not in the right proportion with the map's size and the scale used and this might generate wrong conclusions or make people distrust the information provided (Figure 12.1).

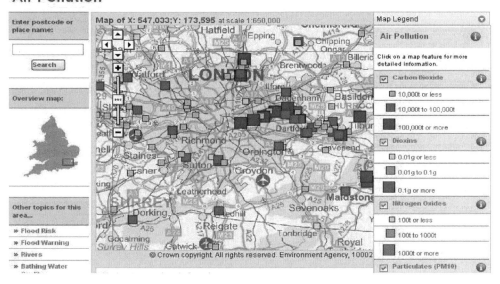

Figure 12.1   Results of air pollution in the 'What's In Your Back Yard' website provided by the Environment Agency (Source: Environment Agency) (a full-colour version of this figure appears in the colour plate section of this book)

As can be seen from Figure 12.1, the air quality results are not proportional to the map scale, which becomes more obvious when the user zooms in, as the size of results remains the same. Another problem is that although air pollutants of carbon dioxide are immediately obvious to the user, other pollutants (i.e. dioxins, nitrogen oxides) are hidden by the top layer. The user can only reveal this information after finding that layers can be selected/deselected, something which is not obvious from the interface. An alternative for this problem is to ensure that when the user enters the site none of the layers is selected, forcing the user to use the function to reveal information, or if the user is informed about the fact that some information is hidden.

Other elements which should be considered during the life-cycle of the development of a Web-mapping application are discussed in the remainder of this chapter. The second part discusses those aspects that the designers of a Web-mapping application should consider and also provides a description of Web-mapping design guidelines suggested in the literature.

12.2   Web-mapping design and HCI considerations

In the first section we were briefly introduced to some of the basic concepts associated with Web-mapping applications. There are several other elements which influence the use of these systems and which if not properly considered can result in significant problems for a Web-mapping system. The context and purpose of the application, the physical environment that the application will be used in and its technological and social contexts should inform the design, development and evaluation processes. The central element in this process is the investigation of the users' needs and expectations, their goals and tasks, spatial and domain knowledge and their mental models. Moreover, recent research on Web-mapping usability can further inform the design and evaluation stages. All these issues are discussed separately in the next sections by providing examples from existing systems.

12.2.1   The purpose and the physical environment

There are thousands of Web-mapping applications, each is used in a different context and for different purposes. The context and the purpose define some of the basic characteristics of a Web-mapping application and determine the services that it provides, and aspects of the desirable resulting user experience.

For example, an interactive Web map designed to describe the migratory history of the human species like the one provided by National Geographic (Figures 12.2 and 12.3). If the user clicks on a point (Figure 12.2) they can view more information about specific historical events, which is provided with textual information, static maps and pictures (Figure 12.3). The purpose of the application is to educate the end-users while its potential user experience attributes include being entertaining and engaging users so that any learning activity is easy and more effective. In such an application accuracy is not a major concern, as it would be with an application that provides environmental information for a specific location or provides a driving directions service.

Notice that while this application does not provide free interactivity as in Google Earth it will still provide the user with a level of freedom that makes the user feel in control of the application and encourages exploration. Such an application demonstrates that the level of interaction should not always be in line with the latest technology, but must match the needs of the users.

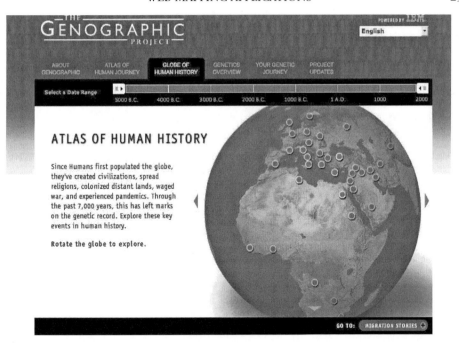

Figure 12.2 Globe of Human History, an interactive Web-mapping application, provided by *National Geographic* and part of the Genographic Project

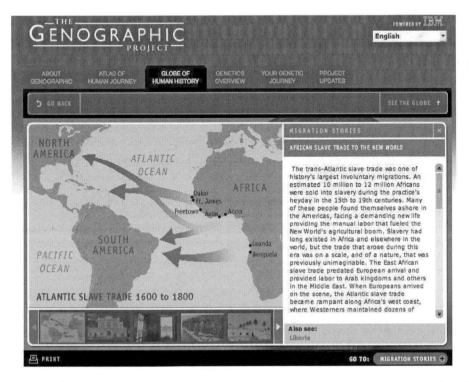

Figure 12.3 Globe of Human History: Information about Migration Stories. *National Geographic*, The Genographic Project

To clearly define the context and the purpose of the application, the designer should answer a series of questions which amongst others involve:

- Is it a standalone Web-mapping application or a Web-mapping system which is integrated into a website? In the first case, the application can be designed with other website convention, in mind, but the level of freedom for the designer is fairly high. In the latter, it is important to consider the constraints and modes of interaction of the rest of the site, and how the application fits within the rest of the user's experience.
- Does it provide only maps or functionality for performing spatial analysis? This consideration can lead to a decision about which architecture should be used and in what ways the best user experience can be achieved.
- Is accuracy the main concern or can accuracy be sacrificed in order to produce easy to perceive geographical representations? In cases where a high level of accuracy is required, more detailed data sets are likely to be used, and this will cause delays in the ability of the system to respond quickly to user requests. It will also mean that much more work will be required to check user's input and validate user's actions. In some cases, such an application will not be suitable for a Web browser, and a dedicated client will be needed.
- Are there any applications which support similar tasks and if so, what are the main difficulties that the users have while interacting with these? These questions can help in understanding the user's context, and providing an application that fits within their workflow. In addition, learning from other applications (and their pros and cons) can assist the designer in developing a more successful application.

Further questions should be developed for the specific application in hand. The context can be further assessed and evaluated by identifying the users, their characteristics, their needs, goals and tasks.

Another important consideration is the physical environment which defines where and how the application will be used. Some important elements involve the monitor size, the lighting conditions and the colour combination as well as the screen's resolution. For Web-mapping applications such as Google Maps, which are usually accessed from different places and different users, the monitor size, the screen resolution and lighting conditions can vary significantly between users. However, for Web-mapping applications that are designed to serve specific user groups in restricted physical environments, for example in a work place, the designers should consider all these characteristics which influence how the application is used. An important consideration that by and large is ignored by most Web-mapping designers is the need of the user for hard copies. Here questions should be asked about the type of printers (colour or black and white) and the use of the printed copies by the user.

12.2.2   The technological and social context

The technological context is very important and most Web-mapping application developers are usually guided by the technological developments. However, although technology should inform the design, it should not be the only driving force in the development of Web-mapping systems. It is essential to take into consideration all the elements which influence the interaction with these systems (i.e. the purpose, physical environment, user needs etc.) to develop new interaction modes and innovative systems which would better suit their users' needs.

Based on the experience from existing Web-mapping applications there are some basic technological principles which should be considered for their successful implementation. As was noted, in the server-side architecture the user sends requests to the server to be processed and the results are sent back to the user. For example, in a task which requires several steps, the user must send each of these steps to the server separately. This communication between the server and the client/user depends highly on the bandwidth and the user's speed in performing each task and sending it to the server. It also depends on the user's speed to cognitively process the information and then request new data from the server, but this is difficult to model and can vary significantly between individuals or groups of users (i.e. experts and novices).

The bandwidth between the client and the server is an important technological consideration as it influences the response and thus the waiting times. Current Web-mapping applications also provide large size maps or maps with high quality graphics (i.e. 3D visualizations) which can further increase the response times. If the waiting time is too long then this can be frustrating for the user and thus the result will be a negative user experience.

Haklay (2005) provides an example about the role of bandwidth in a server-side application which supports the task of area (polygon) digitization in the era of 56kbps modem-based communication (Figure 12.4). As Haklay explains, the application assumed a 56kbps bandwidth based on a dial-up modem and as such the specific task would require about one

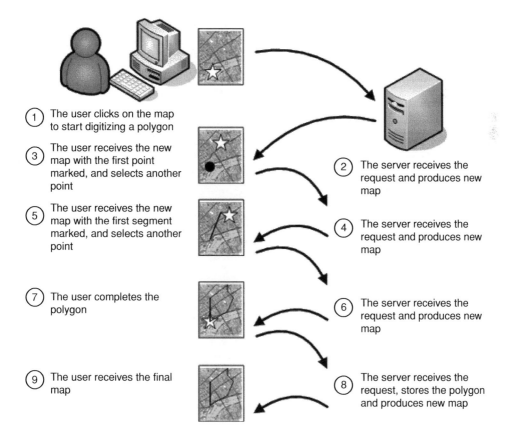

1. The user clicks on the map to start digitizing a polygon

2. The server receives the request and produces new map

3. The user receives the new map with the first point marked, and selects another point

4. The server receives the request and produces new map

5. The user receives the new map with the first segment marked, and selects another point

6. The server receives the request and produces new map

7. The user completes the polygon

8. The server receives the request, stores the polygon and produces new map

9. The user receives the final map

Figure 12.4 Stages of interaction between a user and a server-side Web-mapping application during the task of polygon digitization (Haklay, 2005)

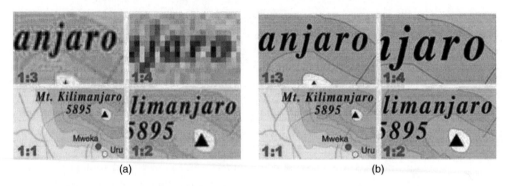

(a)   (b)

Figure 12.5   (a) Raster Zoom-in Function (Shand, 2002); (b) Vector Zoom-in Function (Shand, 2002) (full-colour versions of these figures appear in the colour plate section of this book)

minute to be completed. While this type of interaction is not relevant over fast broadband connection, there are still situations where the bandwidth is limited – for example when the user is using satellite link or when using a mobile application on a moving vehicle. Therefore, the feature, apart from its limited usability, could be extremely annoying for the end-users.

The download times and the file size further influence the data representation that will be used as different data representations result in different file sizes. In Web-mapping applications raster maps are quite popular because of their high availability and their image manipulation but their resolution is a critical concern. Harwell (2004) suggested that by using vector and Scalable Vector Graphics (SVG), analysis and user interaction are better because both the functionality and map quality can be increased. As Figure 12.5 shows, there are clear differences between zooming in a raster and in a vector map.

The majority of Web-mapping applications can and should be accessed through a Web browser, as this is the most available platform to the user. Especially when the lay public are the intended end-users, plug-ins such as Java applets are intimidating due to fear of computer viruses. The exception for this rule is Adobe Flash-based applications, as the Flash plug-in is now widely used, as well as Adobe Acrobat, which increasingly has more geographical ability. There are different Web browsers available, for example, Internet Explorer (IE), Firefox or Safari. They are also available in different versions and thus with different capabilities – for example, only some versions can display SVG. Therefore, it is essential to ensure that a Web-mapping application can be accessed through a variety of Web browsers – at the very least to check what are the most likely browsers that the intended users are likely to use. In case this is not possible, then it is necessary to inform the user which Web browsers can be used and ideally the system should recognize the browser used and report the appropriate error message to the user. An example of a Web-mapping application that can only be accessed with IE is the 'What's In Your Back Yard' website (Figure 12.6) and as can be seen from Figure 12.7, there is no appropriate feedback informing the user about another Web browser's inability to display the website (in this case Mozilla Firefox was used to access the website). As a result it is very likely that some users will leave the website unhappy thinking that it was not working at all.

Moreover, as there are specific Web-mapping sites which are popular for the majority of Web users it is probably wise to use popular interaction conventions. Inventing new interactions is generally a bad idea, as the users cannot use their acquired knowledge. It is

Flooding

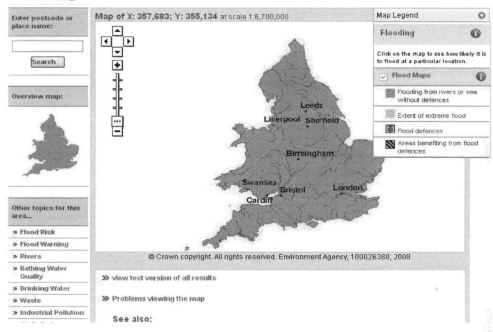

Figure 12.6 The 'What's In Your Back Yard' website, provided by the Environment Agency accessed using Internet Explorer (Source: Environment Agency)

more likely that the majority of users are familiar with common interaction techniques from other, similar sites. However, this depends on the context of the application and the tasks it supports as well as on other user aspects. Therefore, even if a popular interaction technique is to be adopted it would be recommended firstly to test it with real users in the context of the specific application. If a new interaction technique is used, a clear indication on how to use it should be provided to the users, and it should be tested thoroughly.

Apart from the technological context, the social context can further define important aspects and even restrict some of the potential services. For example, the technological developments allow Web-mapping systems to provide not only very accurate maps but also additional services, such as images of specific locations and detailed satellite imagery. There is now a discussion around privacy issues of Internet users. In this context specific Web-mapping applications and some services that these offer have been criticized in some countries or local communities. For example, Google's Street View was stopped in Greece and banned in Minnesota.

Figure 12.7   Using Mozilla Firefox to access the 'What's In Your Back Yard' website (Source: Environment Agency)

Case Study 12.1: Greece's data protection agency has banned Google from expanding its Street View service in the country, pending 'additional information' from the firm

Street View gives users a 360-degree view of a road via Google Maps. Authorities want to know how long the images would be kept on Google's database and what measures it will take to make people aware of privacy rights.

A similar street mapping service, run by local ISP Kapou, was also suspended for the same reason. In a statement, Google said that it had not seen the full details of The Hellenic Data Protection Authority's request, but had taken steps to protect people's privacy. 'Google takes privacy very seriously, and that's why we have put in place a number of features, including the blurring of faces and license plates, to ensure that Street View will respect local norms when it launches in Greece', the statement read.

'We have already spoken with the Hellenic Data Protection Authority to ensure that they understand the importance we place on protecting user privacy. Street View has not been banned in Greece. We have received a request for further information and we are happy to continue discussing these issues with them. We will discuss with them whether it is appropriate for us to continue driving in the meantime'.

'Although that dialogue is ongoing, we believe that launching in Greece will offer enormous benefits to both Greek users and the people elsewhere who are interested in taking a virtual tour of some of its many tourist attractions'.

First launched in the US two years ago, Street View now covers nine countries, including the United Kingdom and Google wants to expand the service to cover all of Europe. Users zoom in to a given location in Google Maps, and then drag the 'Pegman' icon above the zoom bar onto a given street. A picture view of that street appears, which users can control to get a 360-degree view of the area or to progress on the street level, throughout the city. Google says the service shows only imagery already visible from public thoroughfares.

However, it has come in for criticism from some quarters, being accused of an invasion of privacy. While many of these charges have been dismissed, either through the courts or by regional information commissioners, in some cases people have taken a more direct approach. In April, residents near Milton Keynes blocked the driver of a Google Street View car when he started taking photographs of their homes saying the service was 'facilitating crime'. The Pentagon has also banned Google from filming near or inside its military bases, saying it posed a 'potential threat' to security.

The director of the UK-based privacy watchdog Privacy International, Simon Davies, said the Greeks' decision would set a precedent for other nations. 'This is fantastic news. The Greek regulators understand the risks of future technology creep. They have watched what has happened in the US and UK very carefully and will be familiar with the arguments on both sides. This highlights the difference between regulators – some will allow the public space to be exploited, others acknowledge that people's privacy needs to be protected. Now we wait for the domino effect, as the Greek decision sets an example that others may follow – we will see what happens next in Central Europe'.

(adapted from BBC, 2009)

For Web-mapping applications which are open to the public worldwide, it is very difficult to investigate cultural and social obstacles in such scales and it is definitely impossible to capture all the individuals' perceptions. However, it is essential to investigate and make sure that local populations are ready to accept specific services, especially if an application is designed to be used by local communities.

12.2.3   The user context

As with other computerized systems, it is especially important to investigate the user context of a Web-mapping application. Understanding the users and their requirements is central in the design process and should be completed at early stages of the process, especially if a novel system is under development. However, the user context can help designers even during or after the implementation stages of a system in order to improve its usability.

The user context involves an in-depth understanding of the following elements:

- Who are the users: their characteristics and needs.
- The user goals and tasks.
- The spatial and domain knowledge from a user perspective.
- The users' mental models of a system.

These are discussed in the following sections.

12.2.4   Who are the users: their characteristics and needs

Of particular importance in the design of a Web-mapping system are the user demographics which define the potential users of a specific application and further specify some of their basic needs. This involves the age of the people who are going to use the system, their experience levels with similar applications (are they novices or experts?), their educational background (are they people with GIS knowledge and expertise?), their physical capabilities (disabilities) and their gender.

Concerning the age of potential users it is clear that a Web-mapping application designed for kids should probably have a different aesthetics and behaviour compared to one designed for adults. For example, although animated images next to maps can be particularly disturbing as they attract user's attention, they might be effective by attracting kids' attention. Older users respond to websites more slowly and find navigation difficult.

If an application is going to be used by a large population including individuals with colour deficiency, then attention should be paid on the colours used for the base maps and the symbology as well as for the whole website (i.e. colour of menus, links) as was discussed in Chapter 8. Brewer's palette (Chapter 8) provides a basis for the design of effective maps, which can be used even by those with colour deficiency.

An important distinction that determines the behaviour and the interaction levels of a Web-mapping application is between novices and experts. The experience of the users with similar applications will define their needs and expectations from the system as well as how easily they will be able to use the system. Most designers tend to ignore this differentiation choosing to create either simple help documentation for beginners or deep menus with additional information for experts. However, this can be inadequate or confusing for both user groups. Furthermore, the frequency of use is highly important as this will influence how fast a user can move from a beginner to the advanced level. For a simple Web-mapping application the distinction between experts and novices might not be necessary.

However, more often than not, Web-mapping applications are complicated especially to people without any prior GIS experience or knowledge. For example, the term 'buffer', while it might be obvious to a GIS expert, might not be clear to novices. Therefore, how a buffering function will be introduced and explained to a novice is important. Moreover, a stand alone application such as Google Earth, while might be simpler for a person with GIS knowledge as it uses the same metaphor with Desktop GIS, can be challenging to a novice user. The designer should always make sure that functions are clearly communicated, immediately obvious to the user and that the tasks can be completed within a reasonable amount of time.

Although it is obvious that specific user characteristics can determine some of the users' needs and physical interaction with the system, the needs are further influenced by the users' goals and motivations which are discussed in the next section.

12.2.5 User goals and tasks

A goal describes the final expectation that a user has while interacting with a system. A goal is usually broken into smaller steps (tasks) which should be completed in order to accomplish a goal (Cooper *et al.*, 2007 and Chapter 7).

For example, a user who uses a Web-mapping application integrated with a travel information website (i.e. Expedia), will use the application to make sure that they can plan their journey quickly and easily. This described the user's goal. The goal is most likely to include tasks, such as examining different prices and booking the tickets and the hotel. After investigating users' requirements based on their needs and goals, the designers might discover that before the task 'booking the hotel', the user might want to check how close different hotels are to specific places (i.e. a museum, a conference centre). Therefore, the user experience and the system's effectiveness can be enhanced by providing users with the corresponding functionality to help them achieve their goals. In this example the designers could provide the users with a function which gives the option to the user to state the place of interest and show the hotels in relation to this location.

12.2.6 Spatial and domain knowledge

The experience that users have with similar applications can differentiate between novices and experts. The existing knowledge that a user has about the particular domain of the Web-mapping application as well as the existing spatial knowledge for a specific location can further influence the application's usability and the user experience.

It is common to integrate Web-mapping applications in different domains. If, in the case of the travel application that we have discussed earlier, the user is familiar with the place that they are going to visit, they will use the information on the travel website in a different way to a user who is going to visit a place for the first time, and therefore is likely to use other websites to learn about the place, the locations of tourist attractions etc. Therefore, the domain knowledge further influences the goals and the expectations that the users have from the application.

Web maps can support spatial cognition and spatial reasoning but when important decisions are to be made then the existing spatial cognition can influence the user experience in several ways. For example, consider a Web-mapping application designed to support collaborative mapping of the local environment and which will be used by the local community. It is very likely that the designers will not have sufficient spatial local knowledge to put all the necessary information on a map. If important features that the local community is expecting to find on the map are not provided then the resulting user experience might be one of disappointment and distrust. Therefore, in such cases it is important to provide the users with the tools which would allow them to add features or to make sure that it is well understood how people perceive their local environment and that this is reflected on the maps' design.

12.2.7 Users' mental models

As with any other computerized system, the users' mental models can influence the usability of Web-mapping applications. The closer the mental model is to the system's model the more likely it is that the users will be able to easily and efficiently use a Web-mapping system.

To support users' mental models it is necessary to understand what the users expect in terms of the system's behaviour. For example, the designers must understand what people think is happening when they perform specific actions on the interface, to incorporate the corresponding behaviour into the system. In addition, the investigation of users' mental models can further reveal what vocabulary and metaphors the users use to describe specific elements of their world that are reflected in the computerized system. For example, they might refer to the zoom in/out function using the magnifying glass icon. If these metaphors and vocabulary are incorporated into the system, it is more likely that the system will be predictable and easy to use.

12.2.8 Design guidelines

Although HCI research in GIS in the area of Web-based GIS is still limited there are enough studies that provide an insight on how specific elements of Web-mapping applications should be developed. It is important to remember that Web-mapping is used for different purposes and thus applications can differ significantly. Design guidelines can help increase the usability, but not in all circumstances and in specific user, social and technological contexts. Therefore, in parallel to guidelines it is necessary to investigate the various elements of the specific system separately.

Accessing the map and its functions

One of the major problems with Web-mapping applications is that while users access the website with the expectation to find a map, the users cannot access the maps immediately from the website's homepage. In other circumstances users are not aware of the map's complete functionality even after spending hours interacting with the application.

Nivala *et al.* (2008) carried out a usability evaluation of the Web-mapping sites Google Maps, Multimap, Map Quest and MSN Maps & Directions. They explain that people had difficulty finding the map interface as they were confused by maps that looked like links, especially in cases where no feedback was provided to inform the users. Also the same study found that users felt confused by information overload and elements such as advertisements, which further reduced the visibility of the map interface and its functionality.

In a different study of the 'What's In Your Back Yard' website which was mentioned above, accessing the maps from the main page of the Environment Agency's website was difficult, as there was no clear indication of the service (Alsop, 2008).

In Web-mapping applications, the map should be the central element and occupy a large part of the interface. Its functions should be immediately communicated to the users, even if further help is required before using the map tools. For websites which integrate a Web-mapping application with other services, it is essential to help users access the Web maps by providing direct and clear links from the website's homepage, or from the page where they are most likely to need the map – for example, from the list of hotels in a city on a travel website. Moreover, feedback is essential in cases where a specific Web browser is needed in order to access a Web-mapping application.

The search box

The search box is another central element of a Web-mapping application and as such should be easily accessed and used. However, there are cases when the search box is small enough to be left unnoticed or it is located in the wrong place – not to the left and the top of the interface (Nivala *et al.*, 2008).

It is not only important to make the search box easily accessible to the user but to also provide the users with the search options that match to the application's context and meet its users' requirements. For example, Skarlatidou (2005) found that Google Maps search for 'British Museum' was pointing to Crewe, in North West England, and not to London (see Case Study 6.2). This reveals a major usability problem as tourists are one of the most important user groups that access the Web-mapping sites.

Map size

Skarlatidou and Haklay (2006) have shown that the map size is an important factor which influences how easily and efficiently people can perform tasks on the site. In this study it was found that in Google Maps where the map size was considerably larger compared to other sites, the average time for locating a facility (underground station) near a specific location was about three seconds, which was significantly less compared to other websites with smaller map interfaces, which could be up to 40 seconds.

A large map size can significantly improve the spatial navigation of the users as more features are available on the screen and at the same time minimizes the use of tools such as panning, which add to the waiting time for each user's actions.

Back button

An important aspect for Web-mapping design is that in most cases these applications operate on a Web browser and thus the users expect to use the Web browser in a similar way it is used for any other website.

For example, using the back button of the Web browser is popular and consists of an intuitive action for the majority of Web users. Although the mapping application itself might include a similar back button function, which allows users to see the previously accessed map, this is likely to be overlooked by the users, as the browser's back button is the one that will be used. As a result, such considerations should be taken into account, so that the system will recognize the Web browser's functionality and will support instinctive or learned user behaviours.

Map visualization and aesthetics

Map visualization must be effective in order to serve its purpose which is to communicate clearly the spatial information to the user. Different factors influence the effectiveness of map visualization, such as the map's size. However, an important factor is also the aesthetics which refer to beauty and harmony (Franges *et al.*, 2007, as well as Chapters 3 and 8). More specifically aesthetics focus on how different elements and colours are combined together so that the end-user interface is pleasant to use.

A Web map with easy to distinguish colours, clear text and symbology can not only communicate the information to the users more easily, but as MacEachren (1995) pointed out, increases the users' confidence and thus the perceived trustworthiness of the information. In a usability evaluation study of different paper maps Harrower *et al.* (1997) showed that professional maps of high quality were perceived by the participants as the most accurate, while the comments of the users for these maps included amongst others that they were *'easy to read'* and had a *'nice choice of colours'*(Harrower *et al.*, 1997: 33).

Nivala *et al.* (2008) suggested that digital maps which imitate paper map design can frustrate the users and were criticized in usability evaluation. Therefore, attention should be paid on designing Web maps which look well on a computer screen. Furthermore, Nivala and her colleagues suggest that the background colours of the maps should be readable, different colours should be used for different features and the symbols should be easily distinguished and interpreted. For example, using a blue colour for roads would make them look like rivers (Nivala *et al.*, 2008).

When symbols overlap then the designer should consider carefully how to choose easy to distinguish colour combinations with the appropriate transparency levels. Similarly, the legend should include readable text and not too many items in order to avoid confusion.

Aesthetics and visual clarity can further determine the competency of a Web-mapping application. The increased number of available Web-mapping systems allows users to choose between different alternatives. This means that designers should take into account this competition and improve the user experience in their systems by also considering the aesthetics.

Map tools

Web-mapping applications offer a variety of tools depending on their context and the tasks that they were designed to handle, but also to comply with users' expectations. Some suggestions by Nivala *et al.* (2008) for commercial Web-mapping sites include the support of continuous dragging and a measurement tool.

A critical consideration regarding map tools is how these are communicated to the user and where in relation to the map they are located. Nivala *et al.* (2008) suggest that the tools should not hide the map and should be easy to identify. Concerning their visualization, one recommendation would be to use not complex but popular icons describing a tool (i.e. a magnifying glass for zoom in/out, or a ruler for the measurement tool).

Kramers (2008) reports that during the development of the Atlas of Canada *'The general results of usability research revealed that all participants, regardless of user group, responded better to a label accompanying an icon, as icons alone can mean different things to different people'* (Kramers, 2008: 103).

The user context plays a crucial role in both the functionality that a Web-mapping system provides, as well as what are the best visualization options for communicating the map tools. For example, Harrower and Sheesley (2005) carried out a usability evaluation of different panning and zooming options based on criteria of productivity, comfort and efficiency. They have concluded that the best choice depends on the users' characteristics and their tasks. During the investigation of the user context, elements such as the vocabulary the users use to describe specific map elements, including the tools, can help in understanding the best visualization options.

User interface of Web-mapping applications

There are several examples of Web-mapping applications which are not standalone but instead are part of a website (i.e. Expedia). Therefore, the layout and the functionality of the whole website which offers the Web-mapping service should also be considered. It is essential that the website complies with HCI standards and usability principles.

For example, the design should be consistent on all pages of the website. Attention should be paid to how the website and the information on it are organized and structured. Information overload can lead to difficulties in using the site especially if the navigational aids (i.e. links) do not help users to find what they are looking for. Therefore, the menu should be effective and basic elements such as the search box should be immediately clear to the user. The vocabulary and the terminology used should be designed to increase the communication with the end-users, and appropriate feedback should be provided to support the end-users' tasks.

An important aspect that requires special attention is the use of advertisement and animated banners. Studies showed that advertisements can be very distracting for the users of Web-mapping applications especially if these are located close to the maps (Nival *et al.*, 2008; Skarlatidou and Haklay, 2006). In several occasions the number of advertisements cannot be reduced due to operational requirements, and thus it is necessary to avoid placing advertisement boxes next to the map especially if these include animation.

Finally, the Help and Documentation should be designed carefully based on the purposes of the application and the user context. No matter how simple or easy to use a system is it should be always kept in mind that no user is an expert at the beginning. Thus instructions for what tasks the application supports and how these can be achieved are necessary. Moreover, as these applications involve a GIS element and it is most likely that the majority of the users have no prior GIS experience, the vocabulary and instructions should be simple and easy to understand. For example, the term GIS should not be used!

Several Web-mapping application developers seem to think that information concerning the data accuracy and other technical issues on how the maps were constructed are not essential. However, to avoid problems of trust the Help and Documentation section should cover the needs and expectations even of the most demanding user. If the Web-mapping application is open to a wide variety of users, then people with a deep knowledge in these issues might want to find relevant information before they begin using the system, or novices who use the application regularly and thus become experts, might need such information in the future.

Summary

This chapter provided an overview of some of the basic technological principles of Web-mapping, the architectures and the different Web map types that currently exist. Although Web maps have significant advantages over traditional paper maps, there are still some concerns associated with their successful design and implementation. Several of the problematic aspects of existing Web-mapping applications can be eliminated or reduced if a holistic approach, which is not purely technologically-driven, is followed. Elements such as the purpose of the application, the physical environment, the technological and social contexts should be taken into consideration and inform the design and evaluation process. The user context is central in this process and HCI methodology provides the tools and techniques that can be used in order to improve the user experience and the usability of Web-mapping applications.

Further reading

For an in-depth understanding of Web-mapping technologies the readers can refer to Peng, Z. and Tsou, M., 2003, *Internet GIS: Distributed Geographic Information Services for the Internet and Wireless Networks*.

The book describes all the basic issues of Internet GIS which are concerned with techno-logical aspects mainly for their development and implementation.

Peterson, M. ed., 2003, *Maps and the Internet*.

This book provides a collection of different research studies on Internet Cartography, with some of them focusing on technical aspects and others discussing privacy issues or its uses in different contexts such as in education.

The HCI research in the area of Web-mapping is only recent and the number of studies is still limited. However, for a more detailed investigation of the HCI aspects of Web-mapping applications the reader can further review the research papers which are mentioned in this chapter. Indicatively some recent research papers are the following:

Haklay, M., 2006, Usability Dimensions in Collaborative GIS.

Skarlatidou, A. and Haklay, M., 2006, Public Web Mapping: Preliminary Usability Evaluation.

Nivala, A.M., Brewster, S. and Sarjakoski, L.T., 2008, Usability Evaluation on Web Mapping Sites.

Kramers, E., 2008, Interaction with Maps on the Internet – A User Centred Design Approach for the Atlas of Canada.

Revision questions

1. Indicate the names and describe the differences between the two basic Web-mapping architectures.
2. What are the currently available Web-mapping types and their associated properties according to Neumann's classification? Search for some Web-mapping applications and specify their properties.
3. What are the main areas and contexts that a Web-mapping developer should consider during the design life cycle of a Web-mapping application?
4. What are the elements of the user context that should be taken into consideration?
5. Why are aesthetics important in Web-mapping design? Search for some Web-mapping applications and observe their aesthetics. What aspects do you think make the application aesthetically pleasant and how do these affect the perceived overall user experience?

Bibliography

In addition to providing references for the material that was presented in the book, this reference list aims to provide a compendium of sources for the study of HCI and usability studies of geospatial technologies. References that are marked with an asterisk are especially useful or significant.

Abras, C., Maloney-Krichmar, D. and Preece, J. (2004) User-centered design. In Bainbridge, W.S. (ed.) *Berkshire Encyclopedia of Human-Computer Interaction*. Great Barrington, MA: Berkshire Publishing Group.

Aerts, J.C.J.H., Clarke, K.C. and Keuper, A.D. (2003) Testing popular visualization techniques for representing model uncertainty. *Cartography and Geographic Information Science* **30**, 249–261.

Agarwal, P. (2004) Contested nature of place: knowledge mapping for resolving ontological distinctions between geographical concepts. In Egenhofer, M.J., Freksa, C. and Miller, H.J. (eds) *Geographic Information Science, Proceedings*. Berlin: Springer-Verlag, pp. 1–21.

*Albert, W.S. (1997) The role of spatial abilities in the acquisition and representation of geographic space. In Craglia, M. and Couclelis, H. (eds) *Geographic Information Research: Bridging the Atlantic*, London: Taylor & Francis, pp. 320–332.

Albert, W.S. (1999) The Effects of Turn Instruction on Memory for Landmarks During Route Learning. Cambridge Basic Research Technical Report.

Albert, W.S. and Golledge, R.G. (1999) The use of spatial cognitive abilities in geographical information systems: The map overlay operation. *Transactions in GIS* **3**(1), 7–21.

Albert, W.S., Reinitz, M.T., Beusmans, J.M. and Gopal, S. (1999) The role of attention in spatial learning during simulated route navigation. *Environment and Planning* **A 31**, 1459–1472.

Albert, W.S. and Thornton, I.M. (2003) The effect of speed changes on route learning in a desktop virtual environment. *Spatial Cognition* **52**, 127–142.

*Albrecht, J. (1994) Universal elementary GIS tasks – beyond low-level commands. In Waugh, T. and Healey, R. (eds) SDH 94: Sixth International Symposium on Spatial Data Handling, Edinburgh, Scotland, 5–9 September 1994. *Advances in GIS Research Proceedings*, Vol. 1, London: Taylor & Francis, pp. 209–222.

Albrecht, J. (1995) Semantic net of universal elementary GIS functions. *Proceedings, Twelfth International Symposium on Computer-Assisted Cartography (Auto-Carto 12)*, Charlotte, NC, pp. 232–241.

Albrecht, J. (1998) Universal analytical GIS operations – a task-oriented systematisation of datastructure-independent GIS functionality. In Craglia, M. and Onsrud, H. (eds) *Geographic Information Research: transatlantic perspectives*. London: Taylor & Francis, pp. 577–591.

Albrecht, J., Derman, B. and Ramasubramanian, L. (2008) Geo-ontology tools – the missing link. *Transactions in GIS* **12**(4), 409–424.

Albrecht, J., Jung, S. and Mann, S. (1997) VGIS—A GIS shell for the conceptual design of environmental models. In Kemp, Z. (ed.) *Innovations in GIS IV*. London: Taylor & Francis, pp. 154–165.

Albrecht, J. and Pingel, J. (2004) GIS as a communication process. In Wang, F. (ed.) *Geographic Information Systems and Crime Analysis*. Singapore: Idea Group, pp. 2–23.

Alçada-Almeida, L., Tralhão, L., Santos, L. and Coutinho-Rodrigues, J. (2009) A multiobjective approach to locate emergency shelters and identify evacuation routes in urban areas. *Geographical Analysis* **41**(1), 9–29.

Al-Kodmany, K. (1999) Combining artistry and technology in participatory community planning. *Berkeley Planning Journal* **13**, 28–36.

*Al-Kodmany, K. (1999) Using visualization techniques for enhancing public participation in planning and design: Process, implementation, and evaluation. *Landscape and Urban Planning* **45**, 37–45.

Al-Kodmany, K. (2002) GIS and the artist: shaping the image of a neighbourhood in participatory environmental design. In Weiner, D., Harris, T.M. and Craig, W.J. (eds) *Community Participation and Geographic Information Systems*. London: Taylor & Francis, pp. 320–329.

Alonso, G. (1999) OPERA: a design and programming paradigm for heterogeneous distributed applications. Position Paper, International Process Technology Workshop (IPTW'99). September 1–3, (1999). Villars de Lans, France.

Alsop, R. (2008) What's in your back yard? A usability study by persona, MSc Thesis, London: University College London.

Ancona, D., Freeston, M., Smith, T. and Fabrikant, S. (2002) Visual explorations for the Alexandria Digital Earth Prototype. *Visual Interfaces to Digital Libraries*, Berlin: Springer, pp. 199–213.

Andrienko, G.L. and Andrienko, N.V. (1999) Interactive maps for visual data exploration. *International Journal of Geographical Information Systems* **13**, 355–374.

Andrienko, G.L, Andrienko, N.V., Janowski, P., Keim, D., Kraak, M.-J., MacEachren, A. and Wrobel, S. (2007) Geovisual analytics for spatial decision support: Setting the research agenda. *International Journal of Geographical Information Science* **21**(8), 839–857.

*Andrienko, N.V., Andrienko, G.L., Voss, H., Bernardo, F., Hipolito, J. and Kretchmer, U. (2002) Testing the usability of interactive maps in common GIS. *Cartography and Geographic Information Science* **29**, 325–342.

Arctur, D.K. (1996) *Design of an Extensible, Object-Oriented GIS Framework with Reactive Capability*. University of Florida (0070), 110.

Armstrong, M. (1994) Requirements for the development of GIS-based group decision support systems. *Journal of the American Society for Information Science* **45**(9), 669–677.

Bailey, T.C. and Gatrell, A.C. (1995) *Interactive Spatial Data Analysis*. Longman: Harlow.

*Balram, S. and Dragicevic, S. (eds.) (2006) *Collaborative Geographic Information Systems*, Hershey: Idea Group Inc.

Bambacus, M., Yang, P., Evans, J., Cole, M., Alameh, N. and Marley, S. (2007) An interoperable portal supporting prototyping geospatial applications. *URISA Journal* **19**(2), 33–40.

Baranauskas, C., Schimiguel, J., Simoni, C. and Medeiros, C. (2005) Guiding the process of requirements elicitation with a semiotic-based approach. *Proceedings, HCI International – HCI2005*, pp. 100–110.

BBC (2007) PCSOs 'did not watch boy drown'. BBC News website [On-line]. http://news.bbc.co.uk/1/hi/england/manchester/7007081.stm

BBC (2007) Sat Nav driver's car hit by train, http://news.bbc.co.uk/1/hi/wales/southwest/6646331.stm

BBC (2009) Greece put brakes on Street View http://news.bbc.co.uk/1/hi/technology/8045517.stm

Beale, R. and Lonsdale, P. (2004) Mobile context aware systems: the intelligence to support tasks and effectively utilize resources. *Mobile HCI*, Berlin: Springer-Verlag, pp. 240–251.

Bekker, M., Barendregt, W., Crombeen, S. and Biesheuvel, M. (2004) Evaluating usability and challenge during initial and extended use of children's computer games. In Fincher, S., Markopoulos, P., Moore, D. and Ruddle, R. (eds) *HCI*, Vol. 1. Leeds: Springer, pp. 331–346.

Ben-Joseph, E., Ishii, H., Underkoffler, J., Piper, B. and Yeung, L. (2001) Urban simulation and the luminous planning table: bridging the gap between the digital and the tangible. *Journal of Planning Education and Research* **21**, 195–202.

Bennett, D.A., (1997) A framework for the integration of geographical information systems and model base management. *IJGIS* **11**(4), 337–357.

Bennett, D.A. (1997) Managing geographical models as repositories of scientific knowledge. *Geographical and Environmental Modelling* **1**, 115–133.

Bidgoli, H. (2004) *The Internet Encyclopedia*. NY: John Wiley & Sons Ltd.

Bidoshi, K. (2003) *Virtual Reality Visualization for Maps of the Future*. The Ohio State University (0168), 186.

Bishop, I.D. (2002) Determination of thresholds of visual impact: The case of wind turbines. *Environment and Planning B: Planning and Design* **29**, 707–718.

Bishop, I.D. (2003) Assessment of visual qualities, impacts, and behaviours, in the landscape, by using measures of visibility. *Environment and Planning B: Planning and Design* **30**, 677–688.

Bishr, Y.A. (1998) Overcoming the semantic and other barriers to GIS interoperability. *International Journal of Geographical Information Science* **12**, 299–314.

Bishr, Y.A. (1999) A global unique persistent object ID for geospatial information sharing. *Interoperating Geographic Information Systems*, Berlin: Springer, pp. 55–64.

Bishr, Y.A., Pundt, H., Kuhn, W. and Radwan, M. (1999) Probing the concept of information communities – a first step toward semantic interoperability. *Interoperating Geographic Information Systems*, Berlin: Springer, pp. 55–70.

Bishr, Y.A., Pundt, H. and Ruther, C. (1999) Proceeding on the road of semantic interoperability – design of a semantic mapper based on a case study from transportation. *Interoperating Geographic Information Systems*, Berlin: Springer, pp. 203–215.

Bishr, Y.A., Radwan, M.M. and Pandya, J. (1997) SEMWEB – A prototype for seamless sharing of geoinformation on the World Wide Web in a client/server architecture. In Hodgson, S., Rumor, M. and Harts, J.J., (eds) *Geographical Information '97: from Research to Application through Cooperation*, Vols. 1 and 2, pp. 145–154.

Blaser, A.D. (2000) *Sketching Spatial Queries*. University of Maine (0113), 199.

Blaser, A.D., Sester, M. and Egenhofer, M.J. (2000) Visualization in an early stage of the problem solving process in GIS. *Computers & Geosciences* **26**, 57–66.

Blok, C., Kobben, B., Cheng, T. and Kuterema, A.A. (1999) Visualization of relationships between spatial patterns in time by cartographic animation. *Cartography and Geographic Information Science* **26**, 139–151.

Boothby, J. and Dummer, T.J.B. (2003) Facilitating mobility?: The role of GIS. *Geography* **88**, 300–311.

Bornträger, C., Cheverst, K., Davies, N., Dix, A., Friday, A. and Seitz, J. (2003) Experiments with multi-modal interfaces in a context-aware city guide. *Mobile HCI*. Berlin: Springer-Verlag, 116–130.

Bowman, D., Gabbard, J.L. and Hix, D. (2002) A survey of usability evaluation in virtual environments: Classification and comparison of methods. *Presence: Teleoperators and Virtual Environments* **11**(4), 404–424.

Braun, P. (2003) *Primer on Wireless GIS*. Park Ridge, IL. The Urban and Regional Information Systems Association.

*Brewer, C.A. (2003) A transition in improving maps: The ColorBrewer example. *Cartography and Geographic Information Science* **30**, 159–162.

Brewer, C.A. (2004) *Designing Better Maps: a guide for GIS users*. Environmental Systems Research.

*Brewer, C.A. (2008) *Designed Maps: a sourcebook for GIS users*. ESRI Press.

Brewer, C.A. *et al.* (1997) Mapping mortality: Evaluating color schemes for choropleth maps. *Annals of the Association of American Geographers* **87**(3), 411–438.

Brewer, C.A., Hatchard, G.W. and Harrower, M.A. (2003) ColorBrewer in print: A catalog of color schemes for maps. *Cartography and Geographic Information Science* **30**, 5–32.

*Brewer, I. (2002) Cognitive Systems Engineering and GIScience: Lessons learned from a work domain analysis for the design of a collaborative, multimodal emergency management GIS. In Egenhofer, M.J. and Mark, D.M. (eds) *Proceedings of the Second International Conference on GIScience*, Boulder, CO, pp. 22–25.

Brimicombe, A.J. and Li, C. (2009) *Location-Based Services and Geo-Information Engineering*. Chichester: John Wiley & Sons Ltd.

Brimicombe, A.J. and Li, Y. (2006) Mobile space-time envelopes for location-based services. *Transactions in GIS* **10**, 5–23.

Brosnan, M. (1998) *Technophobia: the psychological impact of information technology*. London and New York: Routledge.

Brown, J.S. and Duguid, P. (2000) *The Social Life of Information*. Boston, MA: Harvard Business School Press.

Brunet R. (1986) La carte-modèle et les chorèmes. *Mappemonde* (Apr. 1986), 4–6.

Bruns, H.T. (1994) *Direct Manipulation User Interfaces for GIS Map Algebra*. Master's thesis, University of Maine, Orono ME.

*Bruns, H.T. and Egenhofer, M.J. (1997) User interface for map algebra. *Journal of the Urban and Regional Information Systems Association* **9**(1), 44–54.

Brunswik, E. (1955) Representative design and probabilistic theory. *Psychological Review* **62**, 193–217.

Budic Z.D. (1993) *Human and Institutional Factors in GIS Implementation by Local Governments (Human Factors)*. The University of North Carolina at Chapel Hill (0153), 391.

Bunch, R.L. (2000) *Spatial Cognition and Geographic Information Systems*. University of South Carolina (0202), 134.

Bunch, R. and Lloyd, R. (2006) The cognitive load of geographic information. *The Professional Geographer* **58**(2), 209–220.

Burnett, G. (2000) 'Turn right at the traffic lights': the requirement for landmarks in vehicle navigation systems. *Journal of Navigation* **53**, 499–510.

Busskamp, R. and Liebscher, H.J. (1997) Access to hydrological data from GIS applications by graphical software tools – an example from the Hydrological Atlas of Germany (HAD). In Baumgartner, M.F., Schultz, G.A. and A. I. Johnson (eds) *Remote Sensing and Geographic Information Systems for Design and Operation of Water Resources Systems*. (Proceedings of Rabat Symposium S3, April 1997.) IAHS Publication no 242.

*Buttenfield, B. (1999) Usability evaluation of digital libraries. *Science and Technology Libraries* **17**(3/4), 39–59.

*Buxton, B. (1993) HCI and the inadequacies of direct manipulation systems. *SIGCHI Bulletin* **25**(1), 21–22.

Cai, G. (2007) Contextualization of geospatial database semantics for Human–GIS Interaction. *Geoinformatica* **11**, 217–237.

Cai, G., Sharma, R., MacEachren, A.M. and Brewer, I. (2006) Human-GIS interaction issues in crisis response. *International Journal of Risk Assessment and Management* **6**, 388–407.

CAIN (1996) Map of Belfast divided into Electoral Wards. Maps of Ireland and Northern Ireland. Available at: http://cain.ulst.ac.uk/images/maps/belfast religion.gif [Accessed August 5, (2009)].

Calcinelli, D. and Mainguenaud, M. (1994) Cigales, a visual query language for a geographical information system: the user interface. *Journal of Visual Languages and Computing* **5**, 113–132.

Caplan, L.J. and Lipman, P.D. (1995) Age and gender differences in the effectiveness of map-like learning aids in memory for routes. *Journals of Gerontology Series B: Psychological Sciences and Social Sciences* **50B**(3), 126–133.

Card, S.K., Thomas, T.P. and Newall, A. (1983) *The Psychology of Human-Computer Interaction*. Lawrence Erlbaum Associates.

Carlson-Radvansky, L. and Irwin, D. (1994) Reference frame activation during spatial term assignment. *Journal of Memory and Language* **33**, 646–671.

Carroll, J. (1993) *Human Cognitive Abilities: a survey of factor-analytic studies*. Cambridge: Cambridge University Press.

*Cartwright, W., Crampton, J., Gartner, G., Miller, S., Mitchell, K., Siekierska, E. and Wood, J. (2001) Geospatial information visualization user interface issues. *Cartography and Geographic Information Science* **28**, 45–60.

Carver, S. (2001) Public participation using web-based GIS. Guest editorial. *Environment and Planning B: Planning and Design* **28**(6), 803–804.

Carver, S., Evans, A., Kingston, R. and Turton, I. (2001) Public participation, GIS, and cyberdemocracy: Evaluating on-line spatial decision support systems. *Environment and Planning B: Planning and Design* **28**(6), 907–21.

Castillo, J.C., Hartson, H.R. and Dix, D. (1998) Remote usability evaluation: Can users report their own critical incidents? In Karat, C.M. and Lund, A. (eds) *CHI'98*. Los Angeles: ACM, pp. 253–254.

Cheylan, J., Deffontaines, J., Lardon, S., Thery, H. (1990) Les chorèmes: un outil pour l'étude de l'activité agricole dans l'espace rural. *Mappemonde* **4**, 2–4.

Chittaro, L. (2006) Visualizing information on mobile devices. *Computer* **39** (3 March), 40–45.

Cleveland, W.S. (1993) *Visualizing Data*. New Jersey: Hobart Press.

Cooper, A., Reimann, R. and Cronin, D. (2007) *About Face 3: The Essentials of Interaction Design*. New York: John Wiley & Sons Ltd.

Cordes, R.E. (2001) Task-selection bias: a case for user-defined tasks. *International Journal of Human Computer Interaction* **13**(4), 411–419.

Couclelis, H. (1998) Worlds of information: The geographic metaphor in the visualization of complex information. *Cartography and Geographic Information Systems* **25**, 209–220.

Cowan, N. (2001) The magical number 4 in short-term memory: A reconsideration of mental storage capacity. *Behavioral and Brain Sciences* **24**(1), 87–114.

Cuff, D.J. and Mattson, M.T. (1982) *Thematic Maps: their design and production*. Routledge.

*Davies, C. (1995) Tasks and task descriptions for GIS. In Nyerges, T.L., Mark, D.M., Laurini, R. and Egenhofer, M.J. (eds) *Cognitive Aspects of Human-Computer Interaction for Geographic Information Systems*. Dordrecht, Netherlands: Kluwer, pp. 327–341.

*Davies, C. (1998) Analysing 'work' in complex system tasks: an exploratory study with GIS. *Behavior and Information Technology* **17**(4), 218–230.

Davies, C. (2002) When is a map not a map? Task and language in spatial interpretations with digital map displays. *Applied Cognitive Psychology* **16**, 273–285.

Davies, C., Holt, I., Green, J., Harding, J. and Diamond, L. (2009) User needs and implications for modelling vague named places. *Spatial Cognition and Computation* **9**(3), 174–194.

*Davies, C. and Medyckyj-Scott, D. (1994) GIS usability: recommendations based on the user's view. *International Journal of Geographical Information Systems* **8**, 175–189.

Davies, C. and Medyckyj-Scott, D. (1994) The importance of human factors. In Hearnshaw, H.M. and Unwin, D.J. (eds) *Visualisation in GIS*, Chichester: John Wiley and Sons Ltd.

Davies, C. and Medyckyj-Scott, D. (1995) Feet on the ground: studying user-GIS interaction in the workplace. In Nyerges, T., Mark, D.M., Laurini, R. and Egenhofer, M.J. (eds), *Cognitive Aspects of Human-Computer Interaction for Geographic Information Systems*. Dordrecht: Kluwer Academic Publishers, 123–141.

*Davies, C. and Medyckyj-Scott, D. (1996) GIS users observed. *International Journal of Geographical Information Systems* **10**(4), 363–384.

Davies, C., Wood, L. and Fountain, L. (2005) User-centred GI: hearing the voice of the customer. Paper presented at AGI '05: People, Places and Partnerships, Annual Conference of the Association for Geographic Information, London, UK, 8–10 November 2005. (London: Association for Geographic Information). Available online at: http://www.ordnancesurvey.co.uk/oswebsite/partnerships/research/publications/docs/2005/ClareDavies etal geo.pdf.

De Donatis, M. and Bruciatelli, L. (2006) MAP IT: The GIS software for field mapping with tablet pc. *Computers and Geosciences* **32**, 673–680.

de Oliveira, J.L. and Medeiros, C. (1999) Técnicas, modelos e ferramentas para suporte à construção de interfaces em sistemas de aplicações geográficas. *Proceedings of the XIII Brazilian Software Engineering Symposium*, http://www.inf.ufsc.br/~sbes99/.

de Oliveira, J.L. and Medeiros, C. (2000) A software architecture framework for geographic user interfaces. *Proceedings of the International Workshop on Emerging Technologies for Geo-Based Applications*, Ascona, CH. pp. 233–248.

de Oliveira, J.L., Pires, F. and Medeiros, C.B. (1997) An environment for modeling and design of geographic applications. *GeoInformatica* **1**, 29–58.

De Queiroz, J.E.R. and Ferreira, D.S. (2009). A multidimensional approach for the evaluation of mobile application user interfaces. In Jacko, J.A. (ed.), *Human-Computer Interaction, part I, HCII (2009)*, Berlin, Heidelberg: Springer, pp. 242–251.

de Smith, M.J., Goodchild, M.F., Longley, P.A. (2007) *Geospatial Analysis: a comprehensive guide to principles, techniques and software tools* (2nd edition). Available from http://www.spatialanalysisonline .com/.

Del Fatto, V., Laurini, R., Lopez, K., Loreto, R., Milleret-Raffort, F., Sebillo, M., *et al.* (2007) Potentialities of chorems as visual summaries of spatial databases contents. Proceedings of the 9th International Conference on Visual Information and Information Systems VISUAL (2007), Shanghai (CINA), 28–29 May (2007), in Qui G. et al. (eds) *Lecture Notes in Computer Science*, Springer-Verlag vol. 4781, pp. 537–548.

Densham, P.J. and Armstrong, M.P. (1995) Human-computer interaction: considerations for visual interactive locational analysis. In Nyerges, T.L., Mark, D.M., Laurini, R. and Egenhofer, M.J. (eds) *Cognitive Aspects of Human-Computer Interaction for Geographical Information Systems*, Dordrecht: Kluwer Academic Publishers, pp. 179–196.

Dent, B.D. (1993) *Cartography: Thematic map design*. William C. Brown.

Dillemuth, J. (2005) Map design evaluation for mobile display. *Cartography and Geographic Information Science* **32**, 285–301.

*Dix, A., Finley, J., Abowd, G., and Beale, R. (2004) *Human-Computer Interaction* (3rd edition). Prentice-Hall, Inc.

Dorling, D. (1991) The visualization of spatial structure. PhD Thesis, Department of Geography, University of Newcastle upon Tyne.

Dorling, D. and Fairbairn, D. (1997) *Mapping: Ways of representing the world*. Longman.

Doubleday, A., Ryan, M., Springett, M. and Sutcliffe, A. (1997) A comparison of usability techniques for evaluating design. In *DIS'97 ACM*. Amsterdam, pp. 101–110.

Dourish, P. (2004) What we talk about when we talk about context. *Personal and Ubiquitous Computing Journal* **8**: 19–30.

Downs, R.M. (1997) The geographic eye: seeing through GIS? *Transactions in GIS* **2**, 111–121.

Downton, A. (ed.) (1991) *Engineering the Human-Computer Interface*. New York: McGraw-Hill.

Dray, S. and Siegel, D. (2004) Remote possibilities: international usability testing at a distance. *Interactions* (March and April), 10–17.

Driber, B. and Liles, W. (1983) A Communication model for the design of a computer-assisted cartographic system. *Proceedings of the Fifth International Symposium on Cartography and Computing*, American Congress on Surveying and Mapping. Falls Church, VA, pp. 267–274.

Drummond, J., Joao, E. and Billen, R. (2007) Current and future trends in dynamic and mobile GIS. In Drummond, J., Billen, R., Joao, E. and Forrest, D. (eds) *Dynamic and Mobile GIS: Investing changes in space and time*, Boca Raton, FL: CRC Press, pp. 289–300.

Dubois, E., Truillet, P. and Bach, C. (2007) *Evaluating Advanced Interaction Techniques for Navigating Google Earth*. Volume 2. Proceedings of the 21st BCS HCI Group Conference.

Dumas, J.S. and Redish, J.C. (1999) *A Practical Guide to Usability Testing*. Bristol: Intellect.

Dunbar, K. and Blanchette, I. (2001) The invivo/invitro approach to cognition: the case of analogy. *Trends in Cognitive Sciences*, pp. 5334–5339.

Dunn, C.E. (2007) Participatory GIS – a people's GIS? *Progress in Human Geography* **31**(5), 616–637.

*Dykes, J., MacEachren, A.M. and Kraak, M.-J., (eds.) (2005) *Exploring Geovisualization*. Amsterdam, The Netherlands: Elsevier.

Edsall, R.M. (2003) Design and usability of an enhanced geographic information system for exploration of multivariate health statistics. *Professional Geographer* **55**, 146–160.

Edson, D. (1979) The International Cartographic Association – An overview. In *Proceedings of the International Symposium on Cartography and Computing: Applications in Health and Environment*, Reston, Virginia, pp. 164–167.

Egenhofer, M.J. (1989) *Spatial Query Languages (Query Languages)*. University of Maine (0113), 232.

*Egenhofer, M.J. (1997) Query processing in Spatial-Query-by-Sketch. *Journal of Visual Languages and Computing* **8**, 403–424.

Egenhofer, M.J. (2002) Toward the semantic geospatial web. *Proceedings of the tenth ACM international symposium on Advances in Geographic Information Systems*, McLean, VA: ACM Press, 1–4.

*Egenhofer, M.J. and Bruns, H.T. (1997) Visual map algebra: a direct-manipulation user interface for GIS. Proceedings of the third IFIP WG2. 6 working conference on Visual Database Systems 3 (VDB-3) table of contents, pp. 235–253.

Egenhofer, M.J., Glasgow, J., Gunther, O., Herring, J.R. and Peuquet, D.J. (1999) Progress in computational methods for representing geographical concepts. *International Journal of Geographical Information Science* **13**, 775–796.

Egenhofer, M.J. and Mark, D.M. (1995) Modeling conceptual neighbourhoods of topological line-region relations. *International Journal of Geographical Information Systems* **9**, 555–565.

*Egenhofer, M.J. and Mark, D.M. (1995) *Naive Geography*. Santa Barbara, CA: National Center for Geographic Information and Analysis, Report 95–8. Online from http://www.geog.buffalo.edu/ncgia/i21/ng/NG51.html.

Egenhofer, M.J. and Mark, D.M. (1995) Naive geography. *Spatial Information Theory* 1–15.

*Egenhofer, M.J. and Richards, J.R. (1993) Exploratory access to geographic data based on the map overlay metaphor. *Journal of Visual Languages and Computing* **4**, 105–125.

Einstein, A. (1934) On the method of theoretical physics. *Philosophy of Science* **1**(2), 163–169.

Ekstrom, R.B., French, J.W., Harman, H.H. and Dirmen, D. (1976) *Manual for Kit of Factor-referenced Cognitive Tests*. Princeton NJ: Educational Testing Service.

Ellis, C.A., Gibbs, S.J., and Rein, G. (1991) Groupware: some issues and experiences. *Communications of the ACM* **34**, 1, 39–58.

Ellis, C.D., Quiroga, C., Shin, S.Y. and Pina, R.J. (2003) GIS and human-centered systems design: using ethnographic data collection and analysis methods to design a utility permitting support system. *URISA Journal* **15**.

Ellul, C., Haklay, M., Francis, L. and Rahemtulla, H. (2009) *A Mechanism to Create Community Maps for Non-Technical Users*. The International Conference on Advanced Geographic Information Systems & Web Services – GEOWS (2009), Cancun, Mexico, 1–7 February.

Environment Agency (2009) Environment Agency-What's In Your backyard? [online] Available at: http://maps.environment-agency.gov.uk/wiyby/wiybyController?ep=maptopics&lang=e [Accessed 24 August (2009)]

Evans, I.S. (1977) The selection of class intervals. *Transactions of the Institute of British Geographers* 98–124.

Fabrikant, S.I. (2001) Evaluating the usability of the scale metaphor for querying semantic spaces. In Montello, D.R. (ed.) *Spatial Information Theory: Foundations of Geographic Information Science*. Conference on Spatial Information Theory (COSIT'01), p. 156.

Fabrikant, S.I. and Buttenfield, B.P. (2001) Formalizing semantic spaces for information access. *Annals of the Association of American Geographers* **91**, 263–280.

Fabrikant, S.I., Montello, D.R. and Mark, D.M. (2006) The distance-similarity metaphor in region display spatializations. *IEEE Computer Graphics and Applications* **26**, 34–44.

Fairbairn, D., Andrienko, G.L., Andrienko, N.V., Buziek, G. and Dykes, J. (2001) Representation and its relationship with cartographic visualization. *Cartography and Geographic Information Science* **28**, 13–28.

Finch, H. and Lewis, J. (2003) Focus groups. In Ritchie, J. and Lewis, J. (eds) *Qualitative Research Practice: A Guide for Social Science Students and Researchers*. London: Sage, pp. 170–198.

Finley, D.B. (1997) *Collaborative GIS in a Distributed Work Environment*. The University of New Brunswick Canada (0823), 130.

Friedman, T. (2006) *The World is Flat*, London: Penguin.

Fisher, P.F. (1998) Is GIS hidebound by the legacy of cartography? *Cartographic Journal* **35**, 5–9.

Fonseca, F.T., Agouris, P., Egenhofer, M.J. and Camara, G. (2002) Using ontologies for integrated geographic information systems. *Transactions in GIS* **6**, 231–257.

Fonseca, F.T., Davis, C. and Camara, G. (2003) Bridging ontologies and conceptual schemas in geographic information integration. *GeoInformatica* **7**, 355–378.

Fonseca, F.T. and Martin, J.E. (2005) Toward an alternative notion of information systems ontologies: Information engineering as a hermeneutic enterprise. *Journal of the American Society for Information Science and Technology* **56**, 46–57.

Fotheringham, A.S., Brunsdon, C. and Charlton, M. (2002) *Geographically Weighted Regression: the analysis of spatially varying relationships*. Chichester: John Wiley & Sons Ltd.

Franges, S., Posloncec-Petric, V. and Zupan, R. (2007) Continuous development of cartographic visualisation. In ICA (International Cartographic Association) 23rd International Cartographic Conference, Moscow, Russia 4–10 August. Available at: http://bib.irb.hr/datoteka/358941.3.P.2 CONTINUOUS DEVELOPMENT OF CARTOGRAPHIC VISU.pdf [accessed 13 July (2007)].

*Frank, A.U. (1993) The use of geographical information systems: The user interface is the system. In Medyckyj-Scott, D. and Hearnshaw, H.M. (eds.) *Human Factors in Geographical Information Systems*. London: Belhaven Press, pp. 3–14.

Frank, A.U. (2000) Geographic information science: new methods and technology. *Journal of Geographical Systems* **2**, 99–105.

Frank, A.U. (2001) Tiers of ontology and consistency constraints in geographical information systems. *International Journal of Geographical Information Science* **15**, 667.

Frank, A.U. (2003) Ontology for spatio-temporal databases. In Koubarakis, M. (ed.) *Spatio-Temporal Databases*, Springer, 9–77.

*Frank, A.U. and Mark, D.M. (1991) Language issues for geographical information systems. In Maguire, D., Goodchild, M. and Rhind, D. (eds) *Geographical Information Systems: principles and applications*. London: Longman, pp. 227–237.

Gaffney, G. (2000) Card sorting. http://www.infodesign.com.au/ftp/CardSort.pdf (Accessed: 26/07/2006).

Gahegan, M. (2001) Visual exploration in geography: Analysis with light. In Miller, H.J., Han, J., *Geographic Data Mining and Knowledge Discovery*. London: Taylor and Francis, pp. 260–287.

Gastner, M.T. and Newman, M.E.J. (2004) Diffusion-based method for producing density-equalizing maps. *Proceedings of the National Academy of the United States of America* **101**(20), 7499–7504. Accessible through: http://www.pnas.org/cgi/content/abstract/101/20/7499.

Gerring, J. (2004) What is a case study and what is it good for? *American Political Science Review* **98**(2), 341–354.

Gertler, M. (2003) Tacit knowledge and the economic geography of context. *Journal of Economic Geography* **3**, 75–99.

Geyer, M.W. (1993) *The Effects of a Knowledge-Based System on Organizational Information Input Overload (Organizational Information)*. Texas A and M University (0803), 159.

Gilhooly, K., Wood, M., Kinnear, P. and Green, C. (1988) Skill in map reading and memory for maps. *Quarterly Journal of Experimental Psychology* **40A**(1), 87–107.

Gillespie T.K. (1991) *Mapping Thoughts: Visual Interfaces for Information Retrieval (Interface Design)*. University of California Berkeley (0028), 254.

Gilmartin, P. and Patton, J. (1984) Comparing the sexes on spatial abilities: Map use skills. *Annals of the Association of American Geographers* **74**(4), 605–619.

*Golledge, R.G. (1992) Do people understand spatial concepts: the case of first-order primitives. In Frank, A.U. and Formentini, U. (eds) *Theories and Methods of Spatio-Temporal Reasoning in Geographic Space*. Lecture Notes in Computer Sciences, Vol. 639. New York: Springer, pp. 1–21.

Golledge, R.G. (2003) Reflections on recent cognitive behavioural research with an emphasis on research in the United States of America. *Australian Geographical Studies* **41**, 117–130.

*Golledge, R.G. (2005) Reflections on procedures for learning environments without the use of sight. *Journal of Geography* **104**(3), 95–103.

Golledge, R.G., Dougherty, V. and Bell, S. (1995) Acquiring spatial knowledge: Survey versus route based knowledge in unfamiliar environments. *Annals of the Association of American Geographers* **85**(1), 134–158.

Goodchild, M.F. (1988) Towards an enumeration and classification of GIS functions. *Proceedings, International Geographic Information Systems (IGIS) Symposium*, II. Washington, DC: National Aeronautics and Space Administration, pp. 67–77.

Goodchild, M.F. (1992) Geographical information science. *International Journal of Geographical Information Systems* **6**(1), 31–45.

Gould, M.D. (1989) Human factors research and its value to GIS user interface design. *Proceedings, GIS/LIS'89*, vol. 1, Orlando, Florida, pp. 542–550.

Gould, M.D. (1993) Two views of the User Interface. In Medyckyj-Scott, D. and Hearnshaw, H.M. (eds) *Human Factors in Geographical Information Systems*. London: Belhaven Press, pp. 101–110.

Gould, M.D. (1994) Defining GIS tasks: when you say what you do, do you mean what you say? *GIS Europe* **3**(7).

Gould, M.D. (1994) *Map Use, Spatial Decisions, and Spatial Language in English and Spanish (Puerto Rico)*. State University of New York at Buffalo (0656), 193.

Gould, M.D. and McGranaghan, M. (1990) Metaphor in Geographic Information Systems. In: Brassel, K.E. and H. Kishimoto (eds) *4th International Symposium on Spatial Data Handling*. Zürich, Switzerland (SDH '90). Volume 1/2. Columbus, Ohio: International Geographical Union (IGU), *Commission on Geographic Information Systems*, pp. 433–442.

Gould, M.D., Nunes, J., Comas, D., Egenhofer, M.J., Freundschuh, S. and Mark, D.M. (1996) Formalizing informal geographic information: Cross-cultural human subjects testing. In Rumor, M., McMillan, R. and Ottens, H.F.L., (eds) *Geographical Information – from Research to Application through Cooperation*, Vols. 1 and 2, pp. 285–294.

Gray, W.D. and Salzman, M.C. (1998) Damaged merchandise? A review of experiments that compare usability evaluation methods. *Human-Computer Interaction* **13**(3), 203–261.

Green, D.R. and Horbach, S. (1998) Colour – difficult to both choose and use in practice. *Cartographic Journal* **35**, 169–180.

Green, T.R.G. (1990) The cognitive dimension of viscosity: a sticky problem for HCI. In D. Diaper, D. Gilmore, G. Cockton and B. Shackel (eds.) *Human-Computer Interaction - INTERACT '90*, Amsterdam: North-Holland, pp. 79–86.

Greif, S. (1991) The role of German work psychology in the design of artefacts. In Carroll, J.M. (ed.) *Designing Interaction: Psychology at the Human-Computer Interface*, Cambridge: Cambridge University Press, pp. 203–226.

Griffin, A.L. (2004) Understanding how scientists use data display devices for interactive visual computing with geographical models. Penn State University. PhD.

Gunzelmann, G., Anderson, J.R. and Douglass, S. (2004) Orientation tasks with multiple views of space: strategies and performance. *Spatial Cognition and Computation* **4**, 207–253.

*Guo, D. (2003) A geographic visual query composer (GVQC) for accessing federal databases. In *Proceedings of National Conference for Digital Government Research*, pp. 397–400.

Haklay, M. (2006) Usability dimensions of collaborative GIS. In Dragicevic, S. and Balram, S. (eds) *Collaborative Geographic Information Systems*, Idea Group Publishing, pp. 24–42.

*Haklay, M. and Tobón, C. (2003) Usability evaluation and PPGIS: Towards a user-centred design approach. *International Journal of Geographical Information Science* 17, 577–592.

*Haklay, M. and Zafiri, A. (2008) Usability engineering for GIS: learning from a screenshot. *The Cartographic Journal* 45(2), pp. 87–97.

Haniff, D.J. and Baber, C. (2003) User evaluation of augmented reality systems. In Banissi, E., Borner, K., Chen, C., Clapworthy, G., Maple, C., Lobben, A., Moore, C., Roberts, J., Ursyn, A. and Zhang, J. (eds) *Seventh International Conference on Information Visualization*. London, pp. 505–511.

*Harrower, M. and Sheesley, B. (2005) Designing better map interfaces: a framework for panning and zooming. *Transactions in GIS* 9(2), pp. 77–89.

Harrower, M., Keller, P. and Hocking, D.M. (1997) Cartography on the Internet: thoughts and preliminary user survey. *Cartographic Perspectives* 26, 27–37.

Hartson, H.R., Andre, T.S. and Williges, R.C. (2001) Criteria for evaluating usability evaluation methods. *International Journal of Human Computer Interaction* 13(4), 373–410.

Harvey, F. (1997) Improving multi-purpose GIS design: participative design. In *Proceedings of the international Conference on Spatial Information Theory: A theoretical Basis For GIS (October 15–18)*. S.C. Hirtle and A.U. Frank, (eds) Lecture Notes In Computer Science, vol. 1329. London: Springer-Verlag, 313–328.

Harvey, F. (2008) *A Primer of GIS: fundamental geographic and cartographic concepts*. The Guilford Press.

*Harvey, F., Kuhn, W., Pundt, H., Bishr, Y. and Riedemann, C. (1999) Semantic interoperability: A central issue for sharing geographic information. *Annals of Regional Science* 33, 213–232.

Harwell, R. (2004) Web mapping with SVG. Directions Magazine [internet] 5 November. Available at: http://www.directionsmag.com/article.php?article id=693&trv=1 [Accessed 15 June (2009)].

Haunold, P. and Kuhn, W. (1994) A keystroke level analysis of a graphics application: manual map digitizing. In *CHI '94: Celebrating Interdependence,* Association for Computing Machinery Conference on Human Factors in Computing Systems, Boston, Massachusetts, USA, 24–28 April, New York: Association for Computing Machinery, pp. 337–343.

Hegarty, M., Richardson, A.E., Montello, D.R., Lovelace, K. and Subbiah, I. (2002) Development of a self-report measure of environmental spatial ability. *Intelligence* 30, 425–447.

Heipke, C. (2004) Some requirements for geographic information systems: a photogrammetric point of view. *Photogrammetric Engineering and Remote Sensing* 70, 185–195.

Hertzum, M. and Jacobsen, N.E. (2001) The evaluator effect: a chilling fact about usability evaluation methods. *International Journal of Human Computer Interaction* 13(4), 421–443.

Hirtle, S.C. and Hudson, J. (1991) Acquisition of spatial knowledge for routes. *Journal of Environmental Psychology* 11(4), 335–345.

Hirtle, S.C. and Jonides, J. (1985) Evidence of hierarchies in cognitive maps. *Memory and Cognition* 13(3), 208–217.

Hong Jung, H. (1994) *Qualitative Distance and Direction Reasoning in Geographic Space (Distance Reasoning)*. University of Maine (0113), 188.

Hook, K. (undated) Lab-Studies vs Ecological Validity; What is the Point of Evaluation Studies?; What is Scientific and What is Not?; What is HCI and What is General Emotion Theory? http://www.sics.se/~kia/ideas/evaluation.htm (Accessed: 24/10/2006).

*Hornbaek, K. (2002) Navigation patterns and usability of zoomable user interfaces with and without an overview. *ACM Transactions on Human-Computer Interaction* 9(4), pp. 362–389.

Hutchingson, J.A. and Wittmann, J.H. (1997) Map design and production issues for the Utah GAP Analysis Project. *Cartography and Geographic Information Systems* 24, 91–100.

IIED (2006) Mapping for Change: practice, technologies and communication. *Participatory Learning and Action*, April, IIED.

Iliffe, J. and Lott, R. (2008) *Datums and Map Projections for Remote Sensing, GIS and Surveying*. Dunbeath: Whittles.

Ishikawa, T., Barnston, A.G., Kastens, K.A., Louchouarn, P. and Ropelewski, C.F. (2005) Climate forecast maps as a communication and decision-support tool: An empirical test with prospective policy makers. *Cartography and Geographic Information Science* **32**, 3–16.

ISO 13407 (1999) *Human-Centered Design for Interactive Systems*. Geneva, Switzerland: International Organisation for Standardisation.

Jackson, J. (1990) Visualization of metaphors for interaction with geographic information systems. M.S. thesis, University of Maine, Orono ME.

Jäger, E., Altintas, I., Zhang, J., Ludäscher, B., Pennington, D. and Michener, W. (2005) A scientific workflow approach to distributed geospatial data processing using web services. *Proceedings of the 17th International Conference on Scientific and Statistical Database Management* (SSDBM'05), 27–29 June 2005, Santa Barbara, CA.

Jahn, M. and Frank, A.U. (2004) How to increase usability of spatial data by finding a link between user and data. AGILE Association Geographic Information Laboratories Europe 7th Conference on Geographic Information Science, Heraklion, Crete.

Jakobsson, A. (2002) User requirements for mobile topographic maps. GiMoDig-project, IST-(2000)-30090. At http://gimodig.fgi.fi/deliverables, accessed 5/2005.

Jankowski, P. and Nyerges, T.L. (1989) Design considerations for Mapkbs-map projection knowledge-based system. *American Cartographer* **16**, 85–95.

*Jankowski, P. and Nyerges, T.L. (2001) GIS-supported collaborative decision making: Results of an experiment. *Annals of the Association of American Geographers* **91**, 48–70.

Jankowski, P., Nyerges, T.L., Smith, A., Moore, T.J. and Horvath, E. (1997) Spatial group choice: a SDSS tool for collaborative spatial decision-making. *International Journal of Geographical Information Science* **11**, 577–602.

Jenks, G.F. and Caspall, F.C. (1971) Error on choroplethic maps: definition, measurement, reduction. *Annals of the Association of American Geographers* **61**(2), 217–244.

Jenks, G.F. and Coulson, M.R.C. (1963) Class intervals for statistical maps. *International Yearbook of Cartography* **3**, 119–134.

Johnson, J. (2000) *GUI Bloopers: Don'ts and Do's for Software Developers and Web Designers*. Morgan Kaufmann Publishers.

Johnson, J., Roberts, T., Verplank, W., Smith, D., Irby, C., Beard, M., Mackey, K. (1989) The Xerox Star: A retrospective. *IEEE Computer* **22**(9), September.

Johnston, R.J. *et al.*, (2000) *The Dictionary of Human Geography*. Blackwell Publishers.

Jones, C.E., Haklay, M., Griffiths, S. and Vaughan, L. (2008) Visualising London's suburbs. In Geographical Information Science Research – UK (GISRUK). Manchester Metropolitan University, UK, April. http://eprints.ucl.ac.uk/5184/.

Jones, C.E., Haklay, M., Griffiths, S. and Vaughan, L. (2009) A less-is-more approach to geovisualization – enhancing knowledge construction across multidisciplinary teams, *International Journal of Geographical Information Science* **23**(8), 1077–1093.

Jung, S. and Albrecht, J. (1997) Multi-level comparative analysis of spatial operators in GIS and remote sensing as a foundation for an integrated GIS. In Förstner, W. and L. Plümer (eds.) *Semantic Modeling for the Acquisition of Topographic Information from Images and Maps, SMATI '97*, Basel Boston: Birkhäuser Verlag, pp. 72–88.

Kaasinen, E. (2003) User needs for location aware mobile services. *Personal Ubiquitous Computing* **7**, 70–79.

Kazman, R. and Chen, H. (2009) The metropolis model: a new logic for development of crowdsourced systems. *Communications of the ACM* **52**(7), 76–84.

Kealy, W.A. and Webb, J.M. (1995) Contextual influences of maps and diagrams on learning. *Contemporary Educational Psychology* **20**(3), 340–358.

Kelly, J.E. (1982) *Scientific Management, Job Redesign and Work Performance* (London: Academic Press).

Kingston, R. (2002) Web-based PPGIS in the United Kingdom. In Craig, W.J., Harris, T., and Weiner, D. (eds) *Community Participation and Geographic Information Systems*. London: Taylor & Francis, pp. 101–112.

Kirwan, B. and Ainsworth, L.K. (eds.) (1992) *A Guide to Task Analysis*, Washington DC: Taylor & Francis.

*Kitchin, R.M. (1996) Are there sex differences in geographic knowledge and understanding? *Geographical Journal* **162**(3), 273–286.

Kjeldskov, J. and Graham, C. (2003) A review of mobile HCI research methods. *Mobile HCI* (2003). Springer-Verlag, Berlin, 317–335.

*Klien, E., Lutz, M. and Kuhn, W. (2006) Ontology-based discovery of geographic information services – An application in disaster management. *Computers Environment and Urban Systems* **30**, 102–123.

Klippel A., Richter K.-F., Hansen S. (2005) Wayfinding Choreme Maps in Proceedings of the 8th International Conference on Visual Information and Information Systems VISUAL 2005, Amsterdam, The Netherlands, July 5, 2005, in *Lecture Notes in Computer Science*, Springer-Verlag vol. **3736**, Bres, S. and Laurini, R. (Eds.).

Kösters, G., Pagel, B. and Six, H. (1996a) GeoOOA: Object-oriented analysis for geographic information systems, in *Proceedings of 2nd IEEE International Conference on Requirements Engineering*, pp. 245–253. Los Alamitos: IEEE Press.

Kösters, G., Six, H.W. and Voss, J. (1996b) Combined analysis of user interface and domain requirements. In *Proceedings of 2nd IEEE International Conference on Requirements Engineering*, pp. 199–207. Los Alamitos: IEEE Press.

*Koua, E.L. and Kraak, M.-J. (2004) A Usability Framework for the Design and Evaluation of an Exploratory Geovisualization Environment. Eighth International Conference on Information Visualisation (IV'04) IEEE Computer Society, pp. 153–158.

Koua, E.L., Maceachren, A. and Kraak, M.J. (2006) Evaluating the usability of visualization methods in an exploratory geovisualization environment. *International Journal of Geographical Information Science* **20**, 425–448.

Koussoulakou, A. and Stylianidis, E. (1999) The use of GIS for the visual exploration of archaeological spatio-temporal data. *Cartography and Geographic Information Science* **26**, 152–160.

Kozhevnikov, M. and Hegarty, M. (2001) A dissociation between object manipulation spatial ability and spatial orientation ability. *Memory and Cognition* **29**(5), 745–756.

Kraak, M.J. and Brown, A. (2000) *Web Cartography: Developments and Prospects*, NY: Taylor & Francis.

Kraak, M.J. and Brown, A. (2001) *Web Cartography*. CRC Press.

Kraak, M.J. and Ormeling, F.J. (1996) *Cartography: Visualization of Spatial Data*, Prentice Hall.

Kraak, M.J. and Ormeling, F.J. (2003) *Cartography: visualization of geospatial data*. Addison-Wesley Longman Ltd.

Kramers, E. (2008) Interaction with maps on the internet – a user centred design approach for the atlas of Canada, *The Cartographic Journal*, Use and Users Special Issue, **45**(2), 98–107.

Krueger, R.A. (1994) *Focus Groups: A Practical Guide for Applied Research*. Sage: London.

Krygier, J. and Wood, D. (2005) *Making Maps: a visual guide to map design for GIS*. Guilford Press.

Kuhn, W. (ed.) (2002) *Modeling the Semantics of Geographic Categories through Conceptual Integration*. Berlin/Heidelberg: Springer.

Kuhn, W. (2003) Semantic reference systems. *International Journal of Geographical Information Science* **17**, 405–409.

Kuhn, W. (2005) Geospatial semantics: Why, of what, and how?, *Journal on Data Semantics* **52**, 1–24.

Kuhn,W. and Frank, A.U. (1991) A formalization of metaphors and image schemas in user interfaces. In Mark, D.M. and Frank, A.U. (eds) *Cognitive and Linguistic Aspects of Geographic Space*. Kluwer: Dordrecht.

Kuhn, W. and Raubal, M. (2003) Implementing Semantic Reference Systems. AGILE (2003) – 6th AGILE Conference on Geographic Information Science. Lyon, France: Presses Polytechniques et Universitaires Romandes, 63–72.

*Kuhn, W., Willauer, L., Mark, D.M. and Frank, A.U. (1992) User interfaces for geographic information systems: discussions at the specialist meeting. In *User Interfaces for Geographic Information Systems: Report on the specialist Meeting*, Mark, D.M. and Frank, A.U. (eds). Buffalo, NY: NCGIA.

Lakoff, G. and Johnson, M. (1980) *Metaphors We Live By*. Chicago: University of Chicago Press.

Landauer, T.K. (1995) *The Trouble with Computers: Usefulness, Usability, and Productivity*, MIT Press, Cambridge, MA.

Lanter, D.P. (1994) A Lineage Metadata Approach to Removing Redundancy and Propagating Updates in a GIS Database. *Cartography and Geographic Information Systems* **21**(2), 91–98.

*Lanter, D.P. and Essinger, R. (1991) User-Centered Graphical User Interface Design for GIS. *USCB* **24**.

Lardon, S. (2003) Usage des chorèmes, graphes et jeux dans le diagnostic de territoire. Engref Pop'Ter. Clermont-Ferrand.

Laurini, R., Milleret-Raffort, F. and Lopez, K. (2006) A Primer of Geographic Databases Based on Chorems. In Proceedings of the SebGIS Conference, Montpellier, *Lecture Notes in Computer Science vol. 4278*, Springer Verlag, pp. 1693–1702.

Lawton, C.A. (1994) Gender differences in way-finding strategies – relationship to spatial ability and spatial anxiety. *Sex Roles* **30**(11–12), 765–779.

Lawton, C.A. (2001) Gender and regional differences in spatial referents used in direction giving. *Sex Roles* **44**(5–6), 321–337.

Lawton, C.A. and Kallai, J. (2002) Gender differences in wayfinding strategies and anxiety about wayfinding: A cross-cultural comparison. *Sex Roles* **47**(9–10), 389–401.

Leitner, M. and Buttenfield, B.P. (2000) Guidelines for the display of attribute certainty. *Cartography and Geographic Information Science* **27**, 3–14.

Lembo Jr, A.J. and O'Rourke, T.D. (2003) Software for the user interface of coordinate rubbersheeting in vector-based geographic information systems. *Surveying and Land Information Science* **63**, 155–160.

Lemmens, R. (2006) *Semantic interoperability of distributed geo-services*. Publications on Geodesy, 63. Delft, NL: Netherlands Geodetic Commission.

Levine, M. (1982) You are here maps: psychological considerations. *Environment and Behavior* **14**, 221–237.

Li, C. (2006) User preferences, information transactions and Location-Based Services: a study of urban pedestrian wayfinding. *Computer Environment and Urban Systems* **30**, 726–740.

*Li, C. and Longley, P. (2006) A test environment for location-based services applications. *Transactions in GIS* **10**, 43–61.

Li, C. and Maguire, D. (2003) The handheld revolution: towards ubiquitous GIS. In Longley, P. and Batty, M. (eds). *Advanced Spatial Analysis: The CASA book of GIS*, Redlands, CA: ESRI Press, pp. 193–210.

Li, C. and Willis, K. (2006) Modelling context aware interaction for wayfinding using mobile devices. *Proceedings of Mobile HCI 06*, Espoo, 97–100.

Li, S. and Coleman, D. (2005) Modelling distributed GIS data production workflow. *Computers, Environment and Urban Systems* **29**, 401–424.

Liben, L.S., Myers, L.J. and Kastens, K.A. (2008) Locating oneself on a map in relation to person qualities and map characteristics. In Freksa, C., Newcombe, N., Gardenfors, P. and Hoelscher, C. (eds) *Spatial Cognition VI: Learning, reasoning and talking about space. Lecture Notes in Artificial Intelligence* Vol. **5248**. Berlin: Springer. pp. 171–187

Light, A. (2004) *Design Patterns for Cartography and Data Graphics*. University of Oregon (0171), 97.

Lindgaard, G., Tsuji, B. and Khan, S. (2005) Ecological validity and behavioural measures in the usability testing of new applications: A workshop in reality testing. In *Proceedings of HCI 2005*. Springer-Verlag: Napier University, Edinburgh, Scotland.

Liu, Y., Gao, Y., Wang, Y., Wu, L. and Wang, L. (2005) A component-based geo-workflow framework: A discussion on methodological issues. *Journal of Software* **16**(8), 1395–1406.

Liu, Y., McGrath, R., Myers, J. and Futrelle, J. (2007) Towards a rich-context participatory cyberenvironment. *Proceedings of International Workshop on Grid Computing environments*. November 11–12, Reno, NV. Online resource at http://library.rit.edu/oajournals/index.php/gce/article/view/101/76, last accessed 15 Aug. (2009)

*Lloyd, R. and Bunch, R.L. (2003) Technology and map-learning: Users, methods, and symbols. *Annals of the Association of American Geographers* **93**, 828–850.

*Lobben, A.K. (2004) Tasks, strategies, and cognitive processes associated with navigational map reading: A review perspective. *Professional Geographer* **56**, 270–281.

Lohman, D., Pellegrino, J.W., Alderton, D. and Regian, J.W. (1987) Dimensions and components of individual differences in spatial abilities. In S.H. Irvine and S.E. Newstead (eds) *Intelligence and Cognition: contemporary frames of reference* Dordrecht: Nijhoff, pp. 253–312.

Longley, P.L., Goodchild, M.F., Maguire, D.J. and Rhind, D.W. (2001) *Geographic Information Systems and Science* (1st Edition). Chichester: John Wiley & Sons Ltd.

Longley, P.L., Goodchild, M.F., Maguire, D.J. and Rhind, D.W. (2005) *Geographic Information Systems and Science* (2nd edition). Chichester: John Wiley & Sons Ltd.

Lowe, R. (1999) Extracting information from an animation during complex visual learning. *European Journal of Psychology of Education* **14**, 225–244.

Lu, X.J., Zhou, C.H. and Gong, J.H. (1999) On geographic spatial thinking in images – The development of spatial mental images. *Acta Geographica Sinica* **54**, 401–409.

Lutz, M. (2005) Ontology-based service discovery in spatial data infrastructures. *Proceedings of the 2005 Workshop on Geographic Information Retrieval*, Bremen, Germany: ACM Press, pp. 45–54.

Lutz, M. (2007) Ontology-based descriptions for semantic discovery and composition of geoprocessing services, *GeoInformatica* **11**(1), 1–36.

Lutz, M. and Klien, E. (2006) Ontology-based retrieval of geographic information. *International Journal of Geographical Information Science* **20**, 233–260.

Lutz, M., Riedemann, C. and Probst, F. (2003) A classification framework for approaches to achieving semantic interoperability between GI web services. *Spatial Information Theory, Proceedings*, pp. 186–203.

Lynch, K. (1960) *The Image of the City*. Cambridge, MA: Joint Center for Urban Studies (MIT & Harvard).

Mac Aoidh, E. (2006) Personalised multimodal interfaces for mobile geographic information systems. *Adaptive Hypermedia and Adaptive Web-Based Systems,* Proceedings, pp. 452–456.

MacEachren, A.M. (1992) Visualizing uncertain information. *Cartographic Perspective* **13**(3), 10–19.

MacEachren, A.M. (2000) Cartography and GIS: facilitating collaboration. *Progress in Human Geography* **24**, 445–456.

*MacEachren, A.M. (2004) *How Maps Work: representation, visualization, and design*. The Guilford Press.

MacEachren, A.M. (2004) Visualization for constructing and sharing geo-scientific concepts. *Proceedings of the National Academy of Sciences of the United States of America* **101**, 5279.

MacEachren, A.M. (2005) Enabling collaborative geoinformation access and decision-making through a natural, multimodal interface. *International Journal of Geographical Information Science* **19**, 293.

MacEachren, A.M. and Cai, G. (2006) Supporting group work in crisis management: Visually mediated human – GIS – human dialogue. *Environment and Planning B: Planning and Design* **33**, 435–456.

*MacEachren, A.M. and Kraak, M.J. (eds) (2001) Research Challenges in Geovisualization. *Cartography and Geographic Information Science* **28**(1).

MacEachren, A.M., Cai, G., Sharma, R., Rauschert, I., Brewer, I., Bolelli, L., *et al.* (2005) Enabling collaborative geoinformation access and decision making through a natural, multimodal interface. *International Journal of Geographical Information Science* **19**, 293–317.

MacEachren, A.M., Robinson, A., Hopper, S., Gardner, S., Murray, R., Gahegan, M. and Hetzler, E. (2005) Visualizing geospatial information uncertainty: What we know and what we need to know. *Cartography and Geographic Information Science* **32**, 139–160.

MacEachren, A.M. and Taylor, D.R. (1994) *Visualization in Modern Cartography.* Pergamon.

Maguire, M.C. (1998) *Usability Methods: Diary keeping.* http://www.lboro.ac.uk/eusc/g diary keeping.html (Accessed: 24/03/2005).

Marcus, A. and Gasperini, J. (2006) Almost dead on arrival: a case study of non-user-centered design for a police emergency-response system. *Interactions* **13**(5), 12–18.

Mark, D.M. (1989) Cognitive image-schemata for geographic information: relations to user views and GIS interfaces. *Proceedings, GIS/LIS'89, Orlando, Florida,* vol. 2, pp. 551–560.

Mark, D.M. (1992) Spatial metaphors for human-computer interaction. *Proceedings, Fifth International Symposium on Spatial Data Handling,* Charleston, South Carolina, 1, pp. 104–112.

Mark, D.M. (1993) Human spatial cognition. In Medyckyj-Scott, D. and Hearnshaw H.M. (eds) *Human Factors in Geographic Information Systems,* London: Belhaven, pp. 51–60.

Mark, D.M., Comas, D., Egenhofer, M.J., Freundschuh, S.M., Gould, M.D. and Nunes, J. (1995) Evaluating and refining computational models of spatial relations through cross-linguistic human subjects testing. *Spatial Information Theory: A Theoretical Basis for GIS. Lecture Notes in Computer Science,* Vol. 988. Berlin: Springer-Verlag, pp. 553–568.

Mark, D.M. and Egenhofer, M.J. (1994) Calibrating the meanings of spatial predicates from natural language: Line-region relations. In Waugh, T.C. and Healey, R.G. (eds) *Advances in GIS Research,* Vols. 1 and 2, pp. 538–553.

Mark, D.M. and Egenhofer, M.J. (1994) Modeling spatial relations between lines and regions: combining formal mathematical models and human subjects testing. *Cartography and Geographic Information Systems* **21**, 195–212.

Mark, D.M. and Egenhofer, M.J. (1996) Common-sense geography: Foundations for intuitive geographic information systems. *Gis/Lis '96* – Annual Conference and Exposition Proceedings, pp. 935–941.

Mark, D.M. and Gould, M.D. (1991) Interacting with geographic information. *Photogrammetric Engineering and Remote Sensing* **57**, 1427–1430.

Mark, D.M., Smith, B. and Tversky, B. (1999) Ontology and geographic objects: an empirical study of cognitive categorization. In Freksa, C. and Mark, D.M. (eds) *Spatial Information Theory: A Theoretical Basis for GIS,* Berlin: Springer-Verlag, pp. 283–298.

Mark, D.M., Smith, B., Egenhofer, M.J. and Hirtle, S.C. (2005) Ontological foundations for geographic information science. *Research Agenda for Geographic Information Science* 335–350.

Mark, D.M. and Turk, A.G. (2003) Landscape categories in Yindjibarndi: Ontology, environment, and language. *Spatial Information Theory, Proceedings,* pp. 28–45.

Markopoulos, P., Ruyter, B. and Mackay, W. (2007) Introduction to this special issue on awareness systems design. *Human-Computer Interaction* **22**, 1–6.

Marr, A., Pascoe, R. and G. Benwell (1997) Interoperable GIS and spatial process modelling. *Proceedings, 2nd International Conference on GeoComputation, Dunedin, New Zealand,* pp. 183–90.

Marsh, S.L. (2007) Using and evaluating HCI techniques in geovisualisation: Applying standard and adapted methods in research and education. Unpublished PhD, City University, London.

Martin, D. (1996) *Geographic Information Systems: socioeconomic applications.* Burns & Oates.

Massey, D. (1999) Space-time, 'science' and the relationship between physical geography and human geography. *New Series* **24**(3), 261–276.

May, A., Bayer, S.H. and Ross, T. (2007) A survey of 'young social' and 'professional' users of location-based services in the UK. *Journal of Location Based Services* **1**(2), 112–132.

Mayhew, D.J. (1999) *The Usability Engineering Lifecycle: a Practitioner's Handbook for User Interface Design*. Morgan Kaufmann Publishers Inc.

McFadden, E., Hager, D.R., Elie, C.J. and Blackwell, J.M. (2002) Remote usability evaluation: Overview and case studies. *International Journal of Human Computer Interaction* **14**(3&4), 486–502.

McGee, M. (1979) Human spatial abilities: Psychometric studies and environmental, genetic, hormonal, and neurological influences. *Psychological Bulletin* **86**, 889–918.

McGuinness, C. (1994) Expert/novice use of visualization tools. In A.M. MacEachren and D.R.F. Taylor (eds) *Visualization in Modern Cartography*. Oxford: Pergamon/Elsevier Science Ltd, pp. 185–199.

McHarg, I.L. and American Museum of Natural History (1969) *Design With Nature*. Garden City, NY: Natural History Press.

McMaster, R.B. and Shea, K.S. (1988) Cartographic generalization in a digital environment: a framework for implementation in a Geographic Information System. *Proceedings GIS/LIS'88,* San Antonio, TX, November 30–December 2, 1988, Volume 1, pp. 240–249.

Medeiros, C., Perez-Alcazar, J., Digiampietri, L., Pastorello, G.Z., Santanche, A., Torres, R.S., Madeira, E. and Bacarin, E. (2005) WOODSS and the Web: annotating and reusing scientific workflows. *SIGMOD Record* **34**(3), 18–23.

Medyckyj-Scott, D. (1992) GIS and the concept of usability. In Mark, D.M. and Frank, A.U. (eds) *User Interfaces for Geographic Information Systems: Report on the Specialist Meeting*. Buffalo, NY: NCGIA, pp. 105–110.

*Medyckyj-Scott, D. (1993) Designing geographical information systems for use. In Medyckyj-Scott, D. and Hearnshaw, H.M. (eds) *Human Factors in Geographic Information Systems*, London: Belhaven Press, pp. 87–100.

Medyckyj-Scott, D. (1994) Visualization and Human-Computer Interaction in GIS. In Hearnshaw, H.M. and Unwin, D.J. (eds) *Visualisation in Geographical Information Systems*, Chichester: John Wiley and Sons Ltd, pp. 200–211.

*Medyckyj-Scott, D. and Hearnshaw, H.M. (eds) (1993) *Human Factors in Geographical Information Systems*. London: Belhaven Press.

Meilinger, T. (2008) The network of reference frames theory: a synthesis of graphs and cognitive maps. In Freksa, C., Newcombe, N.S., Gärdenfors, P. and Wölfl, S. (eds) *Spatial Cognition VI. Learning, Reasoning, and Talking about Space*. Lecture Notes in Artificial Intelligence series – Vol. 5248. Berlin, Germany: Springer, pp. 344–360.

Miller, G.A. (1956) The magic number seven plus or minus two: Some limits on our automatization of cognitive skills. *Psychological Review* **63**, 81–97.

Mills, S. (2005) Designing usable marine interfaces: some issues and constraints. *Journal of Navigation* **58**(1), 67–75.

Mitchell, W.J. (1999) The city of bits hypothesis. In Schon, D., Sanyal, B. and Mitchell, W.J. (eds) *High Technology and Low-Income Communities: Prospects for the positive use of advanced information technology*, Cambridge, MA: MIT Press, pp. 105–130.

Monmonier, M. (1996) *How to Lie with Maps*. Chicago: University of Chicago Press.

Montello, D.R. (1992) The geometry of environmental knowledge. In Frank, A.U., Campari, I. and Formentini, U. (eds) *Theories and Methods of Spatio-Temporal Reasoning in Geographic Space* Lecture Notes in Computer Science, Vol. 639. Berlin: Springer-Verlag, pp. 136–152.

Montello, D.R. (1993). Scale and multiple psychologies of space. In Frank, A.U. and Campari, I. (eds) *Spatial Information Theory: a theoretical basis for GIS*. Proceedings of COSIT '93. Lecture Notes in Computer Science, Vol. 716. Berlin: Springer-Verlag, pp. 312–321.

*Montello, D.R. (2001) Spatial cognition. In Smelser, N. and Baltes, P. (eds) *International Encyclopedia of the Social and Behavioural Sciences*. Oxford: Pergamon Press, pp. 14771–14775.

Montello, D.R. (2002) Cognitive map-design research in the twentieth century: Theoretical and empirical approaches. *Cartography and Geographic Information Science* **29**, 283–304.

Montello, D.R., Fabrikant, S.I., Ruocco, M. and Middleton, R.S. (2003) Testing the first law of cognitive geography on point-display spatializations. *Spatial Information Theory, Proceedings*, pp. 316–331.

*Montello, D.R. and Freundschuh, S. (2005) Cognition of geographic information. *Research Agenda for Geographic Information Science* pp. 61–91.

Montello, D.R., Lovelace, K., Golledge, R.G. and Self, C. (1999) Sex-related differences and similarities in geographic and environmental spatial abilities. *Annals of the Association of American Geographers* **89**(3), 515–534.

Mori, M. (2002) *Semantic analysis of spatial expressions in Japanese.* State University of New York at Buffalo (0656), 201.

Morris, K., Hill, D. and Moore, A. (2000) Mapping the environment through three-dimensional space and time. *Computers Environment and Urban Systems* **24**, 435–450.

Mtalo, E.G.B. (1996) *Towards an Extensible GIS/Remote Sensing Knowledge Representation Language (Expert Systems).* The University of New Brunswick Canada (0823), 319.

Muehrcke, P. (1978) *Map Use: reading, analysis and interpretation.* Madison, WI: J.P. Publications.

National Geographic (2009) *The Genographic Project – Human Migration, Population Genetics, Maps, DNA – National Geographic.* [Online]. Available at: https://genographic.nationalgeographic.com/ genographic/lan/en/globe.html [Accessed 10 July (2009)]

Nelson, E.S. (2000) Designing effective bivariate symbols: The influence of perceptual grouping processes. *Cartography and Geographic Information Science* **27**, 261–278.

Neumann, A. (2007) Web mapping and web cartography. In Shekhar, S. and Xiong, H. (eds) *Encyclopedia of GIS*, NY: Springer, pp. 1261–1269.

Newman, M. (2006) Worldmapper: The world as you've never seen it before. Available at: http:// www.worldmapper.org/ [Accessed August 5, (2009)].

Newman, W. and Smith, E.L. (2006) Disruption of meetings by laptop use: is there a 10-second solution? In *CHI '06 Extended Abstracts on Human Factors in Computing Systems* (Montréal, Québec, Canada, April 22–27). New York: ACM, pp. 1145–1150.

Nielsen, J. (1994a) *Usability Engineering.* Morgan-Kaufman.

Nielsen, J. (1994b) Guerrilla HCI: Using discount usability engineering to penetrate the intimidation barrier. http://www.useit.com/papers/guerrilla hci.html.

Nielsen, J. (1996) F-shaped pattern for reading Web content. Usable information technology. http://www.useit.com/alertbox/reading pattern.html.

Nielsen, J. (1999) *Designing Web Usability: The practice of simplicity.* Thousand Oaks, CA: New Riders Publishing.

Nielsen, J. (2005) Usability 101: Introduction to usability. Available on http://www.useit.com/ alertbox/(2003)0825.html accessed 24th May (2009).

Nielsen, J. (2006) Productivity and screen size. Available http://www.useit.com/alertbox/screen-productivity.html, accessed 8th July (2008).

Nivala, A.M. (2005) User-centred design in the development of a mobile map application. Licentiate Thesis, Helsinki University of Technology, Finland. Online at http://www.soberit.hut.fi/t-121/shared/thesis/nivala licentiatethesis.pdf, accessed 25 August (2009).

Nivala, A.M. (2007) *Usability Perspectives for the Design of Interactive Maps.* Publications of the Finnish Geodetic Institute No. 136.

Nivala, A.M., Brewster, S. and Sarjakoski, L.T. (2008) Usability evaluation on web mapping sites. *The Cartographic Journal* **45**(2), 129–138.

Nivala, A.M., Sarjakoski, L.T. and Sarjakoski, T. (2005) User-centred design and development of a mobile map service. In Hauska, H. and Tveite, H. (eds) *Proceedings of the 10th Scandinavian Research Conference on Geographical Information Sciences* (ScanGIS 2005), June 13–15, Stockholm, Sweden, pp. 109–123.

Norman, D.A. (1984) Stages and levels in human-machine interaction. *International Journal of Man-Machine Studies* **21**, 365–375.

Norman, D.A. (1990) *The Design of Everyday Things.* New York: Doubleday.

Nyerges, T.L. (1991) Analytical map use. *Cartography and Geographic Information Systems* **18**, 11–22.

Nyerges, T.L. (1991) Geographic information abstractions – conceptual clarity for geographic modelling. *Environment and Planning A* **23**, 1483–1499.

Nyerges, T.L. (1992) Thinking with geographic information: user interface requirements for GIS. *Position paper for specialist meeting on user interfaces for GIS. NCGIA 92–3.*

Nyerges, T.L. (1993) How do people use geographical information systems? In Medyckyj-Scott, D. and Hearnshaw, H.M. (eds) *Human Factors in Geographical Information Systems*. London: Belhaven Press, pp. 37–50.

Nyerges, T.L. (1995) Cognitive issues in the evolution of GIS user knowledge. In Nyerges, T.L., Laurini, R., Egenhofer, M.J. and Mark, D.M., (eds) *Cognitive Aspects of Human-Computer Interaction for Geographic Information Systems*, Dordrecht: Kluwer Academic Publishers.

Nyerges, T.L. and Jankowski, P. (1989) A knowledge base for map projection selection. *American Cartographer* **16**, 29–38.

*Nyerges, T.L., Mark, D.M., Laurini, R. and Egenhofer, M.J. (eds) (1995a) *Cognitive Aspects of Human-Computer Interaction for Geographic Information Systems*, Dordrecht: Kluwer Academic Publishers.

Nyerges, T.L., Mark, D.M., Laurini, R. and Egenhofer, M.J. (1995b) Cognitive aspects of HCI for GIS: an introduction. In Nyerges, T.L., Mark, D.M., Laurini, R. and Egenhofer, M.J. (eds) *Cognitive Aspects of Human-Computer Interaction for Geographic Information Systems*, Dordrecht: Kluwer Academic Publishers, pp. 1–8.

Nyerges, T.L., Moore, T.J., Montejano, R. and Compton, M. (1998) Developing and using interaction coding systems for studying groupware use. *Human-Computer Interaction* **13**, 127–165.

O'Keefe, J. and Nadel, L. (1978) *The Hippocampus as a Cognitive Map*. Oxford: Oxford University Press.

Oviatt, S. (1997) Multimodal interactive maps: Designing for human performance. *Human-Computer Interaction* 93–129.

Pandey, S., Harbor, J. and Engel, B. (2000) *Internet-based Geographic Information Systems*. Park Ridge, IL: The Urban and Regional Information Systems Association.

Patton, M.Q. (1990) *Qualitative Evaluation and Research Methods*. Sage: London.

Payne, S.J. and Green, T.R.G. (1986) Task-action grammars: a model of the mental representation of task languages. *Human-Computer Interaction* **2**, 93–133.

Pederson, Thomas W. (2000) *The Visual Analysis of Spatial Regression*. University of Pennsylvania (0175), 123.

Peebles, D., Davies, C. and Mora, R. (2007) Effects of geometry, landmarks and orientation strategies in the 'drop-off' orientation task. In Winter, S., Duckham, M., Kulik, L. and Kuipers, B. (eds) *Spatial Information Theory*. Lecture Notes in Computer Science Vol. 4736, Berlin: Springer, pp. 390–405.

Peloux, J.P. and Rigaux, P. (1995) A loosely coupled interface to an object-oriented geographic database. *Spatial Information Theory* 123–137.

Perez Maqueo, O., Equihua, M., Hernandez, A. and Benitez, G. (2001) Visual programming languages as a tool to identify and communicate the effects of a development project evaluated by means of an environmental impact assessment. *Environmental Impact Assessment Review* **21**, 291–306.

Peterson, G.N. (2009) *GIS Cartography*. CRC Press.

Peterson, M.P. (1987) The mental image in cartographic communication. *The Cartographic Journal* **24**(1), 35–41.

Pettersson, M., Randall, D. and Helgeson, B. (2004) Ambiguities, awareness and economy: a study of emergency service work. *Computer Supported Cooperative Work* **13**(2), 125–154.

Philbrick, A.K. (1953) Toward a unity of cartographical forms and geographical content. *The Professional Geographer* **5**(5), 11–15.

Pick, H.L., Heinrichs, M.R., Montello, D.R., Smith, K., Sullivan, C.N. (1995) Topographic map reading. In Hancock, P.A., Flach, J., Caird, J., Vicente, K. (eds) *Local applications of the ecological approach to human-machine systems*. Volume 2, Hillsdale, NJ: Lawrence Erlbaum, pp. 255–284.

Pickles, J. (ed.) (1995) *Ground Truth: The Social Implications of Geographic Information Systems*. New York: Guilford.

Pike, W. and Gahegan, M. (2003) Constructing semantically scalable cognitive spaces. *Spatial Information Theory, Proceedings*, pp. 332–348.

Pires, P., Painho, M. and Kuhn, W. (2005) Measuring semantic differences between conceptualisations: The Portuguese water bodies case – Does education matter? *On the Move to Meaningful Internet Systems*. Otm 2005 Workshops, Proceedings, pp. 1020–1026.

*Pivar, M., Fredkin, E. and Stommel, H. (1963) Computer-compiled oceanographic atlas: An experiment in man-machine interaction. *Proceedings of the National Academy of Sciences* **50**, 396–398.

Plaisant, C. (2004) The challenge of information visualization evaluation. In *Proceedings of Advanced Visual Interfaces, AVI'04. Italy*, pp. 109–116.

Preece, J., Rogers, Y., Sharp, H., Benyon, D., Holland, S. and Carey, T. (1994) *Human-Computer Interaction*. Harlow: Addison-Wesley.

Pundt, H. and Bishr, Y.A. (2002) Domain ontologies for data sharing – an example from environmental monitoring using field GIS. *Computers & Geosciences* **28**, 95–102.

Putz, S. (1994) Interactive information services using World-Wide Web hypertext. In *Proceedings of the First International Conference on World-Wide Web*, Switzerland, Geneva 25–27 May (1994).

Qian, L. (2000) *A Visual Query Language for GIS*. The Pennsylvania State University (0176), 191.

Radwan, M.M., Bishr, Y.A., Friha, N., Driza, O. and Pandya, J. (1997) Guidelines for the development of a geospatial clearing house in a heterogeneous environment. In Hodgson, S., Rumor, M. and Harts, J.J., (eds) *Geographical Information '97: from Research to Application through Cooperation*, Vols. 1 and 2, pp. 277–287.

*Ratti, C. (2004) Tangible user interfaces (TUIs): a novel paradigm for GIS. *Transactions in GIS* **8**, 407.

Rauh, R., Hagen, C., Knauff, M., Kuß, T., Schlieder, C. and Strube, G. (2005) Preferred and alternative mental models in spatial reasoning. *Spatial Cognition and Computation* **5**, 239–269.

Rauschert, I., Agrawal, P., Sharma, R., Fuhrmann, S., Brewer, I. and MacEachren, A. (2002) *Designing a Human-centered, Multimodal GIS Interface to Support Emergency Management*. McLean, VA: ACM Press.

Reichenbacher, T. (2001) Adaptive concepts for a mobile cartography. *Journal of Geographical Sciences* **11**(Supplement 1), 43–53.

Reitsma, F. and Bittner, T. (2003) Scale in object and process ontologies. *Spatial Information Theory, Proceedings*, 13–27.

Reitsma, R., Zigurs, I., Lewis, C., Wilson, V. and Sloane, A. (1996) Experiment with simulation models in water resources negotiation. *Journal of Water Resources Planning and Management* **122**, 64–70.

Renschler, C.C. (2003) Designing geo-spatial interfaces to scale process modes: the GeoWEPP approach. *Hydrological Processes* **17**, 1005–1017.

Rice, M., Jacobson, R.D., Golledge, R.G. and Jones, D. (2005) Design considerations for haptic and auditory map interfaces. *Cartography and Geographic Information Science* **32**, 381–391.

*Richards, J.R. and Egenhofer, M.J. (1995) A comparison of two-directmanipulation GIS user interface for map overlay. *Geographical Systems* **2**(4), 267–290.

Richardson, D.E. and Mackaness, W.A. (1999) Computational processes for map generalization. *Cartography and Geographic Information Science* **26**(1).

Riedemann, C. (2005) Matching names and definitions of topological operators. In Cohn, A.G. and D.M. Mark (eds) *Spatial Information Theory: Cognitive and Computational Foundations*, Proceedings of COSIT '05, Ellicottville, New York, USA (COSIT '05). Berlin: Springer, pp. 165–181.

Riedemann, C. (2005) *Naming Topological Operators at GIS User Interfaces*. 8th AGILE Conference on GIScience, Estoril, Portugal.

Riedemann, C. and Kuhn, W. (1999) What are sports grounds? Or: Why semantics requires interoperability. *Interoperating Geographic Information Systems* 217–229.

Rieger, M. (1999) An analysis of map users' understanding of GIS images. *Geomatica* **53**, 125–137.

Rinner, C. 1999. Argumaps for spatial planning. In *Proceedings of TeleGeo'99, First International Workshop on Telegeoprocessing*, Lyon, France, pp. 95–102. Available at http://set.gmd.de/MS/publications.html.

*Robinson, A.C., Chen, J., Lengerich, E.J., Meyer, H.G. and MacEachren, A.M. (2005) Combining usability techniques to design geovisualization tools for epidemiology. *Cartography and Geographic Information Science* **32**, 243–255.

Robinson, A.H., Morrison, J.L. and Muehrcke, P.C. (1995) *Elements of Cartography*. New York: John Wiley & Sons Inc.

Rodríguez, M.A. and Egenhofer, M.J. (1999) Putting similarity assessments into context: Matching functions with the user's intended operations. In *Proceedings of the Second International and Interdisciplinary Conference on Modeling and Using Context*. Springer-Verlag, 310–323.

Rodriguez, M.A. and Egenhofer, M.J. (2003) Determining semantic similarity among entity classes from different ontologies. *IEEE Transactions on Knowledge and Data Engineering* **15**, 442–456.

Rodriguez, M.A. and Egenhofer, M.J. (2004) Comparing geospatial entity classes: an asymmetric and context-dependent similarity measure. *International Journal of Geographical Information Science* **18**, 229–256.

Rundstrom, R. (1995) GIS, Indigenous peoples, and epistemological diversity. *Cartography and Geographical Information Systems* **22**(1), 45–57.

Sahay, S. (1993) *Social Construction of Geographic Information Systems*. Volumes I and II. Florida International University (1023), 440.

Sandvik, B. (2009) thematicmapping.org. Thematic mapping. Available at: http://thematicmapping.org/ [Accessed August 5, (2009)].

Saraiya, P., North, C. and Duca, K. (2004) An evaluation of microarray visualization tools for biological insight. In *IEEE Symposium on Information Visualization*. IEEE. Austin.

Sauro, J. (2006) Measuring usability: The importance of randomizing tasks. http://www.measuringusability.com/random.htm (Accessed: 01/07/2006).

Scarponcini, P. (1992) *An Inferencing Language System for Automated Graphic Reasoning* (GIS, CAD). University of Missouri Rolla (0135), 231.

Schmiguel, J., Baranauskas, C. and Medeiros, C. (2004) Inspecting user interface quality in Web GIS applications. In *Proceedings, VI Brazilian Symposium on GeoInformatics – GEOINFO(2004)*, pp. 201–219.

Schofield, N.J. and Kirby, J.R. (1994) Position location on topographical maps – effects of task factors, training, and strategies. *Cognition and Instruction* **12**(1), 35–60.

Schuler, D. and Namioka, A., (eds.) (1993) *Participatory Design: Principles and Practices*. Hillsdale, NJ: Lawrence Erlbaum.

Self, C., Gopal, S., Golledge, R. and Fenstermaker, S. (1992) Gender-related differences in spatial abilities. *Progress in Human Geography* **16**(3), 315–342.

Shackel, B. (1997) Human-computer interaction – whence and whither? *Journal of the American Society for Information Science* **48**(11), 970–986.

Shand, M. (2002) Mapping and imaging Africa on the Internet. *The International Archives of the Photogrammetry, Remote Sensing and Spatial Information Sciences* **34**(6/W6), March 25–28, Dar es Salaam, Tanzania, pp. 210–217. Available at: http://www.photogrammetry.ethz.ch/general/persons/jana/daressalaam/papers/shand.pdf.

Shariff, A., Egenhofer, M.J. and Mark, D.M. (1998) Natural-language spatial relations between linear and areal objects: the topology and metric of English-language terms. *International Journal of Geographical Information Science* **12**, 215–245.

Sharma, J. (1996) *Integrated Spatial Reasoning in Geographic Information Systems: Combining Topology and Direction*. University of Maine (0113), 164.

Sharma, R., Yeasin, M., Krahnstoever, N., Rauschert, I., Cai, G., Brewer, I., MacEachren, A.M. and Sengupta, K. (2003) Speech-gesture driven multimodal interfaces for crisis management. *Proceedings of the IEEE* **91**, 1327–1354.

Sharp, H., Rogers, Y. and Preece, J. (2007) *Interaction Design: Beyond Human-Computer Interaction.* 2nd Edition. New York, NY: John Wiley & Sons Inc.

Sharps, M., Welton, A. and Price, J. (1993) Gender and task in the determination of spatial cognitive performance. *Psychology of Women Quarterly* **17**(1), 71–83.

Shea, K.S. and McMaster, R.B. (1989) Cartographic generalization in a digital environment: When and how to generalize. In *Proceedings 9th International Symposium on Computer-Assisted Cartography,* pp. 56–67.

Shepard, R.N. and Metzler, J. (1971) Mental rotation of three-dimensional objects. *Science* **171**, 701–703.

Sheth, A. (1999) Changing focus on interoperability in information systems: From system, syntax, structure to semantics. *Interoperating Geographic Information Systems,* Kluwer, 5–29.

Shiffer, M.J. (1999) Managing public discourse: towards the augmentation of GIS with multimedia. In Longley, P., Goodchild, M.F., Maguire, D.J. and Rhind, D. (eds) *Geographical Information Systems.* New York, NY: John Wiley & Sons Inc, pp. 1101.

Sholl, M. and Egeth, H. (1982) Cognitive correlates of map-reading ability. *Intelligence* **6**, 215–230.

Sidjanin, P. (2001) *A cognitive framework for an urban environment design tool.* Technische Universiteit Delft, The Netherlands (0951), 415.

Sieber, R.E. (2006) Public participation geographic information systems: A literature review and framework. *Annals of the American Association of Geographers* **96**(3), 491–507.

Siegel, A.W. and White, S.H. (1975) The development of spatial representations of large-scale environments. In H.W. Reese (ed.) *Advances in Child Development and Behavior,* Vol. 10. New York: Academic Press, pp. 9–55.

Sillence, E., Briggs, P., Fishwick, L. and Harris, P. (2004) Timeline Analysis: A Tool for Understanding the Selection and Rejection of Health Websites. In Dearden, A. and Watts, L. (eds) *HCI,* Vol. 2. Leeds: Research Press, pp. 113–116.

Silverman, D. (2001) *Interpreting Qualitative Data: Methods for Analysing Talk, Text and Interaction.* London: Sage.

Skarlatidou, A. (2005) The current state of online mapping industry: A usability study of Multimap's public service and that of its competitors. MSc Thesis. London: University College London.

Skarlatidou, A. and Haklay, M. (2006) Public Web mapping: Preliminary usability evaluation. In *Proceedings of 14th Annual GIS Research UK Conference GISRUK (2006),* Nottingham, UK.

Slocum, T.A. (1999) *Thematic Cartography and Visualization.* Upper Saddle River, NJ: Prentice Hall.

Slocum, T.A., Blok, C., Jiang, B., Koussoulakou, A., Montello, D.R., Fuhrmann, S. and Hedley, N.R. (2001) Cognitive and usability issues in geovisualization. *Cartography and Geographic Information Science* **28**, 61–75.

Slocum, T.A., Cliburn, D.C., Feddema, J.J. and Miller, J.R. (2003) Evaluating the usability of a tool for visualizing the uncertainty of the future global water balance. *Cartography and Geographic Information Science* **30**, 299–317.

Smith, T., Su, J., El Abbadi, A., Agrawal, D., Alonso, G. and Saran, A. (1995) Computational modelling systems. *Information Systems* **20**, 127–153.

Sneiderman, B. (1997) *Designing the User Interface.* Third edition. Addison-Wesley Publishing Company.

Sneiderman, B. and Plaisant, C. (2005) *Designing the User Interface: Strategies for Effective Human-Computer Interaction.* 4th Edition. Addison Wesley.

Soon, K. and Kuhn, W. (2004) Formalizing user actions for ontologies. *Geographic Information Science,* Proceedings, 299–312.

Spedding, N. (1997) On growth and form in geomorphology. *Earth Surface Processes and Landforms* **22**, 261–265.

Spencer, C., Blades, M. and Morsley, K. (1989) *The Child in the Physical Environment: the development of spatial knowledge and cognition.* Chichester, UK: John Wiley.

Sperling, G. (1960) The information available in brief visual presentations. *Psychological Monographs: General and Applied* **74**(11), 1–29.

Spinuzzi Clay, I.A.N. (1999) *Designing for Lifeworlds: Genre and Activity in Information Systems Design and Evaluation (Activity Theory, Human-Computer Interaction, Usability).* Iowa State University (0097), 299.

Steves, M.P., Morse, E., Gutwin, C. and Greenberg, S. (2001) A comparison of usage evaluation and inspection methods for assessing groupware usability. In *GROUP'01.* Boulder, CO, pp. 125–134.

*Suchan, T.A. (2002) *Usability Studies of Geovisualization Software in the Workplace.* Los Angeles, CA: Digital Government Research Center.

*Suchan, T.A. and Brewer, C.A. (2000) Qualitative methods for research on mapmaking and map use. *The Professional Geographer* **52**(1), 145–154.

Tang, Q.I.N. (1993) *A Dynamic Visualization Approach to the Exploration of Area-Based, Spatial-Temporal Data (Geographic Information Systems).* The Ohio State University (0168), 225.

The Economist (2003) North Korea: When bluff turns deadly. *The Economist* (May 1st 2003). Available at: http://www.economist.com/displaystory.cfm?story id=1748566 [Accessed August 5, (2009)].

Thurstone, L.L. (1938) *Primary Mental Abilities.* Psychometric Monographs, No. 1. Chicago: University of Chicago Press.

Timpf, S. (2001) The need for task ontologies in interoperable GIS. Online-only document, http://www.geo.uzh.ch/~timpf/docs/Timpf TaskOntologies.pdf.

Timpf, S. (2002) Ontologies of wayfinding: a traveler's perspective. *Networks and Spatial Economics* **2**, 9.

Tobler, W.R. (1961) Map transformations of geographic space. PhD University of Washington. Microfilm.

Tobler, W.R. (1963) Geographic area and map projections. *The Geographical Review* **53**, 59–78.

*Tobón, C. (2005) Evaluating geographic visualization tools and methods: an approach and experiment based upon user tasks. In Dykes, J., MacEachren, A. and Kraak, M.J.) (eds) *Exploring Geovisualization.* Oxford: Elsevier, pp. 645–666.

Tomlinson, R.F. (2007) *Thinking about GIS: Geographic information system planning for managers.* (3rd edition). ESRI Press.

Torres, M., Quintero, R., Moreno, M. and Fonseca, F. (2005) Ontology-driven description of spatial data for their semantic processing. In *Geospatial Semantics, Proceedings.* Berlin: Springer-Verlag, pp. 242–249.

Traynor, C. (1998) *Programming by Demonstration for Geographic Information Systems (GIS, End Users, Interface).* University of Lowell (0111), 164.

*Traynor, C. and Williams, M.G. (1995) Why are geographic information systems hard to use? In Katz, I., Mack, R. and Marks, L. (eds) *Conference Companion on Human Factors in Computing Systems – CHI '95,* Denver, Colorado: ACM Press, pp. 288–289.

Traynor, C. and Williams, M.G. (1997) A study of end-user programming for geographic information systems, empirical studies of programmers: Seventh Workshop, ESP7, Alexandria, VA, ACM Press.

Traynor, C. and Williams, M.G. (1998) Power of persuasion: Non-technical users and geographic information systems. In Chatfield, R.H., Kuhn, S. and Muller, M. (eds) *Pdc 98: Proceedings of the Participatory Design Conference,* pp. 179–180.

Tsou, M.S. and Whittemore, D.O. (2001) User interface for ground-water modeling: Arcview extension. *Journal of Hydrologic Engineering* **6**, 251–257.

Tuckman, B.W. (1965) Developmental sequence in small groups, *Psychological Bulletin* **63**, 384–399.

Tufte, E.R. (2001) *The Visual Display of Quantitative Information.* 2nd edition. Cheshire, CT: Graphics Press.

Tufte, E.R. (2006) *The Cognitive Style of PowerPoint.* 2nd edition. Cheshire, CT: Graphics Press. Available at: http://www.edwardtufte.com/tufte/powerpoint.

Turk, A.G. (1993) The relevance of human factors to geographical information systems. In Medyckyj-Scott, D. and Hearnshaw, H.M. (eds) *Human Factors in Geographical Information Systems*. London: Belhaven, pp. 15–31.

Tversky, B. (1993) Cognitive maps, cognitive collages, and spatial mental models. In Frank, A.U. and Campari, I. (eds) *Spatial Information Theory: a theoretical basis for GIS. Proceedings of COSIT '93*. Berlin: Springer-Verlag, pp. 14–24.

Unwin, D. (2005) Fiddling on a different planet. *Geoforum* **36**, 681–684.

UsabilityNet (2006). Overview of the user centred design process. http://www.usabilitynet.org/management/b_overview.htm (accessed December 2009).

van den Worm, J. (2001) Web map design in practice. In Kraak, M.J. and Brown, A. (eds) *Web Cartography: developments and prospects*, **14**: 87. London: Taylor & Francis.

Van Oosterom, P., Maessen, B. and Quak, W. (2002) Generic query tool for spatio-temporal data. *International Journal of Geographical Information Science* **16**, 713–748.

Vecchi, T. and Girelli, L. (1998) Gender differences in visuo-spatial processing: the importance of distinguishing between passive storage and active manipulation. *Acta Psychologica* **99**(1), 1–16.

Velez, M., Silver, D. and Tremaine, M. (2005) Understanding visualization through spatial ability differences. Paper presented at the IEEE Visualization 2005, Minneapolis, MN. Online available from http://www.caip.rutgers.edu/~mariacv/publications/vis05.pdf, last accessed 22 Mar (2009).

Verbree, E., Van Maren, G., Germs, R., Jansen, F. and Kraak, M.J. (1999) Interaction in virtual world views – linking 3D GIS with VR. *International Journal of Geographical Information Systems* **13**, 385–396.

Volta, G. (1993) *Interaction with Attribute Data in Geographic Information Systems: A Model for Categorical Coverages*. Master Thesis, University of Maine, Orono ME.

Volta, G. and Egenhofer, M.J. (1993) Interaction with GIS attribute data based on categorical coverages. In Frank, A.U. and Campari, I. (eds) *Spatial Information Theory, European Conference COSIT '93*, Marciana Marina, Elba Island, Italy. Lecture Notes in Computer Science 716, New York, NY: Springer-Verlag, pp. 215–233.

Wakabayashi, Y. (2003) Spatial cognition and GIS. *Geographical Review of Japan Series A* **76**, 703–724.

Wardlaw, J., Haklay, M. and Parker, S. (2008) Mapping health information. In *AGI GeoCommunity '08: Shaping a Changing World*. Stratford upon Avon, UK. http://eprints.ucl.ac.uk/13851/.

Weske, M., Vossen, G., Medeiros, C. and Pires, F. (1998) Workflow management in geoprocessing applications. In *Proceedings of the 6th ACM International Symposium on Advances in Geographic Information Systems*. Washington, DC: ACM, pp. 88–93.

Whitefield, A., Esgate, A., Denley, I. and Byerley, P. (1993) On distinguishing work tasks and enabling tasks. *Interacting with Computers* **5**, 333–347.

Whiteside, J., Bennet, J. and Holtzblatt, K. (1988) Usability engineering: our experience and evolution. In Helander, M. (ed.) *Handbook of Human-Computer Interaction*. Amsterdam: North-Holland.

Whitten, J. and Bentley, L. (2005) *Systems Analysis and Design Methods*. New York, NY: McGraw-Hill.

Wiegand, N. and Garcia, C. (2007) A task-based ontology approach to automate geospatial data retrieval. *Transactions in GIS* **11**(3), 355–376.

Wilford, J.N. (2002) *The Mapmakers*. New York: Vintage.

Winograd, T. and Flores, F. (1986) *Understanding Computers and Cognition: A new foundation for design*. Norwood, NJ: Ablex.

Witkin, H.A., Oltman, P.K., Raskin, E. and Karp, S.A. (1971) *A Manual for the Embedded Figures Tests*. Palo Alto CA: Consulting Psychologists Press.

Wood, M. (1993) Interacting with maps. In D. Medyckyj-Scott and H.M. Hearnshaw (eds) *Human Factors in Geographical Information Systems*, London: Belhaven Press, pp. 111–123.

Worboys, M. and Duckham, M. (2004) *GIS – A Computing Perspective* (2nd edition). CRC Press.

Wu, S. (2005) *Information Visualization Methods for GIS, Constraint and Spatiotemporal Databases*. The University of Nebraska Lincoln (0138), 108.

Yank, K. (2002) Interview – Jakob Nielsen, PhD. Available on http://www.sitepoint.com/article/interview-jakob-nielsen-ph-d/ accessed 24th May (2009).

Yuan, M. and Albrecht, J. (1995) Structural analysis of geographic information and GIS operations from a user's perspective. *Spatial Information Theory* 107–122.

Yuan, M., Mark, D.M., Egenhofer, M.J. and Peuquet, D.J. (2005) Extensions to geographic representations. *Research Agenda for Geographic Information Science* 129–156.

Zapf, D., Brodbeck, F.C., Frese, M., Peters, H. and Prumper, J. (1992) Errors in working with office computers: a first validation of a taxonomy for observed errors in a field setting. *International Journal of Human-Computer Interaction* **4**, 311–339.

Zaphiris, P., Gill, K., Ma, T.H.Y., Wilson, S. and Petrie, H. (2004) Exploring the use of information visualization for digital libraries. *The New Review of Information Networking* **10**(1), 51–69.

Zhou, W., Xu, T., Zhang, H. and Zhang, S. (2005) Design and implementation of map annotation object with interoperability in platforms. *Journal of Geomatics* **30**, 28–29.

Zhu, X., Healy, R.C. and Aspinall, R.J. (1998) A knowledge-based systems approach to design of spatial decision support systems for environmental management. *Environmental Management* **22**, 35–48.

Zyszkowska, W. (2000) *Semiotic Aspects of Cartographic Visualization*. Acta Universitatis Wratislaviensis Studia Geograficzne, 130–132.

Index

This index was prepared by Neil Manley.